T0334597

Application of Smart Grid Technologies

Application of Smart Grid Technologies

Case Studies in Saving Electricity in Different Parts of the World

Edited by

Lisa Ann Lamont
Ali Sayigh

ACADEMIC PRESS

An imprint of Elsevier

Academic Press is an imprint of Elsevier
125 London Wall, London EC2Y 5AS, United Kingdom
525 B Street, Suite 1650, San Diego, CA 92101, United States
50 Hampshire Street, 5th Floor, Cambridge, MA 02139, United States
The Boulevard, Langford Lane, Kidlington, Oxford OX5 1GB, United Kingdom

Notices
Knowledge and best practice in this field are constantly changing. As new research and experience broaden our understanding, changes in research methods, professional practices, or medical treatment may become necessary.

Practitioners and researchers must always rely on their own experience and knowledge in evaluating and using any information, methods, compounds, or experiments described herein. In using such information or methods they should be mindful of their own safety and the safety of others, including parties for whom they have a professional responsibility.

To the fullest extent of the law, neither the Publisher nor the authors, contributors, or editors, assume any liability for any injury and/or damage to persons or property as a matter of products liability, negligence or otherwise, or from any use or operation of any methods, products, instructions, or ideas contained in the material herein.

Library of Congress Cataloging-in-Publication Data
A catalog record for this book is available from the Library of Congress

British Library Cataloguing-in-Publication Data
A catalogue record for this book is available from the British Library

ISBN 978-0-12-803128-5

For information on all Academic Press publications
visit our website at https://www.elsevier.com/books-and-journals

Working together
to grow libraries in
developing countries

www.elsevier.com • www.bookaid.org

Publisher: Joe Hayton
Acquisition Editor: Raquel Zanol
Editorial Project Manager: Ana Claudia A. Garcia
Production Project Manager: Kamesh Ramajogi
Cover Designer: Victoria Pearson

Typeset by SPi Global, India

Contents

Part Five   Case Studies   395

List of contributors

Kamal Al Khuffash ADNOC Gas Processing, Abu Dhabi, United Arab Emirates

Lina Bertling Tjernberg KTH Royal Institute of Technology, Stockholm, Sweden

Chen Chen Argonne National Laboratory, Argonne, United States

Guido Coletta University of Sannio, Benevento, Italy

Davide Della Giustina Unareti SpA, Brescia, Italy

Dmitry Efimov Energy Systems Institute, Irkutsk, Russia

Mehdi Ganji Willdan Energy Solutions, Anaheim, CA, United States

Gevork B. Gharehpetian Amirkabir University of Technology, Tehran, Iran

Anna Glazunova Energy Systems Institute, Irkutsk, Russia

Gheorghe Grigoras Gheorghe Asachi Technical University of Iasi, Iasi, Romania

Fengkai Hu University of Tennessee, Knoxville, TN, United States

Norman Ihle OFFIS—Institute for Information Technology, Oldenburg, Germany

Irina Kolosok Energy Systems Institute, Irkutsk, Russia

Elena Korkina Energy Systems Institute, Irkutsk, Russia

Victor Kurbatsky Energy Systems Institute, Irkutsk, Russia

Sebastian Lehnhoff OFFIS—Institute for Information Technology, Oldenburg, Germany

Pierluigi Mancarella The University of Manchester, Manchester, England

Eduardo A. Martínez Ceseña The University of Manchester, Manchester, England

Jürgen Meister OFFIS—Institute for Information Technology, Oldenburg, Germany

Hadi Modaghegh Ministry of Energy, Tehran, Iran

Mehdi Salay Naderi Amirkabir University of Technology, Tehran, Iran

Daniil Panasetsky Energy Systems Institute, Irkutsk, Russia

Ferdinanda Ponci RWTH Aachen University, Aachen, Germany

Sami Repo Tampere University of Technology, Tampere, Finland

Paulo F. Ribeiro Federal University of Itajuba (UNIFEI), Itajuba, Brazil

Yuri R. Rodrigues University of British Columbia (UBC), Vancouver, BC, Canada

Mohammad Shahidehpour Illinois Institute of Technology, Chicago, IL, United States

Kai Sun University of Tennessee, Knoxville, TN, United States

Nikita Tomin Energy Systems Institute, Irkutsk, Russia

Mathias Uslar OFFIS—Institute for Information Technology, Oldenburg, Germany

Alfredo Vaccaro University of Sannio, Benevento, Italy

Domenico Villacci University of Sannio, Benevento, Italy

Nikolai Voropai Energy Systems Institute, Irkutsk, Russia

Carl J. Wallnerström KTH Royal Institute of Technology, Stockholm, Sweden

Ahmad Zahedi James Cook University, Townsville, Australia

Alireza Zakariazadeh University of Science and Technology of Mazandaran, Behshahr, Iran

Ahmed F. Zobaa Brunel University, Uxbridge, United Kingdom

Preface

The subject of this book is smart grids, specifically those related to selected regions of the world, to show how the perceptions of smart grids vary. Although there are several sources that look at smart grids, there is usually no focus on the region where it's installed, how it varies with other locations, or how it is relevant to other locations. This book provides the reader with an international overview of smart grids in one publication.

This book is subdivided into worldwide regions: Asia (Chapters 2 and 3), North America (Chapters 4 and 5), South America (Chapter 6), Europe (Chapters 7–11), and Case Studies (Chapters 12–14).

Chapter 1 provides an introduction to smart grids to give the reader a general understanding of the topic.

Chapter 2 discusses the Iranian smart grid road map, focusing on technology development. The second part of this chapter focuses on the Iranian national advanced metering infrastructure plan and discusses the layer model, communication architecture, and pilot project conclusion.

Chapter 3 discusses intelligent control and protection in the Russian electric power system. This chapter focuses on intelligent electric power systems with active and adaptive networks, state estimation techniques as informational support of intelligent electric power system with an active and adaptive network, intelligent operations and smart emergency protection, and the description of smart grid territorial clusters in the local interconnected power systems.

Chapter 4 discusses the demand response as an enabling technology to achieve energy efficiency in a smart grid. In this chapter, various demand response programs and implementations provided by utilities and independent system operators/regional transmission organizations in the United States will be discussed. Also, a case study of distributed direct load control approach for large-scale residential demand response will be presented.

Chapter 5 discusses microgrids, both for a single-owned building or campus and a community. Microgrids are emerging to eliminate the growth in load, to integrate intermittent renewable energy resources, and to prevent prolonged power outages. In this chapter, an optimal configuration of monitoring and controlling and the communication of such a system is outlined.

Chapter 6 is a smart grid case study in Brazil that focuses on saving electricity. Motivations, projects, thematic coverage, amount of investment, regional distribution, and research and development activities are outlined related to the region. The Brazilian roadmap for smart grid deployment and the actual level of maturity of smart grid concepts there are discussed.

The automation of smart grids in Europe is the focus of Chapter 7, particularly the electricity distribution network as the center of this technology. This technology innovation mainly centered around the improvement of the quality of service through the installation of medium-voltage automation systems, and the remote reading of customer consumption with the use of electronic meters.

Smart distribution networks, demand response, and community energy systems are presented in Chapter 8 via field trial experiences and smart grid modeling advances from the United Kingdom. In the future it is expected that the increased requirements for economic, reliable, sustainable, and socially acceptable energy production will be supported via emerging smart grids. Distributed energy resources are discussed, including their effectiveness. The latest UK research in smart grid applications, distribution networks, distributed energy resources, and relevant results are also documented.

The focus of Chapter 9 is smart metering from all advanced technologies that define the basic architecture of a smart grid in Romania and saving electricity in distribution networks in that country.

Chapter 10 looks at the smart grid digitalization in Germany by a standardized advanced metering infrastructure.

The growth of decentralized energy resources has increased the need for real-time data and telemetry devices and power grids. This information will be used not only to balance out demand and supply but also to mitigate network bottlenecks or react to potential instabilities; this chapter discusses the German initiative toward this.

The analysis of future power systems to enable sustainable energy is presented in Chapter 11. In this chapter the smart grid in Gotland, Sweden, is being used as a case study. This chapter presents an analysis of the integration of wind and solar power, electricity consumption, dynamic rating and energy storage, and solutions for such modernization. The case study is a smart grid demonstration project on the island of Gotland in Sweden. The results provided include how the power systems can handle more electricity consumption and generation and that predominant energy storage will be unused but could support system reliability.

The first case study and Chapter 12 provide details on measurement-based voltage stability monitoring for load areas. This new method can directly calculate the real-time power transfer limit on each tie line. The method is first compared with a Thevenin equivalent-based method using a four-bus test system and then demonstrated by case studies on the Northeast Power Coordinating Council's 48-machine, 140-bus power system.

Chapter 13 discusses the application of cluster analysis for enhancing power consumption awareness in smart grids. This chapter outlines the potential role of self-organizing models based on clustering analysis for classifying the load profiles, correlating them with the endogenous measured variables, and identifying irregularities in energy consumption.

Finally, Chapter 14 highlights how smart grids are capable of handling a large number of generators of different sizes and different technologies and discusses in particular the role of electric vehicles to support the electricity grid during peak demand. It outlines that smart grids will become more efficient and reliable and that the security of the electric grid will improve due to the increased use of digital information and control technologies.

Lisa Ann Lamont, Ali Sayigh

Smart grids—Overview and background information

1

Kamal Al Khuffash
ADNOC Gas Processing, Abu Dhabi, United Arab Emirates

1 Introduction

The power grid is the infrastructure that transports electricity from where it is generated to the consumer [1]. Traditionally, the grid follows a top-down model where electricity is generated in bulk centralized units, then stepped up using a power transformer and a transmission substation. Then, through a distribution substation and power transmission lines, the power reaches the consumer in a one-way flow with no feedback from the consumer side [1–3]. Power utilities have to ensure a sufficient supply of electricity to cope with the demand [2]. In addition, power quality and reliability are managed by utility engineers using meter readings that are distributed in a few critical locations, which supply only limited information about the grid condition [2]. Recently, due to system aging, slow response time, and increased energy demand and load diversity, the rate of system interruptions and blackouts has largely increased [4,5]. Furthermore, the electrical network highly contributes to carbon emissions [4,5]. Therefore, the focus has been shifted toward demand side management, which aims to control demand by educating users to increase their use of energy-efficient products [2]. However, the current measurement procedures do not provide real-time energy consumption [2], which makes it impossible for the consumer to understand how and when they are saving money and energy [2]. Therefore, a new grid infrastructure became essential to overcome all these challenges. Hence, the concept of a smart grid was introduced to present various methods for overcoming those challenges.

2 Definition

Throughout the developed world, the electric utility sector is beginning a fundamental transformation of its infrastructure to overcome the present challenges faced by the sector [6]. These transformations are aiming to making the grid "smarter" and the resulting outcome is referred to as a "smart grid" [4]. Although there are no agreed-upon definitions for the term "smart grid" between organizations [4], there is a common understanding that smart grids should have an information communication structure [4]. Furthermore, the Energy Independence and Security Act of 2007 (EISA 2007) produced what can be considered as the first official definition of a smart grid. The policy states that a smart grid is "the modernization of the Nation's electricity transmission and distribution system to maintain a reliable and secure electricity

Application of Smart Grid Technologies. https://doi.org/10.1016/B978-0-12-803128-5.00001-5

infrastructure that can meet future demand growth" [7]. In addition, the EISA 2007 lists these 10 characteristics of a smart grid:

(1) Increased use of digital information and control technology to improve reliability, security, and efficiency of the electric grid.
(2) Dynamic optimization of grid operations and resources with full cybersecurity.
(3) Deployment and integration of distributed resources and generation, including renewable resources.
(4) Development and incorporation of demand response, demand-side resources, and energy-efficiency resources.
(5) Deployment of "smart" technologies (real-time, automated, interactive technologies that optimize the physical operation of appliances and consumer devices) for metering, communications concerning grid operations and status, and distribution automation.
(6) Integration of "smart" appliances and consumer devices.
(7) Deployment and integration of advanced electricity storage and peak-shaving technologies, including plug-in electric and hybrid electric vehicles as well as thermal-storage air conditioning.
(8) Providing consumers with timely information and control options.
(9) Development of standards for communication and interoperability of appliances and equipment connected to the electric grid, including the infrastructure serving the grid.
(10) Identification and lowering of unreasonable or unnecessary barriers to adoption of smart grid technologies, practices, and services.

The general purpose of a smart grid is to transmit energy in a controlled, smart way from generation units to consumers [8] using a modernized infrastructure that helps improve efficiency, reliability, quality, and safety [2,9]. In addition, smart grids aim to integrate renewable and alternative energy sources [9]. However, to do so requires the implementation of automated control and modern two-way communication technologies [9]. The communication scheme should be able to capture and analyze various data about generation, transmission, and consumption in nearly real time [4].

3 Components

• Smart meters

In a smart grid, timely data is highly important for reliable delivery of electricity [9]. Therefore, smart meters are essential components, considered as the first building block for two-way communication [10]. Smart meters are digital meters with communication capabilities that allow utilities to collect the required information frequently and communicate with devices used by the consumer [2]. A smart meter system consists of the meter, communication capability, and a control device [11]. They are used to assess the health of the equipment and the integrity of the grid [4] as well as execute control commands remotely and locally [11]. In addition, they can be useful to support advanced protection relaying, eliminate meter estimation, and prevent energy theft [4]. Finally, they help relieve power congestion by controlling demand response [4].

Smart meters can distinguish between the energy supplied from the utility and the energy supplied by the distributed generation (DG) units owned by the user.

Therefore, they only bill the energy consumed from the grid, which leads to reducing the energy cost for the consumer [11]. In addition, they can be used by the utility to advise the consumer on the most efficient use of their appliances to reduce the energy cost and the maximum load on the grid [11].

• Transmission and distribution devices

Transmission and distribution devices can also be equipped with communication capabilities to allow the exchange of information as well as the ability to receive commands to modify settings for better grid control [2]. Transformers, voltage regulators, capacitors, and switches are used to enhance the reliability and availability of power to the consumer [2].

• Network

The data collected by the smart meters and transmission and distribution devices require a network to exchange the information between different components or between the utility grid and the consumer [2], all in a timely fashion. In addition, the network can be used for studying the impact of equipment limitations and faults [9]. It also helps to avoid the natural accidents and catastrophes that limit its effect [9]. Furthermore, it helps to maintain the grid safety, reliability, and protection by developing online condition monitoring, diagnostics and protection [9].

• Consumer devices

Currently, there is no way for the consumer to monitor their hourly use of power. This means they only rely on the monthly bill, which makes it difficult for them to understand or feel the effect of installing energy-efficient devices [2]. In addition, such devices can help consumers enhance their energy profiles by disconnecting some loads at peak energy demand times, therefore helping reduce the monthly bill [2]. Such devices can be, but are not limited to, an in-house display or a web portal provided by the utility [2]. Furthermore, installing in-house intelligent devices that can communicate with smart meters to determine the peak demand time and control the energy consumption accordingly [2].

Those components form the new grid infrastructure to achieve grid reliability, flexibility, efficiency, and sustainability. Those features are achieved by the integration of renewable resources, load management, and enhanced energy storage devices [12]. Each one of those methods is discussed next.

4 Renewable energy resources

The integration of renewable energy sources will enhance grid sustainability [10]. However, the high energy fluctuation resulting from the high integration of renewable energy sources might affect the grid's reliability and stability [10,13,14]. In addition, the increased number of DGs increases the complexity of the grid, making it difficult for the operators to handle the system [2]. Therefore, the enhanced communication capabilities of the smart grid can help eliminate those problems because information will be available for balancing supply and demand [2]. Also, because the communication system will be able to send automatic commands, it will respond to the energy fluctuation much faster [2]. Furthermore, the bidirectional energy flow feature of the smart grid will allow the user to supply the grid with energy [14].

Higher integration of renewable energy sources to the grid can represent a way to move toward DG [3]. DG uses small generation units distributed in several locations near the consumer instead of the traditionally bulk units that are usually located far away from cities [3]. Applying DG concepts enhances grid reliability and efficiency by eliminating the high transmission distances [3]. Furthermore, using renewable energy sources in DG reduces carbon emissions produced by the system [3]. However, this high integration of renewable energy sources will cause higher fluctuation in the system, affecting its stability [14]. Several solutions are presented in the literature such as combining the renewable energy sources with combined heat power (CHP) generation [14]; flexible demand such as boilers, heat pumps, and electric vehicles [12,14]; load management [2,10,13]; and energy storage [12,13]. The latter two methods are also viewed as ways for achieving smart grid features discussed in the previous section.

5 Load management

Load management techniques have been implemented since the 1980s as a tool for load shaping and producing a flat load profile [2,12,15]. The main concept of load management is to shift the load from the high demand periods to periods with lower demand [15]. Currently, the load is managed by rejecting loads at high demand periods, using protection relays in a process called "load shedding" to protect the overall grid [12]. However, with the implementation of communication technologies, the load management can be done smarter by responding to the load signal either automatically or manually [12]. The load managers allow the users to program their devices to turn on when the power demand is low and turn off when the grid is congested [13]. To encourage customer participation in load management, utilities have designed different pricing methods such as peak load pricing and adaptive pricing [12]. In the peak load pricing scheme, the utility divides the day into different periods, with each period having a different consumption price announced at the beginning of the day [12]. For, adaptive pricing, the energy price for each period is announced at the beginning of the period, depending on real-time data collected by the smart meters [12]. Consumer participation in load management is essential [15]. Hence, some utility companies offer in-home displays or web portals for their customers so they can access their hourly consumption, their contribution to CO_2 emissions, and the consumption prices [2,15]. This way, the consumer can identify when it is best to use energy devices and be informed about his or her effect on the environment [2]. Also, the consumer can feel the benefits of equal distribution of the load economically [2,15]. Furthermore, with variable pricing schemes, the user can use the time with low energy prices to charge their energy storage devices, then use that power later at high energy price periods [16]. This makes energy storage devices an important part of the smart grid [16]; hence, they are discussed in the next section.

6 Energy storage

Energy storage can play an important role in resolving the previously explained issues with renewable energy sources and load management because it can help balance the load, therefore enhancing the system performance and reliability [16]. Several countries, such as the United States, Germany, and Japan, are planning to expand their energy storage facilities [16]. Currently, most of the storage units exist as large pumped-hydro facilities, or compressed air energy storage (CAES) [16]. Both technologies use the low power demand periods to charge the system (pumping water for the pumped-hydro plants, and compressing air for the CAES), then use it again when the demand is at a critical level [16].

Several energy storage technologies such as flywheel, compressed gas, CHP, super capacitors, and batteries can also be used [14,16]. Due to their fast acting time, batteries, flywheels, and super capacitors have been the main focus for distributed storage systems [16]. However, when considering the growing demand variability imposed by the integration of wind and solar power systems and their economical advantages, batteries appear to be the technology with the highest potential to meet the industry requirement [12,13]. Battery systems have been developed and used for decades in Japan and the United States [16].

As the importance of energy storage is increased to meet the challenges imposed by integrating variable types of energy sources and loads, several concepts and ideas have emerged, such as community energy storage (CES) [16]. The concept is to install relatively small energy storage units at the distribution side to protect the supply at the consumer's side [16]. This is considered as a deviation from the traditional control philosophy [16]. The concept can help meet the customer demand as new electronics loads are added to the load profile [16]. In addition, as more users start installing photovoltaics (PV) panels on the roofs of their houses, CES can handle the energy flow back to the system when the supply exceeds the demand for consumers and use it again when the demand is increased [16]. Positioning the CES units near the consumer side will help capture the excess power and reuse it for the same consumers with minimal transmission losses [16]. Finally, the CES units can help guard against voltage fluctuations due to a cloud passing by Roberts and Sandberg [16]. When the clouds shade a huge number of PV panels, the voltage will rapidly drop, and as the sun reappears the voltage will increase [16]. The fast response of the electronic devices associated with the CES unit will be able to manage those fluctuations without affecting the continuity of supply [16].

Another concept is the utilization of plug-in electric vehicles (PEVs) to support the grid at high demand periods [2]. As the technology develops, the benefits of PEV will outweigh their cost, which will increase the share of PEVs in the market [2]. This increase can either cause an additional stress to the grid or can be useful in supporting the grid [12,16]. If the charging and discharging patterns are well managed, the PEV can be charged when the demand is low [2,12], and then supplied back to the system when the power demand is at peak value [2]. However, this concept needs further improvement on the power storage elements used for the car [16].

So far, the smart grid definition has been introduced and the different elements of a smart grid are explained. In the following sections, the characteristics of the smart grid are introduced and discussed.

7 Self-healing

The current grid philosophy only responds to limit the effect of any fault cascading through the grid and focus on protecting the assets [17]. In a smart grid, the system is expected to perform online assessments and respond to signals from its sensors [17]. The assessment is used to automatically restore the affected components or sectors [17]. The system will use its communication capabilities to analyze undesirable conditions and take the appropriate action [17]. The self-healing feature can help identify the issue before it escalates, therefore maintaining the system stability [17]. It can also help with handling problems too large or too fast-moving for human intervention [17]. To equip the system with self-healing capabilities, the system should be supported with technologies that help stabilize it, such as flexible alternating current transmission system and wide area measurement system [17]. In addition, new protections, communications, and control elements are required to sense the circuit parameters, isolate faults, and automatically restore the service [17]. Implementing the self-healing abilities will help not only with supporting the grid reliability, security, and power quality but it will also significantly reduce the average outage duration, saving costs for the industry and customers [17].

8 Customer active participation

Currently, the customers are uninformed about their hourly consumption trends [17] as the only information they receive is the monthly electricity bill [2]. As a month period is too long, it is almost impossible for the customer to understand how their consumption affects the grid conditions or the electricity cost [2]. Implementing real-time devices and information elements will help enhance the customer's understanding and participation [2,17]. An informed customer can use the information to modify their consumption pattern to minimize their cost and maximize their benefits [17]. In a smart grid, the customer's active participation will help relieve the grid from congestion and reduce the net cost to the customer [17]. As the utility uses the enhanced measurement equipment to provide real-time pricing algorithms, the customer can utilize this data to modify their consumption either manually or automatically by implementing computer programs to act accordingly [17]. The customer's response to the pricing signals can help provide a flatter energy consumption profile and hence reduce their cost along with their environmental impact [17]. Finally, the utility can benefit from the flat consumption profile to use its assets more efficiently [17].

9 Security

The electric grid should incorporate solutions to guard it against physical and cyber threats [17]. The current infrastructure is highly vulnerable to terror attacks and natural disasters [17]. For example, in 2008, severe damage was caused to the power system in China due to continuous snowfall. This resulted in the loss of the power supply to 170 cities, paralyzing 13 provinces and leaving 2018 substations dysfunctional [18]. Therefore, the smart grid design and operation should minimize the consequences of such disasters to speed up service restoration [17]. In addition, the increased integration of communication technologies has raised concerns about privacy, and cyber-security [19]. Such concerns include, but are not limited to, hackers attacking the network to either manipulate the consumption data and prices, fake consumption data and overload the grid, manipulate the electric devices of the consumer, or use the energy consumption data to learn the victim's daily activities [19]. To guard against natural disasters or terrorist attacks, designers should anticipate the system's weak points and study different scenarios during system planning [17]. In addition, cyber-security can be enhanced by utilizing technologies such as authorization, authentication, encryption, and intrusion detection [17]. Ensuring the grid security enhances its reliability by reducing system vulnerability, and minimizes the consequences of any disturbance [17].

10 Power quality

The power quality is affected mainly by the harmonics, imbalance, sags, and spikes introduced by the distribution system [17]. Therefore, electronic devices connected to the system can highly damage the grid's power quality [17]. Low power quality severely damages equipment life expectancy and grid reliability [17]. It also causes financial distress to the utility companies by increasing the losses, which need to be compensated by increasing generation [17]. The current electric grid utilizes a great number of devices to maintain power quality, such as voltage regulators, capacitor banks, and transformers [2]. However, those devices are not equipped with two-way communication facilities, and hence act in an isolated manner [2]. By adding the communication capabilities to those devices, the operators can adjust the settings of the devices to enhance the power quality of the overall system [2]. Furthermore, the hourly information measured by the smart sensor can help route the electricity based on system conditions [2]. Enhancing the power quality will reduce the operation costs and system downtime, resulting in increased customer satisfaction [17].

11 DG and storage

The current power plants are bulk units located far from consumption points [3]. In addition, the distribution grid is designed for one-way power flow [2]. The smart grid aims to shift the generation philosophy to a more decentralized model by including a variety of small distribution sources closer to the consumer [17]. Integration of renewable energy sources in DG units offers an environmentally friendly alternative to fossil fuel-based generation units [3]. This integration along with the various energy storage units can be located almost anywhere [12]. The DG and storage units offer many benefits to the system, including cost reductions, increased system capacity, and increased tolerance to threats and attacks [17]. However, implementing the DG concept requires enhanced communication capabilities [13]. Therefore, the smart grid's two-way communication enables monitoring and controlling the energy flow in a timely manner [2].

12   Efficient operation

It has always been challenging to design and operate electricity networks capable of matching the supply and demand [20], due to the limited integration of operational data [17]. The current grid infrastructure is designed based on the peak load requirement [20]. Therefore, a considerable amount of the substation equipment is only used for a few hours [20]. Hence, the data obtained from the implemented smart meters in the smart grid can be utilized for better grid operation [17,20]. The obtained data along with demand-side management techniques will help balance the supply and demand [21]. Furthermore, the real-time data collected helps create a predictive maintenance program where the equipment is only maintained when needed, hence removing the extra cost of implementing a preventive maintenance strategy and reducing the cost [17]. Implementing software to manage the system and send alerts to the operators or commands to the different equipment to automatically modify its settings based on the network conditions will reduce the frequency of human errors and enhance the efficiency and decision-making process [17].

13   Summary

The increased energy demand and environmental concerns in addition to the aging grid infrastructure have been the driving force toward a new grid philosophy called "the smart grid" [9]. The new infrastructure needs to be efficient, reliable, economic, secure, and sustainable [17]. To achieve the mentioned goals, smart meters equipped with advanced two-way communication technology are essential for providing real-time data. The information received from the smart meters can be utilized to enhance the reliability and security of the grid by allowing the integration of renewable energy sources and distributed storage units into the system, forming a DG philosophy. Furthermore, several software programs can benefit from the gathered data either to automatically modify the settings of the grid equipment to meet the demand requirements or to identify weak points in the system. Furthermore, the equipment maintenance cost can be significantly reduced by developing a predictive maintenance program to reduce the routine maintenance frequency. Finally, by the implementation of demand-side management and pricing programs, the consumer can actively participate in producing a flat energy consumption profile to reduce his or her total electricity bill.

The customer's active participation in controlling the energy demand does not necessary mean they have to worry about when to use their electric appliances [10]. Instead, software programs can be implemented to receive the data and act accordingly. In addition, the grid hardware and software will be designed as plug-and-play devices [17]. In addition, Rahman [10] is suggesting a scenario where companies will offer a service to the consumer to sell and buy energy on their behalf, providing them with the benefits of smart grids without having to deal with the associated issues.

References

[1] J. Berst, Smart grid 101, Smart Grid News (2011). Available from: https://www.smartgrid. gov/files/Smart_grid_101_Everything_you_wanted_to_know_about_grid_mode_201112.pdf.
[2] CADMUS group, Smart Grid 101 for Local Governments, Public Technology Institute, 2011.
[3] M. Wolsink, The research agenda on social acceptance of distributed generation in smart grids: renewable as common pool resources, Renew. Sust. Energ. Rev. 1364-0321,16 (1) (2012) 822–835, https://doi.org/10.1016/j.rser.2011.09.006. Available from: http://www.sciencedirect.com/science/article/pii/S1364032111004564.
[4] J. Gao, Y. Xiao, J. Liu, W. Liang, C.L.P. Chen, A survey of communication/networking in Smart Grids, Future Gener. Comput. Syst. 0167-739X, 28 (2) (2012) 391–404, https://doi.org/10.1016/j.future.2011.04.014. Available from: http://www.sciencedirect.com/science/article/pii/S0167739X11000653.
[5] V.C. Gungor, et al., Smart grid technologies: communication technologies and standards, IEEE Trans. Ind. Inf. 7 (4) (2011) 529–539, https://doi.org/10.1109/TII.2011.2166794 URL. http://ieeexplore.ieee.org/stamp/stamp.jsp?tp=&arnumber=6011696&isnumber=6056501.
[6] Americans for a Clean Energy Grid, Utility modernization, in: Energy Future Coalition, 2009. Available from: http://energyfuturecoalition.org/our-campaigns/utility-modernization/.
[7] 110th Congress, "Energy Independence and Security Act of 2007", U.S. Government Printing Office, December, 2007. Available from: https://www.gpo.gov/fdsys/pkg/PLAW-110publ140/html/PLAW-110publ140.htm.
[8] P. Siano, Demand response and smart grids—a survey, Renew. Sust. Energy Rev. 1364-0321, 30 (2014) 461–478, https://doi.org/10.1016/j.rser.2013.10.022. Available from: http://www.sciencedirect.com/science/article/pii/S1364032113007211.
[9] V.C. Gungor, B. Lu, G.P. Hancke, Opportunities and challenges of wireless sensor networks in smart grid, IEEE Trans. Ind. Electron. 57 (10) (2010) 3557–3564, https://doi.org/10.1109/TIE.2009.2039455. Available from: http://ieeexplore.ieee.org/stamp/stamp.jsp?tp=&arnumber=5406152&isnumber=5567234.
[10] S. Rahman, Smart grid expectations [In my view], IEEE Power Energy Mag. 7 (5) (2009). 88, 84–85, https://doi.org/10.1109/MPE.2009.933415. Available from: http://ieeexplore.ieee.org/stamp/stamp.jsp?tp=&arnumber=5208439&isnumber=5208405.
[11] S.S.S.R. Depuru, L. Wang, V. Devabhaktuni, N. Gudi, in: Smart meters for power grid—challenges, issues, advantages and status, 2011 IEEE/PES Power Systems Conference and Exposition, Phoenix, AZ, 2011, pp. 1–7, https://doi.org/10.1109/PSCE.2011.5772451.

Available from: http://ieeexplore.ieee.org/stamp/stamp.jsp?tp=&arnumber=5772451&
isnumber=5772439.

[12] K. Moslehi, R. Kumar, in: Smart grid—a reliability perspective, *2010 Innovative Smart Grid Technologies (ISGT)*, Gaithersburg, MD, 2010, pp. 1–8, https://doi.org/10.1109/ISGT.2010.5434765. Available from: http://ieeexplore.ieee.org/stamp/stamp.jsp?tp=&arnumber=5434765&isnumber=5434721.

[13] H. Kanchev, D. Lu, F. Colas, V. Lazarov, B. Francois, Energy management and operational planning of a microgrid with a PV-based active generator for smart grid applications, IEEE Trans. Ind. Electron. 58 (10) (2011) 4583–4592, https://doi.org/10.1109/TIE.2011.2119451. Available from: http://ieeexplore.ieee.org/stamp/stamp.jsp?tp=&arnumber=5720519&isnumber=6005674.

[14] H. Lund, A.N. Andersen, P.A. Østergaard, B.V. Mathiesen, D. Connolly, From electricity smart grids to smart energy systems—a market operation based approach and understanding, Energy 0360-5442, 42 (1) (2012) 96–102, https://doi.org/10.1016/j.energy.2012.04.003. Available from: http://www.sciencedirect.com/science/article/pii/S0360544212002836.

[15] P. Samadi, A.H. Mohsenian-Rad, R. Schober, V.W.S. Wong, J. Jatskevich, in: Optimal real-time pricing algorithm based on utility maximization for smart grid, 2010 First IEEE International Conference on Smart Grid Communications, Gaithersburg, MD, 2010, pp. 415–420, https://doi.org/10.1109/SMARTGRID.2010.5622077. Available from: http://ieeexplore.ieee.org/stamp/stamp.jsp?tp=&arnumber=5622077&isnumber=5621989.

[16] B.P. Roberts, C. Sandberg, The role of energy storage in development of smart grids, Proc. IEEE 99 (6) (2011) 1139–1144, https://doi.org/10.1109/JPROC.2011.2116752. Available from: http://ieeexplore.ieee.org/stamp/stamp.jsp?tp=&arnumber=5768106&isnumber=5768087.

[17] National Energy Technology Laboratory (NETL), A System View of the Modern Grid, vol. 2, U.S. Department of Energy, 2006. Available from: https://www.smartgrid.gov/files/Systems_View_Modern_Grid_200712.pdf.

[18] Y. Yu, J. Yang, B. Chen, The smart grids in China—a review, Energies 5 (2012) 1321–1338, https://doi.org/10.3390/en5051321. Available from: http://www.mdpi.com/journal/energies.

[19] F. Li, B. Luo, P. Liu, in: Secure information aggregation for smart grids using homomorphic encryption, 2010 First IEEE International Conference on Smart Grid Communications, Gaithersburg, MD, 2010, pp. 327–332, https://doi.org/10.1109/SMARTGRID.2010.5622064. Available from: http://ieeexplore.ieee.org/stamp/stamp.jsp?tp=&arnumber=5622064&isnumber=5621989.

[20] C. Chen, S. Kishore, L.V. Snyder, in: An innovative RTP-based residential power scheduling scheme for smart grids, *2011 IEEE International Conference on Acoustics,* Speech and Signal Processing (ICASSP), Prague, 2011, pp. 5956–5959, https://doi.org/10.1109/ICASSP.2011.5947718. Available from: http://ieeexplore.ieee.org/stamp/stamp.jsp?tp=&arnumber=5947718&isnumber=5946226.

[21] F. Li, et al., Smart transmission grid: vision and framework, IEEE Trans. Smart Grid 1 (2) (2010) 168–177, https://doi.org/10.1109/TSG.2010.2053726. Available from: http://ieeexplore.ieee.org/stamp/stamp.jsp?tp=&arnumber=5535240&isnumber=5552162.

Part One

Asia

Iranian smart grid: road map and metering program

2

Gevork B. Gharehpetian*, Mehdi Salay Naderi*, Hadi Modaghegh[†],
Alireza Zakariazadeh[‡]
*Amirkabir University of Technology, Tehran, Iran, [†]Ministry of Energy, Tehran, Iran,
[‡]University of Science and Technology of Mazandaran, Behshahr, Iran

Abbreviations

AMI	advanced metering infrastructure
CAS	central access system
CIS	Customer Information System
DA	distribution automation
DC	data concentrator
DER	distributed energy resources
DMS	distribution management system
DR	demand response
EC	European Commission
EMS	energy management system
FA	feeder automation
FAHAM	National Smart Metering Program in Iran
GIS	geographic information system
GPRS	General Packet Radio Service
HVDC	high voltage direct current
IEA	International Energy Agency
IEC	International Electro Technical Commission
ISGC	Iran Smart Grid Company
JRC	Joint Research Center (of EU Commission)
MDM	meter data management
NOC	Network Operation Center
OMS	outage management system
PLC	power line carrier
SCADA	supervisory control and data acquisition
SG	smart grid
SOC	Security Operation Center
VVC	volt-var control
WAMS	wide area monitoring systems
UML	unified modeling language

Application of Smart Grid Technologies. https://doi.org/10.1016/B978-0-12-803128-5.00002-7

1 Smart grid technology roadmap in Iran

1.1 Introduction

Smart grid development in different countries is affected by six major drivers:

* Economic competitiveness
* Knowledge-based development
* Enhancement of interaction with customers
* Energy security
* Enhancement of grid performance
* Increasing grid reliability and sustainable development

In Iran, the mentioned drivers motivate technology development and the implementation of smart grids with different degrees of importance. To develop the technology development roadmap of the Iran smart grid, we first need to recognize the details of smart grid technology and then select a reliable reference among numerous references to maintain the consistency and integrity of the programs and actions in the roadmap.

For this purpose, the European Union (EU) has defined a smart grid platform. Most European countries have used this reference for their smart grid development programs. Also, it is possible for different countries with diverse economies, technology, and social circumstances to use this methodology. Its information is available and provides the ability to compare the countries.

This document is prepared to accomplish the Iran smart grid roadmap project, which is one of the projects of the Iran Smart Grid National Grand Project. This roadmap focuses on the technology development; the deployment roadmap should be written as well.

To prepare the technology development roadmap of the Iran smart grid, two recommended documents of the EU commission have been used:

– The Smart Grid Architecture Model (SGAM), which introduces interoperability aspects and how they are taken into account via a domain, zone, and layer-based approach. The SGAM is a method to fully assign and categorize processes, products, and utility operations and align standards to them.
– The JRC (Joint Research Center) method, which has a management-technical concept. The major purpose of the JRC method is presenting eight steps to evaluate the cost-benefit of smart grid deployment.

To prepare the technology development roadmap of the Iran smart grid, the first smart grid and its technologies and areas are investigated. Then, the Iran smart grid vision was outlined according to expert opinions and upstream national documents in the smart grid field as well as comparative studies. Based on this vision, policies, actions, strategies and, finally, the technology development roadmap are extracted.

The technology development roadmap contains three parts: the development approach, prioritizing technologies, and determining the technology acquisition method. Considering the variety of smart grid technologies, the International Energy Agency (IEA) standard has been used to categorize smart grid technologies. Based on this standard, smart grid technologies are categorized into eight groups. Fig. 1 demonstrates the technology groups of smart grids in different domains of the SGAM

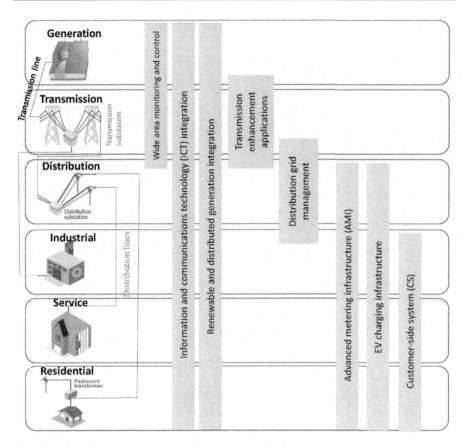

Fig. 1 Smart grid technology groups in the IEA standard.

standard. For each IEA category, the development approach and method of technology acquisition have been extracted separately.

Subgroups of each technology area in the IEA standard in hardware, software, and systems sections are expressed in Table 1 in detail.

Prioritization of technologies for each IEA group has been performed based on an attractiveness-capability matrix as well as the opinions of experts. To obtain the attractiveness-capability matrix for all smart grid technologies, a structured questionnaire form has been prepared with two sections. In the first section, the maturity level of each technology in nine levels was asked; in the second section, the attractiveness of technologies considering seven criteria was investigated. These criteria are employment, number of competitors, new markets, added value, a key technology, application in other industries, and the position in the life cycle of technology. Then the attractiveness-capability matrix was extracted.

In terms of time, the attractiveness-capability matrix is divided into three groups. According to the position of each technology in the attractiveness-capability matrix,

Table 1 **Subgroups of the IEA standard for smart grid technologies**

Technology area	Hardware	Systems and software
Wide-area monitoring and control	Sensor equipment, phasor measurement units (PMU)	Supervisory control and data acquisition (SCADA), wide area monitoring systems (WAMS), control and automation (WAAPCA), wide area adaptive protection, wide area situational awareness (WASA)
Information and communication technology integration	Communication equipment (power line carrier, WIMAX, LTE, RF mesh network, cellular), routers, relays, switches, gateway, computers (servers)	Enterprise resource planning software (ERP), customer information system (CIS)
Renewable and distributed generation integration	Power conditioning equipment for bulk power and grid support, communication and control hardware for generation and enabling storage technology	SCADA, geographic information system (GIS), energy management system (EMS), distribution management system (DMS)
Transmission enhancement	Superconductors, FACTS, high-voltage direct current (HVDC)	Network stability analysis, automatic recovery systems
Distribution grid management	Automated reclosers, switches and capacitors, remote-controlled distributed generation and storage, transformer sensors, wire and cable sensors	Geographic information system (GIS), DMS, outage management system (OMS), workforce management system (WMS)
Advanced metering infrastructure	Smart meter, in-home displays, servers, relays	Meter data management system (MDMS)
Electric vehicle charging infrastructure	Charging infrastructure, batteries, inverters	Smart grid-to-vehicle charging (G2V) and discharging vehicle-to-grid (V2G) methodologies, energy billing
Customer-side systems	Smart appliances, routers, in-home display, building automation systems, thermal accumulators, smart thermostat	Energy dashboards, EMS, energy applications for smart phones and tablets

the priority of development in short, mid and long-term periods can be determined in Fig. 2 for each technology. Fig. 3 shows the IEA standard groups position in the attractiveness-capability matrix for Iran.

Based on the basic features of the Iran smart grid technologies groups (according to the IEA standard and SGAM method), the appropriate method of technology acquisition for each group has been determined. Among a number of valid traditional methods for technology acquisition, the Chiesa (1998) and Narula (2001) methods have been selected. The accuracy and widespread use of these methods were the reasons of this selection. To apply these methods, expert opinions have been collected and analyzed by using a detailed questionnaire form. Comparing the results of these two methods makes the technology acquisition method reliable for each technology.

To obtain a technology development policy, Technology Innovation System (TIS) and Multilevel Perspective models have been employed. Using event logging, measures in the eight IEA standard groups have been analyzed from the TIS perspective. In order to develop more effective actions and strategies in the smart grid roadmap, it is necessary to investigate and classify the players of the Iran smart grid. Fig. 4 represents Iran smart grid players and their relationships in the smart grid ecosystem.

To prepare the roadmap, all the lessons learned in comparative studies, background or benchmarking studies, frameworks, validated methodologies, and upstream national documents and drivers have been used to compile the Iran smart grid road map. Fig. 5 shows the stages of technology and the business development road map of the Iran smart grid.

Fig. 6 shows the fundamental components of the Iran smart grid technology development roadmap. The methodology and components of the roadmap are based on legislation approved in the 18th Supreme Council for Science Research and Technology meeting in January 2016. In the next sections, details of the Iran smart grid technology development roadmap will be described.

1.2 Economic, social, and environmental requirements of smart grid development

I. Enhancement of performance and reliability of the transmission and distribution network.
II. Reduction of electrical losses in transmission and distribution systems.
III. Best usage of existing capacities and, as a result, investment postponement for new generation, transmission, and distribution system development.
IV. Optimal asset management and operational cost savings.
V. Reduction of peak demand.
VI. Optimal management of energy consumption.
VII. Development of DGS Infrastructure, including CHP systems and renewable resources.
VIII. Enhancement of stability and security of energy.
IX. Offering better services and raising consumer satisfaction.
X. ICT technologies and infrastructure development.
XI. Reducing environmental pollution emissions.

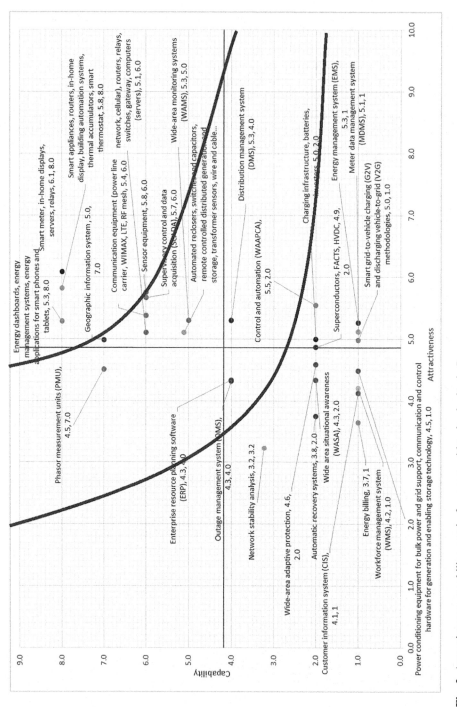

Fig. 2 Attractiveness-capability matrix for smart grid technologies in Iran.

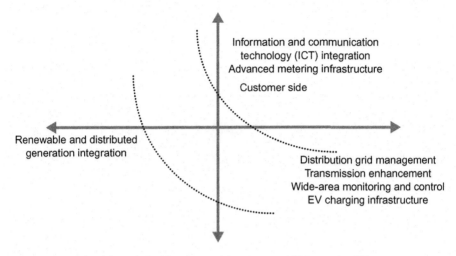

Fig. 3 IEA standard groups position in attractiveness-capability matrix of Iran.

1.3 Values

Iran smart grid policy-making values are based on the following principles:

   I. The belief in the values of science and knowledge and the utilization of human resources.
  II. Economic use of resources and preventing their waste.
 III. Optimal use of resources and facilities.
 IV. Put effort into foresight and risk acceptance.
  V. Observance of fairness in development of all districts of the country.
 VI. Service delivery and raising social satisfaction.
VII. Preserving the rights of the next generation.
VIII. Protecting and preventing degradation of the environment.
 IX. Protecting the private sanctum of citizens.
  X. Boosting mentality of social partnerships and accepting the responsibility of people.

1.4 Vision of Iran smart grid

By 2025, the Islamic Republic of Iran aims to develop an electric smart grid as an efficient, secure, flexible and stable grid that delivers required high quality and reliable power to consumers and stakeholders. ICT, smart management systems, new technologies in the area of smart grids, IoT and integration of DGs, CHP systems, renewable energy resources, and energy storage systems cause dynamic interactions between stakeholders of the whole energy system. The smart grid provides optimal management of demand and supply in a competitive electricity market. Iran seeks to elevate and consolidate their position as the first country in the Middle East in technology development and the implementation of a smart grid.

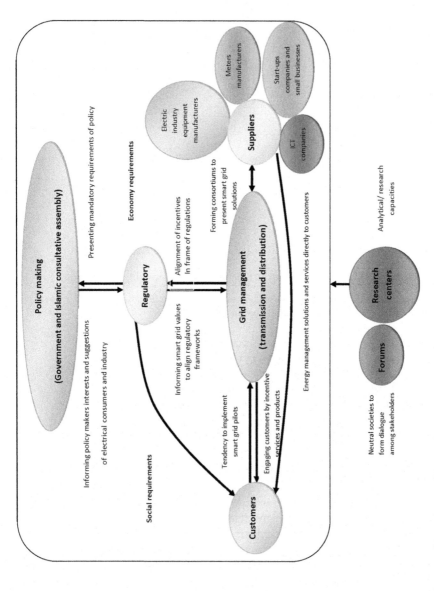

Fig. 4 Ecosystem of Iran smart grid players.

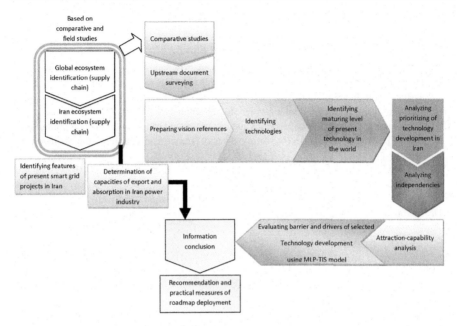

Fig. 5 Stages of technology and business development of the road map of the Iran smart grid.

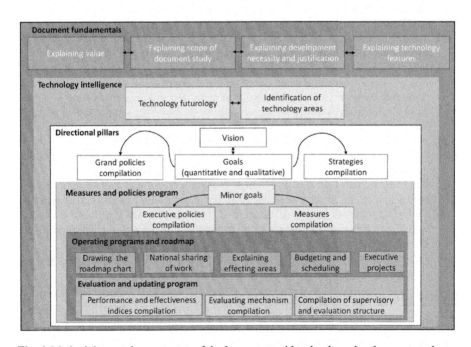

Fig. 6 Methodology and components of the Iran smart grid technology development roadmap.

1.5   Grand policies

The macro policies of the Iran smart grid technology development are as follows:

   I. Attention to human capital as a source of competitive advantage and value added.

   II. Focus on policy-making with a national and multisectorial approach and coordinated and systematic implementation.

   III. Attention to maximize the added value due to smart grid technology development.

   IV. Maximum use of national capabilities and capacities and support of domestic production.

   V. Giving priority to private-sector participation and maximizing participation of citizens to achieve goals.

   VI. Emphasizing international cooperation, engagement, and participation in the systematic development of smart grid technology.

   VII. Building a public culture to expand the application of smart grid technology.

   VIII. Enabling consumers to efficiently manage energy consumption.

1.6   Grand goals

According to the vision of the Iran smart grid and its technology development and considering upstream documents, the goals of the Iran smart grid technology development are listed below:

   I. Achieving the technology cycle of smart grid equipment manufacturing.

   II. Presence among well-known manufacturers and achieving the first position of the region in domestic, competitive industrial products of smart grid equipment, especially smart meters, smart energy management systems (MDMS), building management systems (BMS), and grid monitoring, control and protection systems.

   III. Meeting the domestic needs of smart meter production by 2025. About 70% of installed smart meters must be made in Iran until that time.

   IV. Increasing export capacity of engineering services and smart grid equipment. At least 20% of domestic smart meter products must be exported.

   V. Achieving domestic technical knowledge of meter data management system (MDMS).

   VI. Achieving technical knowledge and self-reliance on the design and production of hardware and software of energy management systems, BMS, smart homes, and IoT.

   VII. Self-reliance on smart grid ICT and data center infrastructure.

   VIII. Self-reliance on design and production of software and implementation of grid monitoring and control center.

   IX. Achieving technical knowledge of design and production of local and wide area monitoring, control, and protection systems, distribution automation, and SACADA systems.

   X. Technology development of load management to reduce electricity cost and damage due to grid contingencies and improve grid radiance to supply sensitive consumers and export commitments.

   XI. Achieving first position in the Middle East and being among the top five Asian countries in science and paper publications.

1.7   Technology development, strategies, and measures

Table 2 presents actions and strategies to achieve expected deliveries and outputs corresponding to the funds allocated to different areas of smart grid technology development. It has two columns representing deliveries and outputs corresponding to the funds supported by Iran's Electric Smart Grid National Grand Project (IESGNGP) and other resources.

Table 2 Deliveries and outputs of goals and actions

No.	Areas of measures	Outputs expected in national smart grid document and financially supported by:	
		IESGNGP	**Other resources**
1	Smart meter development and related technologies	– Compilation of smart meter domestic comprehensive plan – Compilation of national standard of smart meters and related systems – Conducting research project in the field of smart meters and communications systems security – Supporting academic projects and theses in the field of smart meter data utilization	– Development and commercialization of smart meter technology and its related technologies – Development of smart meter data management systems (MDMS) – Development of meter local data concentrators hardware and local and regional area networks – Establishment of smart meter research centers (data generation and utilization technology development) – Establishment of smart meters quality test labs – Definition of projects for utilizing smart meter data
2	Customer side technology development	– Supporting customer side technology development – Supporting smart home appliances technology development – Project definition for classification and analysis of customer group behavior – Designing a variety of tariffs based on customer group lifestyles – Supporting start-up companies in the field of customer side technologies	– Commercialization and development of customer side technologies: smart homes systems, building energy management, energy consumption management, energy management portals and dashboards – Manufacturing key hardware of smart homes and EMS including routers, data concentrators, metering equipment, and control devices such as smart thermostats – Development of interaction tools with customers

Continued

Table 2 Continued

No.	Areas of measures	Outputs expected in national smart grid document and financially supported by:	
		IESGNGP	**Other resources**
		– Promotion programs: introducing businesses in customer side section and encouraging hardware and software companies to enter the market of smart homes and customer side technologies – Compilation of obligatory regulations for governmental organizations, companies, and offices to implement building EMS	– Development of system for customer services – Development of home appliance technologies – Offering facilities to venture capital companies to encourage them to enter customer side technologies – Providing incentives for application of smart homes and building EMS
3	Development of ICT technologies and communications infrastructures	– Compilation of rules and standards for smart grid communication and interoperability – Compilation of rules and standards for smart grid software and hardware – Research project definition on smart grid data security and cyber attacks – Promotion of business opportunities and encouraging communication operators to: (1) produce mobile software in the field of smart grids, (2) manufacture equipment installed in smart grids, (3) develop IOT technologies, and (4) attend smart grid communications market	– Producing domestic software and systems either on the customer side or in grid management – Producing mobile software, devices, and equipment as well as devices connected to grid and IOT – Implementation of pilot projects of data transfer access network using available and diverse technologies
4	Smart grid technology development in distribution and transmission systems	– Development of monitoring and control systems in microgrids – Research projects on effects of home appliance remote control	– Development of distribution automation systems – Development of DMS

5	WAMS development	– Describing smart cities/grids infrastructure standards – Studies on demand response (DS) advanced technologies – Technology development of domestic WAMS – Technology development of sensors and measuring devices – Development of domestic SCADA systems – Technology development and fulfillment of devices and hardware for domestic PMU production	– Knowledge and technology development of FACTS devices – Development and commercialization of WAMS – Commercialization and manufacturing of equipment and hardware for measuring devices and sensors – Development and commercialization of domestic SCADA systems – Commercialization and manufacturing equipment and hardware of PMUs
6	Grid monitoring and control technology development	– Compilation of strategic plan for technology development of control and management of grid – Compilation of technical specifications for grid domestic monitoring and control center – Designing software for grid monitoring and control center – Designing practical power system software	– Producing monitoring software – Producing practical power system and control software – Implementation of domestic center for grid monitoring and control
7	Grid protection technology development	– Supporting domestic local load protection systems technology development – Supporting domestic wide area protection systems technology development – Supporting establishment of domestic wide area protection systems	– Development and commercialization of domestic local protection systems at national and international levels – Producing and commercialization of network simulation and analyzing tools on national and international levels

Continued

Table 2 Continued

| No. | Areas of measures | Outputs expected in national smart grid document and financially supported by: | |
		IESGNGP	Other resources
		– Supporting development of grid events simulation and analysis tools – Establishment of national supervision structure for protection knowledge and technology development – Supporting domestic network protection systems export – Supporting projects and theses in the field of protection systems design and implementation	– Development and commercialization of domestic wide area protection systems
8	Infrastructure for development of DGs and renewable energy resources integration	– Supporting plans and implementation of technology development projects for DGs and renewable energy resources integration – Implementation of MV and LV small-scale pilot projects for integration of DGs	– Implementation of large-scale MV and LV pilot projects for integration of DGs
9	Electric vehicle infrastructure and novel technology development	– Technology development and manufacturing electric vehicle infrastructure equipment and devices – Implementation of small-scale electric vehicle charging pilot project	– Technology development and commercialization of manufacturing electric vehicle infrastructure equipment and devices – Implementation of large-scale electric vehicle charging pilot project

Continued

10	Support of knowledge based start-up businesses	– Establishment and support of smart grid incubators – Compilation of comprehensive plan of support of knowledge-based start-up businesses (venture capital companies and business accelerators)	– Implementation of comprehensive plan of support of knowledge-based start-up businesses (venture capital companies and business accelerators) – Offering facilities to venture capital companies to attend smart grid business market – Supporting export of technical and engineering services domestic fabrication – Supporting projects, plans, theses, and academic papers in priority areas of smart grids – Equipping active education centers education of smart grid skills
11	Human resources education and training	– Allocation of research grants to faculties and postgraduate students working on smart grids – Supporting projects, plans, theses, and academic papers in priorities of smart grid – Designing educational and training courses in different academic levels and fields of smart grids (among electrical, computer, industrial engineering, management, etc.) – Supporting universities to mobilize smart grid labs – Need assessment and defining vocational and technical courses as well as skills teaching	
12	Obtaining technology development infrastructure and smart grid lab development	– Compilation of legal requirements for generation, transmission, and distribution companies to use domestic products of smart grid technologies – Providing rules and procedures to support intellectual property rights in research and development projects	– Establishment of smart grid labs – Implementation of practical mechanisms to cover technology development risks and production and consumption of domestic technologies

Table 2 Continued

No.	Areas of measures	Outputs expected in national smart grid document and financially supported by:	
		IESGNGP	**Other resources**
		– Compilation of comprehensive and consistent smart grid standards	
		– Compilation of procedures and supporting establishment of executive mechanisms to cover risks of technology development, production, and consumption of domestic technologies	
		– Supporting establishment of smart grid labs in universities	
		– Design and implementation of reference smart grid labs in generation, transmission, distribution, and consumer levels	
13	Culture making and promotion of smart grid applications	– Introducing opportunities in smart grids and governmental support to electric energy suppliers and actors	– Culture making and introducing smart grid systems and their benefits to users and encouraging them to apply for them
		– Programs to introduce smart grid systems, opportunities, plans, and national incentives for domestic products to power industry producers	
		– Introducing opportunities and country energy manager programs to ICT sector players	

| 14 | Implementation of pilot projects | – Introducing programs to support knowledge-based start-up companies and cooperating with main new business development players such as incubators, business accelerators, venture capital companies, etc., in order to develop and support ideas and businesses in smart grids, especially on customer side
– Introducing benefits of smart grid and IOT employment to electricity consumers
– Design and implementation of a number of pilot plans and technical and financial support for some of them to demonstrate performance problems of smart grids in generation, transmission, distribution, and consumer sectors
– Holding smart grid expos to introduce domestic products and developed technologies
– Presentation of pilot project results to researchers and authorities
– Allocation of place for implementation of selected post graduate theses and dissertations | – Implementing plans and pilot projects in different areas of smart grids |

1.8   Financing and resource allocation

Total financial resources required for implementing smart grid technology development in a short period (3 years) is $155 million. This budget is allocated to achieve the goals expressed in the national Iranian smart grid roadmap. The budget resources are:

– $25 million will jointly be allocated by the Ministry of Energy (MOE) and the Supreme Council for Science Research and Technology to the national master plan for Iran smart grid technology development.
– $130 million will be financed by the private sector, which provides different services to customers.

Table 3 presents the various areas of the Iran smart grid technology development road map and the funds allocated to each area. The table lists the allocated funds supplied from each resource mentioned above. In order to meet the goals and vision of the plan in 10 years, the resources for mid- and long-term periods should be revised considering the short-term results.

1.9   Updating and evaluation of the road map

In this section, the updating and evaluation mechanisms for smart grid technology development are presented. To do this, two evaluation methods are used: evaluation indices and evaluation reports, which are described.

1.9.1   Evaluation indices

In Table 4, evaluation indices of research and development programs have been collected based on similar domestic and international literature and documents. In this table, evaluation indices are divided into several groups considering their domain. The name of the group and the indices of each group are defined in Table 4.

1.9.2   Evaluation reports

The other evaluation method of smart grid technology development is based on reports that are generated by different committees and organizations. In this section, the subject of the reports and their descriptions are presented.

(a) *Annual report for assessing the progress of development of technologies based on the stage-gate model*

   The responsible organization to prepare this report is the Research and Technology Committee (RTC). The RTC updates the list of prioritized smart grid technologies according to the technology development process and the country's research and development portfolio results. Reports include:
   –   Assessment of technology research and development portfolio.
   –   Precise evaluation and updating of prioritized technologies (for each subgroup of technology groups).
   –   Forecasting future opportunities and financing needs.

Table 3 Funds allocated to different areas of the Iran smart grid technology development road map

No.	Areas of measure for Iran smart grid technology development	Iran's Electric Smart Grid National Grand Project (IESGNGP) budget ($ million)	Other resources ($ million)
1	Smart meter development and related technologies	0.6	6.5
2	Customer side technology development	1.4	12
3	ICT infrastructures technology development	1.1	3.5
4	Smart grid technology development (distribution and transmission systems)	2.7	3.5
5	WAMS development	1.4	5.5
6	Grid monitoring and control technology development	2.1	9.2
7	Grid protection systems development	1.4	2.5
8	DGs and renewable energy resources integration infrastructure development	1.4	6.5
9	Electric vehicle infrastructure and new technology development	1.4	5.3
10	Knowledge-based start-up businesses development support	5.2	17
11	Human resources education and training	1.4	4
12	Obtaining technology development infrastructure and smart grid lab development	1.4	6.5
13	Culture making and promotion of smart grid applications	0.8	6
14	Implementation of pilot projects	2.7	42
Total budget		25	130

(b) *Roadmap updating report at the end of the year (based on the results of the annual report of technology progress)*

The responsible organization to prepare this report is the RTC. This report will be published at the end of short-, mid-, and long-term periods. The report contents include:

- Evaluation of program deviations.
- Funds allocation mechanism.
- Coordination of commercialization and technology development considering the Iran MOE needs in the implementation phase.
- Performance analysis of smart grid innovation systems.
- Condition assessment and supporting mechanisms for industrial and research groups related to the smart grid.
- Determination of research needs according to the technology development progress reports in the Stage-Gate model.

Table 4 **Evaluation indices of the road map**

No.	Index	Domain
1	Percentage of academic staff working as leaders of the smart grid research project	Human resource
2	Percentage of postgraduate students studying in the smart grid field	
3	Number of visiting professors participating in conferences and workshops held in Iran or publishing joint papers	
4	Number of universities or research centers that are active in at least five smart grid projects	Research centers
5	Number of universities and research centers that are ranked among the top 10% best research centers or universities	
6	Number of ISI-ranked published papers	Scientific
7	Mean citation of each ISI-ranked paper	productions
8	Iranian researchers' share from total published papers worldwide	
9	Number of highly cited papers	
10	Number of patents registered within Iran	Technology
11	Number of patents registered outside Iran	development
12	Commercialization of plans	Commercialization
13	Percentage of yearly growth of GDP per capita due to smart grid science and technology	and industry
14	Share of generation of products and services in GDP obtained from domestic knowledge and technology	
15	Share of value added by industrial products from national production value added	
16	Amount of domestic technology production	
17	Number of businesses with at least 1 product/service	
18	Key player interactions (cofinancing, research and development by public and private sectors, cooperation in R&D projects, participating in local and national scientific and technology programs, outsourcing, participation in registration of patents and publications)	Increasing key player interactions
19	Signing contracts	
20	Share of nongovernmental sector in financing research costs	Financial costs
21	Research expenditures share in GDP	
22	Education costs share in GDP	
23	Income absorption	Financial
24	Organizational incentives (e.g., sharing commercialization income among research groups and universities, a credit and promotion system, industrial incentives for innovative plans)	advantageous
25	Ability to attract research funds, venture capital, and university capital as well as university-based companies	
26	License income, copyright and registration rights, research income, tax income, legal costs	
27	Earned value of development projects	

(c) *Evaluation report of MSC and PhD experts' contributions in R&D and rollout projects at the end of each year*

This report is prepared by RTC and includes:

- Comparison of the planned and allocated budget and assessing goal realization.
- Number of PhD and MSC researchers who contribute in smart grid projects (part-time or full-time).
- Monitoring of PhD and MSc experts' activities in different sections of projects in the Stage-Gate procedure.
- Monitoring of PhD and MSC experts' activities in different levels of smart grid deployment.
- The number and size of projects based on the allocated budget.
- Realization level of objectives of projects.
- The number of registered and commercialized patents.

(d) *Triple reports of project managers (charter, mid-term, and final reports)*

The project managers need to prepare the charter report, the mid-term report, and the final report of implemented projects (based on the PMBOK standard). RTC is the responsible organization receiving these reports. These reports are published in RTC portal of knowledge by the project managers with public or limited access.

- The project charter: This is a brief report, a maximum of one page that contains the project definition, budget, manager, scope, necessary equipment, key stakeholders, and schedules.
- The mid-term report: This report contains a maximum two or three pages in the middle of the project period. The report states the project's progress, existing challenges, risks, and needs of the project. This report also states the interaction status of the project with other smart grid players. The aim of this report is to facilitate the flow of information among key players of the smart grid. This report can be employed as the reference of necessary research titles in the future.
- The final report and lessons learned: This report is a maximum of four pages and contains a real-time schedule of projects related to the initial programs, challenges passed, risks, and measures taken to fix them; it also includes deliverables and outputs identified. This report is an important report for future projects and studies.

(e) *Annual report of research studies and applied technology development projects*

This report lists the studies and projects in the field of smart grids. This list can automatically be extracted from the portal of knowledge or generated separately. The aim of this report is to give information to industry sector players and researchers about applied research activities in Iran.

(f) *Annual accumulated and detailed report of R&D and rollout project budget*

This report is prepared by the smart grid policy and supervision committee. The aim of this report is to publish approved, allocated, and operating budgets of government and organizations that can potentially be spent in smart grid R&D projects.

(g) *Evaluation report of achievements of high-tech pilot and future needs at the end of each year*

The Niroo Research Institute (NRI), which is the research arm of the MOE, prepares this report. The purpose of this report is to inform researchers and orient and prioritize studies. Also, this report aims to coordinate different research institutes and administrative organizations based on the results of the implemented pilot project. This report also specifies the projects that should be carried out in this pilot at the next period.

1.10   Deployment strategy

In the development of the Iran smart grid road map, the JRC methodology has been selected as the analyzing method. Based on the JRC methodology, countries that intend to implement smart grids are required to adopt 10 functionalities in their smart grid. To achieve the stated functionalities, it is necessary to determine the assets of the present electricity grid. To identify the priorities and assets of the Iran power electricity network, experts' viewpoints from industries and universities have been collected using designed forms. Fig. 7 demonstrates the diagram of the Iran electric network priorities, which have been obtained by analyzing the received forms.

Considering the EU requirements and assets and priorities of the Iran smart grid, the deployment of the smart grid is divided into four pillars: customer empowerment, market development, grid development, and governmental institutions. Table 5 presents the pillars of smart grid development, the baseline of the Iran grid and the measures that will be taken through 2020 and 2025.

2   National smart meter program

Iran is located in the Middle East and, as of 2016, has a total population of more than 80 million. More than 99% of the country's population has access to electricity. The power-generation capacity of the Iran grid is more than 76 GW.

In Fig. 8, the percentage of electricity customers in each section has been illustrated. From the 34 million customers of the Iran grid, 35% and 32% are industrial and residential users, respectively. Also, the remaining 16%, 9%, 6%, and 2% are agricultural, public loads, commercial, and lighting, respectively.

Fig. 7 Priorities of the Iran grid for smart grid deployment.

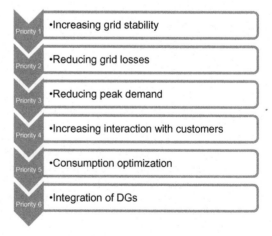

Priority 1 •Increasing grid stability

Priority 2 •Reducing grid losses

Priority 3 •Reducing peak demand

Priority 4 •Increasing interaction with customers

Priority 5 •Consumption optimization

Priority 6 •Integration of DGs

Table 5 **Iran smart grid deployment measures through 2020 and 2025**

Pillar	Base line	Up to 2020	Up to 2025
Customer empowerment	AMI Telecommunication CIS DR	AMI CIS Smart grid (pilot) Smart home (pilot) Storage systems	Smart city Smart home
Market development	Tariff definition Tariff diversification Electricity retailers	Privatization sector Smart operators	
Grid development	WAMS SCADA (transmission)	WAMS Street lightening OMS GIS Volt-Var control (VVC) Asset management system Distribution automation (DA)/feeder automation (FA) Security Operation Center (SOC)/Network Operation Center (NOC) Microgrid (pilot) Integration of distributed energy resources (DER)	HVDC DMS EMS Development and integration of microgrids EV charging infrastructure
Governmental institutions	SG Steering Committee Renewable energy feed in tariff SG regulatory body AMI mandate	Standard definition Iran Smart Grid Company (ISGC) Instructions and regulations completion	

Advanced metering infrastructure (AMI) systems are comprised of state-of-the-art electronic/digital hardware and software that combine interval data measurement with continuously available remote communications. AMI gives the system operator and consumers the information they need to make smart decisions, and also the ability to execute those decisions that they are not currently able to do. Implementation and deployment of the National Smart Metering System project (called FAHAM) in Iran was begun in 2009. The FAHAM project follows promoting energy efficiency and load management, improving system reliability, and reducing operational costs by

Fig. 8 The electricity consumption share per consumer type.

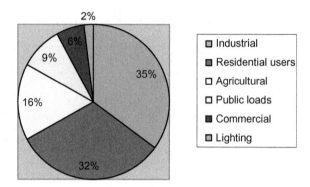

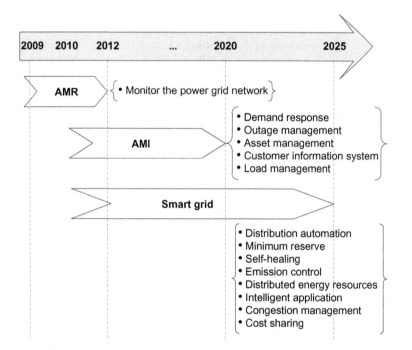

Fig. 9 Iran roadmap for smart grid roll-out.

implementing the smart meter project. Moreover, the FAHAM project plans to replace conventional customer meters with smart meters in order to give consumers greater control over their energy use. Smart meters enable a utility to provide customers with detailed information about their energy usage at different times of the day, which in turn enables customers to manage their energy use more proactively.

In 2009, the Iran Energy Efficiency Organization (IEEO) published the roadmap for smart grid roll-out in Iran, illustrated in Fig. 9. As shown, the first steps were scheduled for implementation and deployment of the AMI.

2.1 Pilot project

To rollout the pilot project, five areas were selected as the main priority. Zanjan, Bushehr, Mashhad, Ahwaz, and Tehran+Alborz are the first distribution companies that will execute bulk rollout. Considering FAHAM action plans for Tehran+Alborz, it should be noted that distribution substations, lighting feeders, loss monitoring purposes, and large and demand customers have been included in the plan. For the other areas, the residential customers have also been included in FAHAM.

Moreover, 300,000 m for large customers (demand customers) in all 39 distribution companies (all around the country) will be connected to FAHAM in parallel with the aforementioned rollout plans. More details about the FAHAM pilot action plan are shown in Fig. 10.

At the conclusion of FAHAM, all 34 million customers of the Iran grid, all distribution substations, and all feeders will be included in the metering infrastructure.

2.2 Goals and benefits of AMI implementation in Iran

The aims of AMI implementation in Iran are given below; they should include a plan for achieving each goal.

- Correcting customer consumption patterns.
- Preparing for complete elimination of subsidies.
- Applying energy management by the network operator in normal and critical conditions.
- Improving meter reading and billing processes.
- Monitoring technical losses as well as reducing nontechnical losses in distribution networks.
- Improving the quality of service and reducing the duration of power interruptions.
- Developing distributed generation and clean energy usage.
- Possibility of electricity presale and establishing electricity retail markets.

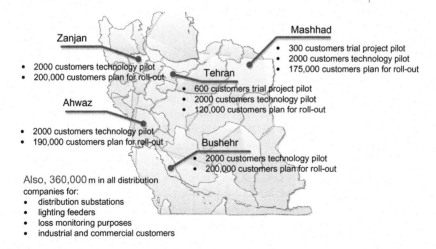

Fig. 10 Selected area for the pilot project.

- Optimizing operation and maintenance costs.
- Providing appropriate management of water and gas meters.

Smart grid development in Iran has several advantages, which are summarized in three areas: economic, social, and environmental benefits:

Economic benefits
- Reducing nontechnical losses.
- Demand side management and diversification of tariffs.
- Modifying electricity consumption patterns by sharing information with customers.
- Improving the billing and payment systems.
- Reducing total costs of meter reading, operation, and maintenance as well as customer disconnection and reconnection.
- Preparing the necessary infrastructure for the development of retail markets.

Social benefits
- No need for periodic visits to each physical location to read the meters.
- Establishment of appropriate services for developing the electronic government.
- Increasing electricity sale options with different prices.
- Power delivery with higher quality and reliability.
- Allows for faster outage detection and restoration of service by a utility when an outage occurs.
- Increasing billing accuracy and speed by eliminating the human error factor.
- Providing better customer service.
- Allows customers to make informed decisions by providing highly detailed information.

Environmental benefits
- Reducing polluted gas and CO_2 emissions.
- Reducing electricity consumption through energy management and reducing network losses.
- Demand-side management through sharing information with customers.

2.3   System components and interfaces

The AMI system in the FAHAM project typically refers to the full measurement and collection system that includes meters at the customer site; communication networks between the customer and a service provider, such as an electric, gas, or water utility; and data reception and management systems that make the information available to the service provider. Fig. 11 shows a typical distribution system including AMI. Data can be provided at the customer level and for other enterprise-level systems either on a scheduled basis or on demand. FAHAM will communicate this data to a central location, sorting and analyzing it for a variety of purposes such as customer billing, outage response, system loading conditions, and demand-side management. FAHAM as a two-way communication network will also send this information to other systems, customers, and third parties as well as send information back through the network and meters to capture additional data, control equipment, and update the configuration and software of equipment. Components and interfaces forming the AMI system in the FAHAM project are illustrated in Fig. 12. The main components are described in the following:

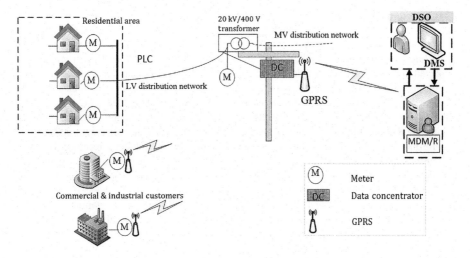

Fig. 11 A typical AMI system.

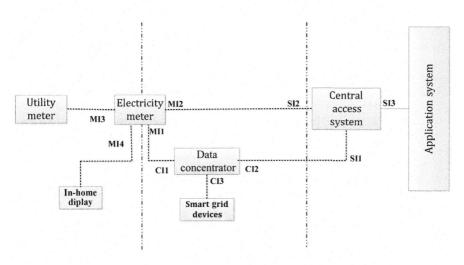

Fig. 12 FAHAM system and interface architecture.

- The *electricity meter/communication hub* is an electronic smart meter device for measuring electricity, that is, an electricity meter. In addition to being a communication hub, it also incorporates additional processing capacities, memory, and communications for the storage and transmission of data received from multiutility meters, if the interface MI3 is present, and the end customer device, if the interface MI4 is present. The central system, either directly through the wide-area communication interface MI2, if present, or indirectly via a concentrator using interface MI1, remotely manages the electricity meter/communication hub.

- The *multiutility meter* is an electronic smart meter device for measuring water or gas with a general communication link to an electricity meter/communication hub using the local interface MI3.
- *End customer devices* are ancillary equipment such as an in-home display, which can be connected to the customer installation in order to permit an interaction and/or display consumption or other information to the customer. The end customer devices communicate to the electricity meter/communication hub using the optional meter interface MI4.
- The *data concentrator (DC)* is an intermediate element between the electricity meter/communication hub and the central system. Its main purpose is to collect and manage the information directly received from the electricity meters and indirectly received from the multiutility meters, if present, as well as also indirectly from the end-customer devices, if present. This information, consisting of measurement registers, alarms, etc., which is collected through interface CI1, is then sent to the central system via the wide-area interface CI2. For control commands, programming, reconfigurations, etc., coming from the central system, the data flow is the opposite direction and reaches the electricity meter/communication hub via CI1-MI1. In addition, there might exist the option of having external devices connected via the local interface CI3.
- *External devices* are other types of equipment that can be connected to the concentrator using the optional interface CI3. They can be used, for example, to permit future smart grid functionality that needs control, monitoring, or sensor elements at the transformer station, which is typically the location where the concentrator is installed.
- The *central access system (CAS)* interfaces with application systems through the interface SI3 and is responsible for the management of all information and data related to smart metering as well as the configuration, control, and operation of all system components using communication via the wide-area interfaces SI1 and SI2. This functionality also includes the treatment of events and alarms and the management and operation of all system communications. In most cases, the central system will receive orders from the application system and has to assure their correct and timely execution, and then returns the result of the operation to the application system. These orders can include reading of different parameters, reconfiguration of field components, remote disconnect of supply, etc. The CAS might delegate parts of its operation to the concentrators, if present, such that certain operations can be performed locally without the need for continuous wide-area communications.
- *Application systems* or legacy systems are the existing commercial or technical systems that manage the business processes of the utility. They communicate with the CAS using interface SI3.

2.4 Communication profile

The communications solutions deployed in the FAHAM project have to fulfill a set of requirements in order to confirm the economic viability as well as functional reliability of the whole system. The selected communication systems between AMI components shown in Fig. 12 are described as follows:

2.4.1 MI1-CI1 (electricity meter-concentrator)

This interface is between the electricity meter/communication hub, and the concentrator, located, for example, in an electrical MV/LV substation. The protocol architecture is shown in Fig. 13.

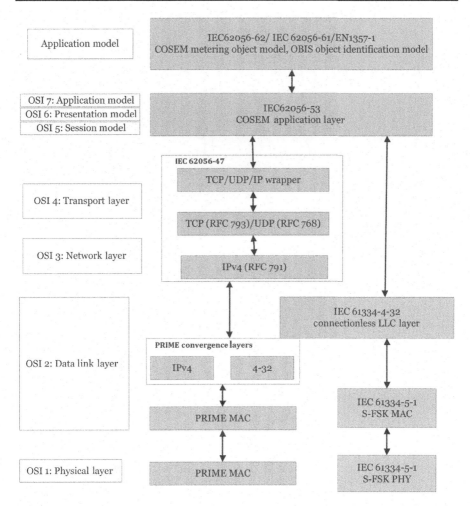

Fig. 13 MI1-CI1 communication architecture.

For this interface, two power line carrier (PLC) technologies, both working in the CENELEC A band, have been selected:

- IEC 61334-5-1 S-FSK
- PRIME

2.4.2 MI2-SI2 (electricity meter-CAS)

MI2-SI2 is the interface directly linking the electricity meter with the CAS. Due to cost constraints, this will usually be cellular. If a large number of nodes are installed using the network of a telecom provider, operation costs may become significant. In this case, data exchange with the meter has to be kept to a minimum.

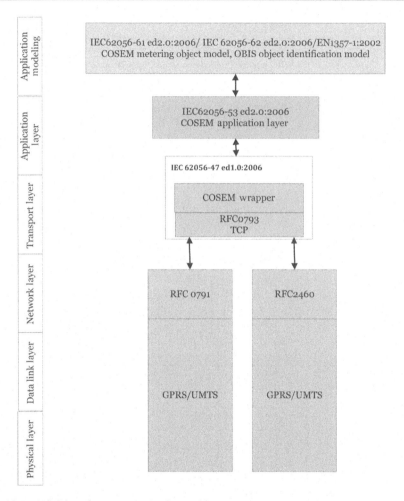

Fig. 14 MI2-SI2 interface communication architecture.

A large number of nodes to be managed by the CAS imply this interface does not usually have high bandwidths from the electricity meter perspective. GPRS (General Packet Radio Service)/UMTS (Universal Mobile Telecommunications System) wireless technology was considered to be the most suitable technology for this interface. A protocol architecture will be studied for DLMS/COSEM. Fig. 14 illustrates this interface communication architecture.

GPRS is a mobile data service offered in GSM systems, in addition to GSM service (it is integrated into GSM Release 97 and newer releases). It was originally standardized by the European Telecommunications Standards Institute (ETSI) and now by the 3rd Generation Partnership Project (3GPP). It is nowadays globally available in nearly all countries (except South Korea and Japan). In general terms, GPRS coverage is readily available in populated areas in most countries.

GPRS is widely used in IP networks today as a WAN wireless (cellular) technology. Each GPRS subscriber obtains an IP address, which can be public or private and at the same time fixed or dynamic, depending on the contracted service features and operator capabilities. GPRS service is provided in the GSM licensed frequency bands of 800, 900, 1800, and 1900 MHz.

The GPRS Access network is divided into two sections:

- GERAN (for GSM Radio Access Network) which comprises all the layers 1, 2, and 3 for the radio access network, plus all the normalized interfaces between them.
- The GSM core network, comprising the GSM network switching subsystem, the Gateway GSM support nodes (GGSN), and the Service GSM support nodes (SGSN).

An electricity meter would function as an MS (Mobile Station) from a GPRS perspective.

UMTS, also known as 3G or third generation mobile technology, is an evolution of existing 2G/GPRS networks using WCDMA modulation techniques in the air interface. It is specified by 3GPP and is part of the global ITU IMT-2000 standard. There have been different releases of UMTS issued by 3GPP. The UMTS lower layer networks are owned by the mobile operator. This network can be further subdivided into two different sections:

- UTRAN (UMTS Terrestrial Radio Access Network), which contains the nodes B (Base Stations) and RNCs that comprise the Radio Access Network. This network provides the coverage area, linking the electricity meter with the UMTS core network. Each of the different network sections in UTRAN contains standard interfaces to the others, so the technology can be easily upgraded while maintaining full compatibility. All this is specified in the 25 series of 3GPP specifications.
- The Core Network, linking all the RNCs. The UMTS core network is an evolution of the previous 2G (GSM) core networks. It is an access-agnostic network where some services can be directly connected to the core network.

2.4.3 CI2-SI1 (concentrator-CAS)

Two protocol profiles are proposed: A new highly scalable solution based on SNMPv3 plus secure file transfer protocol and a web services-based profile.

2.4.4 CI3 (data concentrator to the smart grid devices)

The interface CI3 is used to connect the concentrator to external devices (e.g., sensors that are located in a relatively short radius around the concentrator, power meters, etc.). PLC is not considered as an optimal solution from both installation and component perspectives. Some studies propose wireless as a good alternative; the preferred data model and application layer (IEC 61850) is currently extensively used over wired ethernet, and this is currently considered a good compromise solution for most applications. Given that the concept of an external device is not fully defined in the scope of AMI projects, the profile proposed here should not be considered as exhaustive.

2.4.5 MI3 (multiutility meter-electricity meter/ communication hub)

In Table 6, profiles for the multiutility meter communications interface MI3 are proposed.

The twisted pair profiles are proposed additionally to suggest communication assessment because of the following advantages for multiutility meters:

- Twisted pair bus can provide power;
- "Paired" installations of electricity and for example, natural gas in close proximity are not uncommon in many countries;
- Hardwired pairing of a multiutility meter and communication hub, therefore no special pairing procedures are needed—no data security and integrity issues.

There is a suggestion for the use of IEEE 802.11 (WiFi); this is not considered here. The main reason for this is the fundamental requirement of "low-power" for multiutility meters, which is deemed impossible to fulfill with the current IEEE 802.11.

List of standards that must be taking into account for FAHAM deployment are given in reference [1–31].

2.5 Layer model of AMI

In the layer model, five layers are defined within the AMI system, illustrated in Fig. 15. These layers are described as follows:

Table 6 MI4 interface communication architecture choices

Short name	Physical medium	Physical layer	Link layer	Application layer
M-bus TP	Twisted pair base band signaling	EN 13757-2	EN 13757-2	M-bus dedicated application layer EN13757-3+IEC 62056-53 DLMS/COSEM
M-bus TP DLMS/COSEM	Twisted pair	EN 13757-2	IEC 62056-46 HDLC	IEC 62056-53 DLMS/COSEM
Wireless M-bus	Radio 886 MHz	EN 13757-4 various modes	EN 13757-4	M-bus dedicated application layer EN13757-3+IEC 62056-53 DLMS/COSEM
IEEE 802.15.4 radio	Radio 886 MHz or 2, 4 GHz	IEEE 802.15.4	IEEE 802.15.4	IEC 62056-53 DLMS/COSEM
ZigBee DLMS/COSEM tunneling	Radio 886 MHz or 2, 4 GHz	IEEE 802.15.4	IEEE 802.15.4	IEC 62056-53 DLMS/COSEM

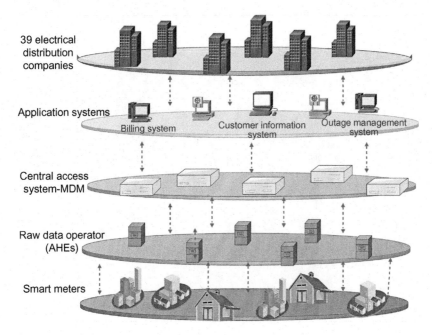

39 electrical
distribution
companies

Application systems

Billing system   Customer information   Outage management
                              system   system

Central access
system-MDM

Raw data operator
(AHEs)

Smart meters

Fig. 15 Layer model in FAHAM.

The first layer is the physical layer related to smart meters that are located in customers' places.

The second layer is AHE software as communications servers that are charged with the task of communicating with smart meters. These devices are different for different meter manufacturers. The diversity of AHE is dependent on a variety of meter manufacturers.

The third layer, meter data management (MDM), is responsible for collecting and managing data from AHE. The task of the layer is to match information in the same format used for other operational surfaces.

The fourth layer pertains to operational and commercial software that provides required statistical data and information for users by received information from MDM.

The fifth layer is distribution companies, which are responsible for information management in order to service the customers based on the business process.

2.6   ICT architecture and CAS communications

ICT architecture in FAHAM project is depicted in Fig. 16. The communications and information transactions between different parts are described as follows:

1. All the measured data from meters and concentrators are saved via GPRS links in one or more centralized MDM (five MDM centers).
2. Collected data is locally placed in information banks of legacy systems such as billing and outage management system (OMS) in the distribution company as the current format.

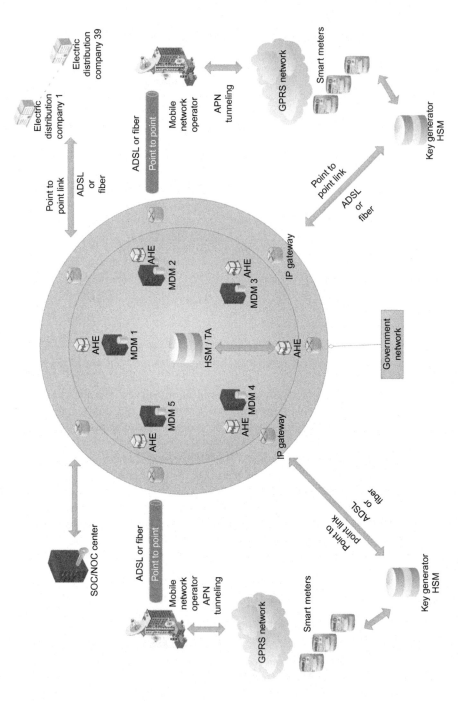

Fig. 16 ICT architecture in FAHAM.

3. Depending on the requirements of application systems, they extract information of smart meters form corresponding MDMs by adding web service call mechanisms.
4. After receiving data from MDM, it is stored in the local information bank of application systems such as billing and OMS.
5. Relations between the distribution company and MDM are limited to sectional transactions of information extraction or sending commands. In other situations, there are no communication and transactions.

2.7 Interoperability

Interoperability can be defined as the ability of systems, components, or equipment to provide services to and accept services from other systems, components, or equipment and to use the services exchanged to enable them to operate effectively together. With respect to software, the term is also used to describe the capability of different programs to exchange data via a common set of exchange formats, to read and write the same file formats, and to use the same protocols.

If two or more systems are capable of communicating and exchanging data, they are exhibiting syntactic interoperability.

Interoperability is the ability to automatically interpret the information exchanged meaningfully and accurately in order to produce useful results as defined by the end users of both systems. To achieve semantic interoperability, both sides must defer to a common information exchange reference model. The content of the information exchange requests is unambiguously defined: what is sent is the same as what is understood.

Interoperability can have important economic consequences. If competitors' products are not interoperable (due to causes such as patents, trade secrets, or coordination failures), the result may well be monopoly or market failure.

Smart metering communication systems should be based on standard metering protocols to confirm interoperability with changing energy supplier equipment and/or consumer equipment over the life of the meter.

Interoperability in the FAHAM system means that meters from different manufacturers should be able to work with all various types of concentrators made by other manufacturers. Every operation and maintenance device can connect to different types of meters and concentrators and CAS can manage all FAHAM devices regardless of their manufacturers. All these mentioned items shall be fulfilled without any additional devices or protocol convertors and without interfering in system online operation.

2.8 Security

When two-way command and control systems are embedded into power systems, several security threats must be addressed:

- To guarantee the confidentiality of data, which only allows authorized entities to access the data;
- To guarantee the integrity of data, which assures that data is not manipulated;
- To guarantee the authenticity of data, which confirms that the data is sent by a dedicated entity;
- To guarantee the availability of data, confirming that data is available when needed.

Security is everywhere in the metering process, from the meter and the DC to the back-office information system, including each network and media used to communicate (home network, public network, and enterprise network). Also, all components are concerned and we need to handle global security. All partners, from manufacturers to suppliers and regulators have to work together to raise awareness and secure future metering systems.

2.8.1 Security assumptions

- If a physical intrusion of a meter or DC happens, the compromising of one device should not permit the compromising of the entire system.
- Sensitive information and commands will have to be strongly protected.
- Hardware devices will support cryptographic algorithms. Whenever possible, association of symmetric and asymmetric algorithms should be used in accordance with standard implementation. A combination of both algorithms will be used when equipment characteristics allow it, otherwise symmetric mechanisms will be deployed or mixed with other technical methods. Usage of trusted equipment, such as a cryptographic processor embedded in smart cards, should be considered because they are tamper-resistant.
- Because security standards are available for IT systems, industrial automation, and control systems, they should be applied from the very conception of the systems for the deployment of devices and global systems.

2.8.2 Foundational security requirements

The FAHAM system should prevent:

- Unauthorized access, theft, or misuse of confidential information (data cannot be read or altered on a meter or in transit across all networks).
- Loss of integrity or reliability of process data and production information.
- Loss of system availability (back-office and data processing is secured).
- Invasions and illicit changes—for example, illicit firmware upgrades.
- Process upsets leading to compromising of process functionality or loss of system capacity (separation of responsibilities for appropriate actions).

Identified requirements to complete these needs are:

- Access and use control
- Data integrity
- Data confidentiality
- Resource availability

2.9 Use cases

A use case is a term in software and system engineering that defines how a user uses a system to accomplish a particular goal. Use cases describe interactions among external actors and the system to attain particular goals. Use cases are modeled by means of a unified modeling language (UML) and are represented by ovals containing the names of the use case.

This part provides the business use cases for electricity meters installed at the premises of domestic customers. In this section, "meter" refers to the electricity meter/communication hub. The procedure of each desired application of the AMI system is defined as a use case. These use cases are as follows:

- Provide periodic meter reads.
- Provide on-demand meter reads.
- Provide information through a local customer interface.
- Provide load profile.
- Provide power quality information.
- Provide interruption information.
- Provide tamper history (tamper detection).
- Disconnect/connect electricity.
- Apply electricity threshold and load management.
- Send long messages to end customer device.
- Shift tariff times electricity.
- Clock synchronization.
- Remote firmware upgrade.
- Planned on-site maintenance.
- Adjust equipment.
- Equipment installation.
- Uninstall electricity meter/communication hub.
- Retrieve electricity meter/communication hub state.
- Perform self-check.
- Verify topology.
- Communication use cases.
- Event logging and error reporting requirements.

Some of these use cases are described in the following:

2.9.1 Use case 1: Provide periodic meter reads

This use case describes the process of gathering and providing periodic meter reads. This process is triggered after the installation of the electricity meter. Periodic meter readings are daily and monthly meter readings. Daily meter readings are used in power market issues such as real-time pricing. The trigger description and block diagram are depicted in Fig. 17.

Legacy systems assign the task of obtaining cycle meter readings to the central system. Meter reading requests are linked to a deadline date for acquiring the reading of the meters involved. Deadline dates are fixed by legacy systems according to legal, technical, or business requirements.

In order to acquire the most recent energy values, the central system could request the direct reading of the meters involved via the concentrator (communication via SI2-MI2). For those meters whose reading is not available in time, the last value stored on the system could be provided. Thus, the central system or concentrators should periodically retrieve and store meter readings.

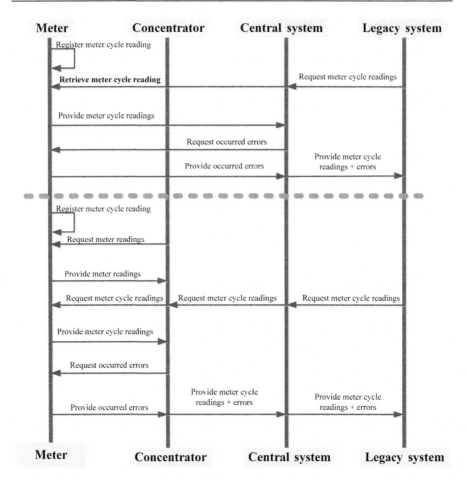

Fig. 17 Provide periodic meter read—UML.

2.9.2 Use case 2: Provide load profile

This use case provides the description of the process of making the load profile available to the CAS. The load profile is made available through the electricity meter (both load profiles for electricity and gas). The process of registering the load profile (after sending the activation comment from the application layer) is an uninterrupted process that runs throughout the lifecycle of the metering equipment. This process is hence triggered after the installation of the electricity meter. The trigger description, block diagram, and UML sequence diagram are depicted in Fig. 18.

The load profile provides a measurement of the variation in the electrical load versus time. It represents the pattern of electricity usage of a customer. This information can also be used for billing.

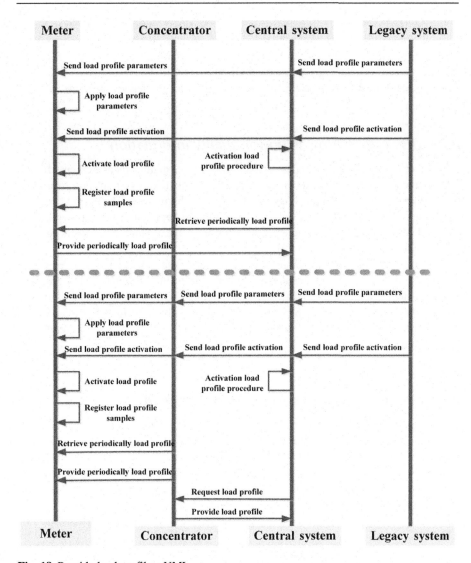

Fig. 18 Provide load profile—UML.

Meters shall have the capability of registering load profiles. The number and length of the load profiles registered by the meter shall accomplish the current legal directives.

The CAS shall be able to order the activation of the load profile storage in the meter (if it is not active by default) as well as the programming of the parameters that define the load profiles (i.e., magnitude and periodicity). These orders shall be remotely managed by the CAS, which communicates with meters directly (communication via SI2-MI2) or through a concentrator communication via SI1-CI2 and CI1-MI2).

In addition, the CAS shall periodically gather and store these load profiles in order to provide this information to the legacy systems when required. The gathering of the load profile data by the CAS can be done directly (communication via SI2-MI2) or aided by a concentrator (communication via SI1-CI1-CI2-MI1).

The CAS shall deliver load profiles at the request of the legacy systems.

2.9.3   Use case 3: Provide power quality information

This use case describes the process of gathering power quality measurements in the CAS. Some indices are defined to measure the power quality of systems ; for example, the number of occurred over-voltage, dip voltage and harmonics. The trigger description, block diagram, and UML sequence diagram are depicted in the following diagrams. The power quality parameters in addition to the voltage parameters for the CT/PT meter are harmonic and THD.

The grid company should guarantee its customers a user-specified quality electricity supply, based on current legal regulations. Due to the aforementioned requirements, meters should register measurements related to interruption and variation voltage. Subsequently, the CAS periodically retrieves this information by means of an on-demand reading to send it to the legacy systems responsible for managing this information.

2.9.4   Use case 4: Provide interruption information

This use case describes the process to register and provide the customer with standard information related to the interruptions. In this standard, the duration (T) for short and long interruptions has been defined as 3 min, to differentiate between short and long interruptions. In the future, this definition might change. Therefore, it is required that T is configurable. The meter will be able to detect any interruptions and register the related information to be shown to the customer on the meter display or to be sent to end customer devices. The meter will be able to collect this set of information for different interruption events.

2.9.5   Use case 5: Provide tamper history (tamper detection)

This use case describes the activities associated with tampering. Attempts to violate (parts of) the metering installation or the removal of the meter cover must be detected and registered with a time stamp; this detection applies for both the electricity meter and the gas and water meter. Further, fraud attempts using magnetic fields must be registered in the metering equipment. The metering installation must be able to register at least the last 10 fraud attempts for each tampering. Tamper detection (fraud and violation) is always active on all equipment (even during interruptions).

2.9.6 Use case 6: Apply electricity threshold and load management

This use case describes the process for applying a threshold on the supply of electrical power. It must be possible to set two different threshold values simultaneously, one for the normal contractual value of the electricity connection and one to be used in case a shortage of electricity is anticipated. The electricity thresholds can be set remotely.

It should be possible to apply demand management settings locally and remotely. Demand limitation for normal and emergency situations should be adjusted either when energy flows from the grid to the customer or from the customer to the grid. When demand limitation in emergency situations is activated it will have the priority to the demand limitation in normal situations.

The activation of the demanded power control mode in the meter allows legacy systems to control the excesses of consumption of electrical power by customers. Meter disconnection elements shall act as a programmable power control switch complying with the applicable legal requirements established. If the power control mode is active, the meter disconnection element interrupts the power supply when the power demanded is higher that a programmable threshold. To restore the power supply, the customer shall, manually or by means of an external domestic device (if such a device exists), close the disconnection device.

Therefore, the legacy systems shall order the activation or deactivation of the power control mode in meters as well as the programming of the power threshold. Those orders should be remotely managed by the CAS, which communicates with meters directly (communication via SI2-MI2) or through a concentrator (communication via SI1-CI2-CI1-MI1).

2.10 Application systems

The smart management of electric distribution grids is one of the key success factors to reach ambitious smart grid goals. Application systems such as an OMS, customer information system, and demand response management system(DRMS) are software that act as decision support systems to assist the distribution system operator with the monitoring and control of the distribution system. The foundation of an application system is the data that is received from AMI. Fig. 19 illustrates the FAHAM system equipped with application systems.

3 Conclusions

This chapter has been composed of two major sections: the Smart Grid Technology Roadmap in IRAN and the FAHAM. The roadmap focuses on technology development and has been approved in the 147th meeting of the Supreme Council for Science Research and Technology of Iran. The vision of this roadmap says, the Islamic Republic of Iran, by 2025, aims to develop an electric smart grid as an efficient, secure,

FAHAM distribution management system

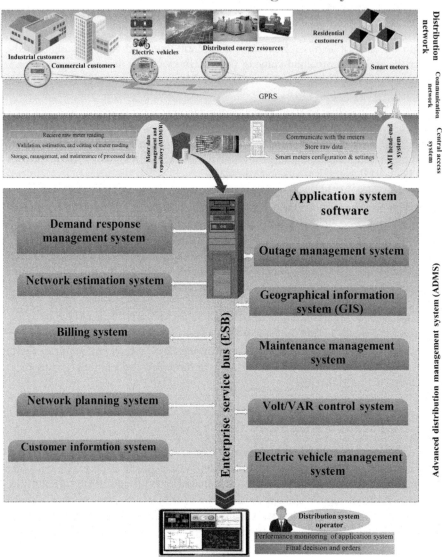

Fig. 19 AMI and application systems.

flexible, and stable grid that delivers the required high quality and reliable power to consumers and stakeholders. ICT, smart management systems, new technologies in the area of smart grids, IoT and integration of DGs, CHP systems, renewable energy resources, and energy storage systems cause dynamic interactions among stakeholders of the whole energy system. A smart grid should provide optimal management of demand and supply in competitive electricity market. Iran seeks to elevate and consolidate the country's position as a first country in the Middle East region in technology development and implementation of smart grid.

FAHAM is an important foundational step in the modernization of the Iran power system. The program involves replacing existing customer meters, now becoming obsolete, with a comprehensive smart meter system. This system includes the technology and telecommunications infrastructure needed for the Iranian power system to continue to manage the electricity system in a reliable, safe, and cost-effective manner. The following key learnings taken from this project have been summarized as follows:

- The smart meter system helps utilities with demand-side management programs.
- The security of the AMI system is the most important issue in the project that should be continuously upgraded.
- Without an application system, the benefits of the AMI system will not be fully revealed. Application systems such as the Volt/Var optimization system, a DRMS, and an OMS use collected data from smart meters to optimally operate and manage the electricity network (Boxes 1 and 2).

Box 1 Technological opportunities

Areas of Iran smart grid technology development that are deducted from the Iran smart grid roadmap vision are:

- Smart meter development and related technologies.
- Customer side technology development.
- ICT infrastructures technology development.
- Smart grid technology development (distribution and transmission systems).
- Wide area monitoring system development.
- Grid monitoring and control technology development.
- Grid protection systems development.
- DGs and renewable energy resources integration infrastructure development.
- Electric vehicle infrastructure and new technology development.
- Knowledge-based start-up businesses development support.
- Human resources education and training.
- Obtaining technology development infrastructure and smart grid lab development.
- Culture making and promotion of smart grid applications.
- Implementation of pilot projects.

Continued

Box 1 Continued

For updating and evaluation of smart grid technology development, two evaluation methods are designed: indices and reports. The domain of evaluation indices are:

- Human resources.
- Research centers.
- Scientific production.
- Technology development.
- Commercialization and industry.
- Increasing key player interactions.
- Financial costs.
- Financially advantageous.

and the subject of the evaluation reports is:

- Annual report for assessing progress of development of technologies based on the Stage-Gate model.
- Roadmap updating report at the end of year (based on the results of the annual report of technology progress).
- Evaluation report of MSC and PhD experts' contribution in R&D and rollout projects at the end of each year.
- Triple reports of project managers (charter, mid-term, and final reports).
- Annual report of research studies and applied technology development projects.
- Annual accumulated and detailed report of R&D and rollout project budget.

Evaluation report of achievements of high-tech pilot and future needs at the end of each year (Table 7).

Table 7 Iran smart grid deployment measures through 2020 and 2025

Pillar	Base line	Up to 2020	Up to 2025
Customer empowerment	AMI Telecommunication CIS DR	AMI CIS Smart grid (pilot) Smart home (pilot) Storage systems	Smart city Smart home
Market development	Tariff definition Tariff diversification Electricity retailers	Privatization sector Smart operators	
Grid development	WAMS SCADA (transmission)	WAMS Street lighting OMS GIS	HVDC DMS EMS Development and

Table 7 Continued

Pillar	Base line	Up to 2020	Up to 2025
		VVC Asset management system DA/FA SOC/NOC Microgrid (pilot) Integration of DER	integration of microgrids EV charging infrastructure
Governmental institutions	SG Steering Committee Renewable energy feed in tariff SG regulatory body AMI mandate	Standard definition ISGC Instructions and regulations completion	

Box 2 Financial allocation (Table 8)

Table 8 Funds allocated to different areas of Iran smart grid technologies development road map

No.	Areas of measure for Iran smart grid technology development	Iran's Electric Smart Grid National Grand Project (IESGNGP) budget ($ million)	Other resources ($ million)
1	Smart meter development and related technologies	0.6	6.5
2	Customer side technology development	1.4	12
3	ICT infrastructures technology development	1.1	3.5
4	Smart grid technology development (distribution and transmission systems)	2.7	3.5
5	WAMS development	1.4	5.5
6	Grid monitoring and control technology development	2.1	9.2

Continued

Table 8 Continued

No.	Areas of measure for Iran smart grid technology development	Iran's Electric Smart Grid National Grand Project (IESGNGP) budget ($ million)	Other resources ($ million)
7	Grid protection systems development	1.4	2.5
8	DGs and renewable energy resources integration infrastructure development	1.4	6.5
9	Electric vehicle infrastructure and new technology development	1.4	5.3
10	Knowledge based start-up businesses development support	5.2	17
11	Human resources education and training	1.4	4
12	Obtaining technology development infrastructure and smart grid lab development	1.4	6.5
13	Culture making and promotion of smart grid applications	0.8	6
14	Implementation of pilot projects	2.7	42
Total budget		25	130

References

[1] IEC 62051 (1999) Electricity metering—glossary of terms.
[2] IEC 62051-1 (2004) Electricity metering—data exchange for meter reading, tariff and load control—glossary of terms—Part 1: terms related to data exchange with metering equipment using DLMS/COSEM.
[3] IEC 62052-11 (2003) (Electricity metering equipment (AC)—general requirements, test and tests conditions—Part 11: metering equipment).
[4] IEC62053-21 (2003) Electricity metering equipment (a.c.)—particular requirements—Part 21: static meters for active energy (classes 1 and 2).
[5] IEC 62052-21 (2004) Electricity metering equipment (AC)—general requirements, tests and test conditions—Part 21: tariff and load control equipment.
[6] IEC 62053-22 (2003) Electricity metering equipment (a.c.)—particular requirements—Part 22: static meters for active energy (classes 0, 2 S and 0, 5 S).
[7] IEC 62053-23 Electricity metering equipment (a.c.)—particular requirements—Part 23: static meters for reactive energy (classes 2 and 3).
[8] IEC 62053-31 (1998) Electricity metering—particular requirements—Part 31: pulse output devices for electromechanical and electronic meters (two wires only).

[9] IEC 62054-21 (2004) Electricity metering—tariff and load control—Part 21: particular requirements for time switches.

[10] IEC 62055-21 (2005) Electricity metering—payment systems—Part 21: framework for standardization.

[11] IEC 62055-31 (2007) Electricity metering—payment systems—Part 31: particular requirements—static payment meters for active energy (classes 1 and 2).

[12] IEC 62056-21 Electricity metering—data exchange for meter reading, tariff and load control—Part 21: direct local data exchange.

[13] IEC 62056-46 Electricity metering—data exchange for meter reading, tariff and load control—Part 46: data link layer using HDLC protocol.

[14] IEC 62056-47 Electricity metering—data exchange for meter reading, tariff and load control—Part 47: COSEM transport layers for IPv4 networks.

[15] IEC 62056-53 Electricity metering—data exchange for meter reading, tariff and load control—Part 53: COSEM application layer.

[16] IEC 62056-61 Electricity metering—data exchange for meter reading, tariff and load control—Part 61: object identification system (OBIS).

[17] IEC 62056-62 Electricity metering—data exchange for meter reading, tariff and load control—Part 62: interface classes.

[18] IEC 62058-11 Electricity metering equipment (AC)—acceptance inspection—general acceptance inspection methods.

[19] IEC 62058-31 Electricity metering equipment (AC)—acceptance inspection—Part 31: particular requirements for static meters for active energy (classes 0, 2 S, 0, 5 S, 1 and 2).

[20] IEC 62059-11 Electricity metering equipment—dependability—Part 11: general concepts.

[21] IEC 61334 Standard for low-speed reliable power line communications by electricity meters. It is also known as S-FSK, for "spread frequency shift keying." It is actually a series of standards describing the researched physical environment of power lines, a well-adapted physical layer, a workable low-power media access layer, and a management interface.

[22] IEC 62059-21 Electricity metering equipment—dependability—Part 11: collection of meter dependability data from the field.

[23] IEC 62059-31-1 Electricity metering equipment—dependability—Part 31-1: accelerated reliability testing—elevated temperature and humidity.

[24] IEC 62059-41 Electricity metering equipment—dependability—Part 41: reliability prediction.

[25] IEC60950 Information technology equipment—safety—Part 1: general requirements (2005); Information technology equipment—safety—Part 21: remote power feeding (2002); Information technology equipment—safety—Part 22: equipment to be installed outdoors (2005); Information technology equipment—safety—Part 23: large data storage equipment (2005).

[26] EN 41003 (2009) Particular safety requirements for equipment to be connected to telecommunication networks and/or a cable distribution system.

[27] EN 50065-1 (2010) Signaling on low-voltage electrical installations in the frequency range 3 kHz to 148.5 kHz. General requirements, frequency bands and electromagnetic disturbances.

[28] EN 50360 (2001) Product standard to demonstrate the compliance of mobile phones with the basic restrictions related to human exposure to electromagnetic fields (300 MHz–3 GHz).

[29] EN 50371 (2002) Generic standard to demonstrate the compliance of low power electronic and electrical apparatus with the basic restrictions related to human exposure to electromagnetic fields (10 MHz–300 GHz). General public.

[30] EN 50385 (2002) Product standard to demonstrate the compliances of radio base stations and fixed terminal stations for wireless telecommunication systems with the basic restrictions or the reference levels related to human exposure to radio frequency electromagnetic fields (110 MHz–40 GHz). General public.

[31] EN 50401 (2006) Product standard to demonstrate the compliance of fixed equipment for radio transmission (110 MHz–40 GHz) intended for use in wireless telecommunication networks with the basic restrictions or the reference levels related to general public exposure to radio frequency electromagnetic fields, when put into service.

Further reading

[32] Specification of general, economical, functional, technical and communicational requirements for the advanced metering infrastructure (AMI). IEEO; 2009.

Intelligent control and protection in the Russian electric power system

Nikolai Voropai, Dmitry Efimov, Irina Kolosok, Victor Kurbatsky, Anna Glazunova,
Elena Korkina, Nikita Tomin, Daniil Panasetsky
Energy Systems Institute, Irkutsk, Russia

Acronyms

AAN	active and adaptive electric network
ANN	artificial neural network
APEC	Asia Pacific Economic Cooperation
ARIMA	autoregressive integrated moving average
BDD	bad data detection
BESS	battery energy storage systems
CACS	conventional automatic control system
CHPP	combined heat and power plant
DC	direct current
DSE	dynamic state estimation
ECS	emergency control system
EMD	empirical mode decomposition
EMS	energy management system
ENTSO-E	European Network of Transmission System Operators for Electricity
EPS	electric power system
ESS	electricity supply system
FACTS	flexible alternating current transmission systems
FGC UES	JSC "Federal Grid Company of Unified Energy System"
GARCH	generalized autoregressive conditional heteroscedasticity
GLONASS	Global Navigation Satellite System in Russia
GPS	global positioning system
HHT	two-step algorithm, combining EMD and HT
HPP	hydro power plant
HT	Hilbert transform
HTSC	high-temperature superconductivity
HVDC	high-voltage direct current
ICOEUR	intelligent coordination of operation and emergency control of European Union and Russian power grids, the international European/Russian FP7 project
IED	intelligent electronic device
IEEE	Institute of Electrical and Electronics Engineers
IESAAN	intelligent electric power system with active and adaptive network

Application of Smart Grid Technologies. https://doi.org/10.1016/B978-0-12-803128-5.00003-9

IMF	intrinsic mode functions
IPS	interconnected power system
MAAS	multiagent automation system
MAE	mean absolute error
MAPE	mean absolute percent error
MLP	multilayer perceptron
OLTC	on load tap changer
OTL	overhead transmission line
PDC	phasor data concentrator
PDSRF	algorithm of proximity-driven streaming random forest
PM	pseudomeasurement
PMU	phasor measurement unit
RMSE	root mean square error
RTU	remote terminal unit
SC	series capacitor
SCADA	supervisory control and data acquisition
SE	state estimation
SO UES	JSC "System Operator of Unified Energy System"
SS	substation
STATCOM	static synchronous compensator
STC	static thyristor compensator
SVC	static var. compensator
SVM	support vector machine
TCR	thyristor-controlled reactor
TCSC	thyristor-controlled series capacitor
TE	test equation
TPP	thermal power plant
TSC	thyristor switched capacitor
TTC	total transfer capability
UES	unified energy system of Russia
UNEG	unified national electric grid
UVLS	under voltage load shedding
VSC	voltage source converter
WACS	wide area control system
WAMS	wide area measurement system
WAPS	wide area protection system

1 Summary

The unified energy system (UES) of Russia is a power interconnection where seven interconnected power systems (IPSs) are combined by weak ties. Under emergency conditions, the Russian UES is able to disintegrate into autonomously operating self-balanced IPSs without grave consequences.

The trends in expansion of electric power systems (EPSs) and changes in the conditions of their operation have complicated EPS operations as well as increased its changeability and unpredictability that call for a more prompt and more adequate response of control systems.

A totally novel approach was suggested to make the transition to a new level of technology and control of the Russian UES. The main goal of this approach is the implementation of intelligent technologies in the Russian power industry to ensure an innovative breakthrough in the development of that industry and to increase the efficiency, reliability, and security of its operation.

The chapter presents the following contributions: intelligent energy system as a Russian vision of a smart grid; informational support of an active and adaptive network (IESAAN) control problems; intelligent operation and smart emergency protection; smart grid clusters in Russia.

1.1 Intelligent energy system as Russian vision of smart grid

In 2010–12, the concept of an intelligent EPS with IESAAN was developed in Russia. The concept stipulates that all subjects of the electricity market (generation, grid, and consumers) take an active part in the processes of electric power transmission and distribution.

The problem of efficient control of IESAAN is among the most important scientific and technical problems. With the availability of an effective control system, IESAAN can provide reliable interaction between consumer grid units of different functions and generators, using uniform principles and the common information-technological platform.

The section presents:

–A general overview of IESAAN as the Russian vision of a smart grid.
–Its technological platform and control system.
–The problems to be solved by the control system online under the conditions of incomplete information on the parameters of EPS and disturbances, means, and systems for the solution of these problems.
–The uniform principles of control and new kinds of techniques and technologies.

1.2 Informational support of IESAAN control problems

State estimation (SE) is the most important procedure that provides EPS control with reliable and quality information. The result of SE is a model of an EPS steady state based on measured parameters of the system and data on the state of network components. IESAAN involves new technologies and systems for measurement of state parameters which form new conditions for SE of EPS, and put forward new objectives to be accomplished on the way of modernization of the data computing system.

The section discusses the following issues:

–*SCADA and WAMS applications.* Large system blackouts that have occurred in different countries in the beginning of the new millennium prove the need for updating and improving the software for EPS monitoring and control (SCADA/EMS applications). The use of phasor measurement unit (PMU) measurements improves the observability of the calculated network, enhances the efficiency of methods for bad data detection (BDD), increases the accuracy and reliability of the obtained estimates, and offers new possibilities in the decomposition of the SE problem.

−*The EPS SE problem formulation and its specificity in the control of IESAAN.* The SE problem consists of calculating the EPS steady-state conditions by measurements. The measurements applied in EPS SE are mostly the measurements obtained from the SCADA system and PMUs. The existing SE methods and approaches that are applied today were developed in the conditions of insufficient telemetry data, low redundancy of measurements, incomplete visibility of the network, asynchronous measurements and considerable delays in their transfer, and low accuracy and reliability of analog measurements and remote signals. IESAAN suggests a wide use of new technologies and systems for measurement of state parameters (such as digital instrument transformers and transducers, PMUs, etc.), as well as modern communication systems. All of this radically change the processes of data collection, processing, and transfer, digitalization of substations. A change in the properties of an observed EPS itself due to saturation of the network with active components (FACTS, HVDC) forms new conditions for SE of EPS and puts forward new objectives to be accomplished in modernization of the data computing system of the IESAAN.

−*The main directions in the development of SE methods and technologies in their application to control of IESAAN* include:

- *Phasor measurements in the SE problem,* including the use of test equations (TEs) for validation of measurements, systematic errors in PMU measurements, and decomposition of the problem.

- *SE of EPS involving FACTS models.* Today, FACTS are among the most promising electrical network technologies, which make it possible for the electrical network to turn from a passive facility for electricity transportation into the facility that takes an active part in the control of the network operation. Until recently, the models for the present-day FACTS devices based on power electronics were not involved in the design diagram when solving the SE problem. The section presents the models for the main FACTS devices and the algorithms to involve these models in the problem.

- *Dynamic state estimation (DSE) and its application,* including criteria for the estimate accuracy, detection of bad data in measurements, and description of the devised method. The DSE in the section is based on the extended Kalman filter and deals with testing the validity of measurements in the areas of low information redundancy, which contain critical measurements and critical sets.

- *Supporting cyber-physical security of EPS by the SE technique,* including the cybersecurity of SCADA systems and wide area measurement system (WAMS). Malicious activities or cyberattacks should be considered as disturbances that may lead to a large-scale outage. The section presents the analysis of possible consequences of cyberattacks against SCADA and WAMS and discusses the accuracy of solutions to the SE problem. Also, consideration is given to the methods applied to detect bad data in measurements, identify the consequences, and mitigate them. The analysis technique of the cybersecurity of SCADA and WAMS in two-level SE and the methodology of cyberattack identification are discussed.

1.3   Intelligent operation and smart emergency protection

This section discusses intelligent operation and smart emergency protection in Russia, including a description of an emergency control system (ECS), requirements to new operation and emergency protection, and a novel system of monitoring, forecasting, and control of EPS while also presenting several recent applications of artificial intelligence in EPS:

—*ECS in Russia and requirements to new emergency protection and operation systems.* Emergency control in the EPS is performed by the technological (dispatching and automatic) control systems. Emergency control in the Russian UES is structured hierarchically. It is realized by the coordinated operation of many control devices and with joint participation from generators, networks, and consumers, which maintains EPS stability and interrupts the expansion of an emergency situation in the case of stability violation and a threat of undesirable cascade emergency development. Nevertheless, existing electricity grids are still vulnerable to large-scale blackouts. The existing ECSs and the actions of transmission system operators may prove to be ineffective to prevent the subsequent catastrophic development of the emergency. The section lists drawbacks of the existing ECSs and ways of improving the ECS operation. Besides, specific requirements for next-generation intelligent ECSs are formulated.

—*The system of monitoring, forecasting, and control of EPSs.* Advanced smart devices and technologies—PMU, artificial intelligence, and so on—promise new possibilities for solving the complex problem of EPS control in normal, preemergency, emergency, and postemergency operating conditions. From the viewpoint of this system efficiency, it is essential to increase the adaptability of control and improve coordination of the control stages, means, and systems. The blocks of the system solve the following problems: EPS SE, forecasting the parameters, detection of weak points in the EPS, determination of margins for transfer capabilities of ties in the expected operating conditions, and visualization of the expected conditions. When information about some EPS characteristics is either unavailable or limited and contradictory, a forecasting performance can be improved by using a transition from a time series task to a regression task when an initial time series is decomposed into some components. Moreover, the online control of EPSs deals with the notion of "prediction of an emergency situation" (probability of its occurrence).

—*Artificial intelligence application* is an advanced way to carry out smart emergency control in EPS. Several specific possibilities of artificial intelligence applications to the development of principles and tools of automatic emergency control in EPS are considered, such as:

■ *Forecasts of state variables based on the DSE method.* The DSE is used to forecast the components of state vectors some time ahead and to calculate all state variables. This technique uses the archives of measurements. The estimates of state variables are obtained based on the formed snapshots using DSE on the basis of the extended Kalman filters. Such a DSE technique can be used to forecast all the EPS state variables for a short period of time.

■ *Forecast of power system parameters based on a hybrid data-driven approach.* This approach is developed for forecasting EPS parameters (such as load, distributed embedded generation, wind energy, and distributed storage) to improve grid operation efficiency as well as reduce fossil fuel consumption and carbon emissions. The main goal of the approach development was to increase the efficiency of forecasting studies when dealing with limited retrospective data and a sharply variable, nonstationary time series of EPS parameters. The approach is based on mode decomposition and a feature analysis of initial retrospective data using the Hilbert-Huang transform and machine learning algorithms. The decision tree techniques are used to rank the importance of variables employed in the forecasting models. The resulting hybrid forecasting models employ the radial basis function neural network and support vector regression. The efficiency of the developed approach was demonstrated with a real-time series in such problems as active power flows, electricity prices, and wind speed forecasting.

- *Total transfer capability (TTC) estimation method.* The method calculates a steady state using TTC for the controlled lines and the parameters of current conditions for the remaining part of EPS. The values of power flows in the controlled lines change through the adjustment of EPS state variables determined in advance. The Kohonen ANNs are applied to adjust the TTC estimation method in real time.
- *Automatic decision tree-based system for online voltage security control of EPSs.* An online method instead of conventional offline methods is proposed to solve the problems of EPS security on the basis of decision trees. The online method updates the existing model, using new data without its total restructuring. A proposed proximity-driven, streaming random forest algorithm can serve as one of such techniques. The operation of the developed intelligent system is mostly automatic, and the major part of control actions is generated with the minimum involvement of the operator.
- *Multiagent coordination of emergency control devices.* The absence of the control device coordination during the postdisturbance period is one of the main causes of the voltage instability that permanently occurs in EPS. A decentralized agent-based system for operation and emergency control (a multiagent automation system, MAAS) is proposed that provides control of reactive power, prevents emergency tripping of generators, and maintains voltage at load buses within acceptable limits. The coordination of local devices through the implementation of additional control actions from the proposed MAAS makes it possible to effectively cope with the equipment overload and voltage collapse.
- *Intelligent system for preventing large-scale emergencies in a power system.* Supplementing the existing principles of EPS control is proposed with the method of "control with prediction," which suggests early detection and prevention of dangerous states and emergency situations before they lead to a voltage collapse. The control system based on this approach adjusts the operation of local emergency control devices, both in a steady state and in postemergency conditions to reduce the probability of a severe emergency. An architecture of the intelligent system, a structure, and a database of emergency disturbances are presented.

1.4   Smart grid clusters in Russia

The process of IESAAN formation suggests the implementation of pilot projects and the creation of territorial smart grid clusters. These clusters are supposed to use information, technology, and control systems providing adaptive control of network parameters, remote control of switching devices, and real-time estimation of the technical state of the network. Currently, along with implementation of the pilot projects and creation of smart grid clusters, new equipment is being installed at the energy facilities of the unified national electric grid (UNEG) of Russia. This section describes several examples of smart grid clusters and pilot projects in Russia's EPSs:

- Smart grid clusters in the East IPS (Amur region, Sakha Republic, Primorye, and Khabarovsk Territories).
- Pilot project for creation of a territorial smart grid cluster on the Russky and Popov Islands. The 220/35 kV substation "Russkaya" and 220 kV cable—overhead line "Zeleny ugol-Russkaya" with a submarine bridge through the Eastern Bosforus strait is under construction to provide a centralized electricity supply to Russky Island.

—Smart grid clusters in the Northwest IPS (Karelskaya power system, power systems of Komi Republic and Arkhangelsk, "Big ring" and "Small ring" of electric networks in St. Petersburg).

—Pilot project on smart electricity supply to the Skolkovo innovation center in the eastern part of the Odintsovo district of the Moscow region. Construction of the center is a far-reaching project for creation of a high-tech research and technological complex for design and commercialization of new energy, information, telecommunication, biomedical, and nuclear technologies.

2 Intelligent energy system as Russian vision of smart grid

A totally novel approach was suggested to make the transition to a qualitatively new level of technology and control of the UES of Russia. The key role in the approach belongs to the core of EPSs, that is, the electrical network with active-adaptive properties that provides reliable and efficient connection between generation and consumers. The main goal of this approach is the implementation of intelligent technology in the Russian power industry to ensure an innovative breakthrough in the development of the industry, and to increase the efficiency, reliability, and security of its operation.

In 2010–12, the concept of an intelligent EPS with an active and adaptive network (IESAAN) [1] was developed by the JSC R&D Centre for Power Engineering. The concept of IESAAN stipulates that all subjects of the electricity market (generation, grid, consumers) take an active part in the processes of electric power transmission and distribution. Electric power consumers as a part of IESAAN should become its leading component. Transmission and distribution networks, which traditionally were passive, will turn into active components whose parameters and characteristics will become flexible according to the operation requirements of the entire system.

2.1 Technological platform, intelligent energy system of Russia

The decision on development of concepts of technological platforms for Russia was made in August 2010 by the Governmental Commission for High Technologies and Innovations [2]. A technological platform is a form of public-private partnership in the field of innovation, a way of uniting the efforts of all parties concerned (various departments, businesses, and the scientific community) to accomplish the final goals of certain strategic priority areas.

In 2010 the JSC Federal Grid Company of Unified Energy System (FGC UES) and Federal State Institution Russian Energy Agency supported the development of an application for creation of the technological platform, "Intelligent energy system of Russia," and its submission to the Ministry of Economic Development of Russia (Fig. 1). The main mission of the technological platform is implementation of intelligent technologies in the Russian power industry to ensure an innovative breakthrough in development of the industry, and to increase the efficiency, reliability, and security of its operation [3].

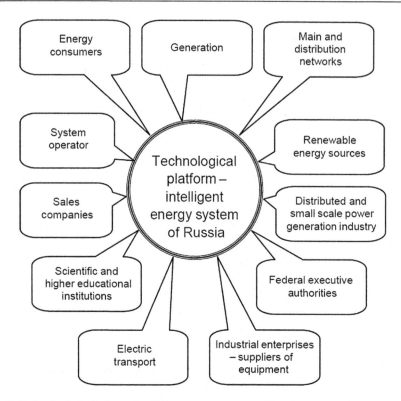

Fig. 1 Technological platform, intelligent energy system of Russia.

2.2 Intelligent electric power system with an active and adaptive network (IESAAN)

An intelligent EPS (Fig. 2) is a customer-oriented EPS of a new generation that should provide high-quality, reliable, and efficient services for electricity supply through flexible interaction in all types of generation, electric networks, and consumers based on cutting-edge technologies and a common hierarchical system of control. An important role in the intelligent EPS is assigned to the active and adaptive electric network (AAN). Radically new properties imparted to the power system by ANN are [4,5]:

- Standardized high-technology flexible interface "Network-Network user."
- Efficient use of electricity through adaptive regulation of the load with the maximal consideration of consumer requirements (including economic ones).
- Network topology providing control of power exchanges with an appropriate system for control of active network components and generation facilities.
- Real-time adaptive responses based on the combination of centralized and local operation and emergency control under normal and emergency conditions.
- Information resources and technologies for assessment of situations in order to find solutions and make prompt and long-term decisions.
- Expansion of market capabilities of the infrastructure through a wide range of services to be mutually rendered by market participants and the infrastructure.

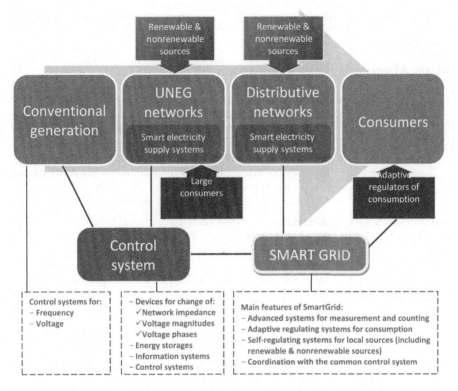

Fig. 2 IESAAN technological infrastructure.

The UNEG carries out backbone functions and includes power segment grids, electricity supply systems (ESSs), and the bulk transmission and international grids.

The goals of an intelligent EPS [6,7] are presented in Table 1.

The following technologies are used to accomplish the above goals [4,5,8,9]:

—Digital substations.
—Battery energy storage systems (BESS).
—Smart meters.
—Systems for power quality improvement (filter-compensating devices).
—Smart grid components (automated dispatching control system, system of energy distribution control, algorithms of energy system control).
—WAMS/WAPS/WACS.
—FACTS.
—Gas-insulated cable lines.
—Direct current lines.
—Cogeneration.
—Wind turbines.
—Alternative domestic mini-generation.

At the formation of IESAAN, the UNEG components of different voltage levels should contain the devices changing the impedance of the grid elements and voltage

Table 1 **The goals of intelligent electric power system**

Energy sector	Goals
Electric networks	• Power quality control • Peak load shaving in electric network • Creation of an alternative to reconstruction of the network infrastructure in the network "bottlenecks" • Creation of an alternative to expansion of the network infrastructure for electricity supply to remote and isolated areas • Reducing the duration of interruptions in the electricity supply • Reduction of expenses for the network infrastructure maintenance and its automation • Operation and emergency control of the grid
Electricity consumption	• Reduction of electricity costs • Provision of continuous energy supply • Improvement of power quality • Use of infrastructure for smart houses and electric cars • Integration of consumers with their own microgeneration
Power generation	• Maintenance of frequency • Increase in available capacity • Integration of variable generation sources, including wind turbines and mini-cogeneration power plants (mini-CPP) • Creation of an alternative to construction of generation capacities for electricity supply to remote and isolated areas

(both magnitude and phase) at various points of the grid. Possibilities of combining alternating and direct (lines and back-to-back stations) currents as well as modern devices of short circuit current limitation in powerful switching equipment will be widely used.

The ESS structure can include sources of electric power, electric power transmission, distribution, transformation devices, and various auxiliary devices and constructions (power plants and feeding lines of regional power systems, transmission lines, substations, and distribution devices). One of the smart ESS's key functional characteristics is the motivation of active behavior of the end consumer. Active behavior implies consumer possibility to change independently the received electricity amount and functional properties (level of reliability, quality, etc.), using the information about prices, electricity supply volumes, reliability, quality, etc. The aim of the motivation is steering those changes according to the balance of the requirements and possibilities of ESS.

2.3   Control system of IESAAN

The problem of efficient control of IESAAN, whose urgency and importance have been confirmed by recent large, human-caused failures, is among the most important scientific and technical problems. With the availability of an effective control system,

IESAAN can provide reliable interaction between consumer grid units of different functions and generators, using uniform principles and the common information-technological platform [6].

The IESAAN control system should provide the solution for the following problems:

−control of IESAAN operation conditions and parameters,
−control of settings, parameters, and conditions of active-adaptive devices and technological objects of IESAAN,
−control of routing and transmission of electricity and power on the basis of active-adaptive algorithms,
−monitoring of the technical state, operation, and reliability of IESAAN basic objects,
−monitoring of a technical state of auxiliary technological systems,
−planning and development of an IESAAN technological complex,
−power quality control,
−provision of technological safety and control of IESAAN and its objects in emergency situations,
−estimation of situations and forecasting of their development for decision-making,
−technical support and operation of technical objects and facilities of IESAAN,
−the commercial and technical accounting of the electricity and power,
−information interaction in a hierarchical network of IESAAN dispatching and operating structures, and
−information security of a control system.

For the effective solution of IESAAN control problems online in the conditions of incomplete information on parameters of EPS and disturbances, the use of uniform principles of control and qualitatively new kinds of techniques and technologies is necessary, including means and systems for:

−control and regulation of active and reactive power with the application of power electronics,
−limitation of short circuit currents,
−electricity storage,
−forecasting and intelligent analysis of emergencies,
−support of operative decisions, generation of recommendations, and control actions on localization and liquidation of failures,
−control and analysis of a technical state and a residual resource of the technological equipment,
−availability of high-speed, completely integrated, bilateral technology of communication and commutations between EPS subjects for interactive interchange of information, energy, and monetary flows between them in real time, and
−intelligent accounting of the electricity and control of power consumption.

Thus the structure of IESAAN control (Fig. 3) should be formed on the basis of the following principles:

(1) A control system realizing operating functions with a high degree of automation is a key link at the construction of IESAAN.
(2) IESAAN control systems should have the uniform (coordinated) principles of management regardless of the form of property and the hierarchical level.
(3) When combining the local and system functions of management, the relations between controls of generation, networks, and load-controlled consumers should be traced explicitly.

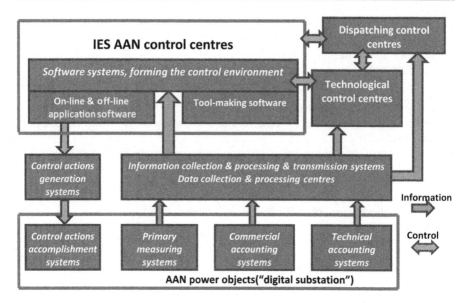

Fig. 3 The functional structure of IESAAN control.

(4) Information support of control systems should provide transmission of controlled parameters and signals into control loops in real time.

(5) The software of control systems should be realized in online and offline modes to trace and control both normal (optimization of conditions) and abnormal (prevention of overloaded and liquidation of preemergency conditions) EPS operation, and should be adaptive to changing situations.

(6) The communication systems included in the control loop should provide high speed and reliability of information and control signals transmission.

3   Informational support of IESAAN control problems

3.1   SCADA and WAMS

Large system blackouts that have occurred in different countries in the beginning of the new millennium prove the need for updating and improving of software for EPS monitoring and control (SCADA/EMS-applications). It supposes the use of

−advanced technical systems for collection,
−transfer and processing of information on state variables of the network and its components,
−high-speed software for calculation of the current state of EPS, and
−forecast of state variables to detect bottlenecks in the system and the creation of a flexible system for control (combination of centralized and local control) of all network components.

Creation of satellite communication systems provided a new generation of measurement equipment—PMUs [10]. Integrated into the WAMS, the PMU sensors give a real picture of the EPS state. WAMS technology is intended for the PMU-based calculation of angles between voltage and current vectors synchronized with accuracy up to 1 μs.

Each PMU installed at the ith node provides direct measurements of voltage magnitude U_i and phase angle δ_i at buses, magnitude of current I_{ij} along the outgoing line, and angle φ_{ij} between current and voltage as well as the calculated flows of active P_{ij} and reactive Q_{ij} power in a line. GPS-synchronized equipment is capable of measuring voltage magnitude with an accuracy of 0.1% and a phase angle with accuracy of 0.2°.

The updating of software for EPS monitoring and control (SCADA/EMS-applications) has become possible on a qualitatively new level owing to (WAMS) allowing the EPS state to be controlled synchronously and with high accuracy. The PMUs are the main measurement equipment in these systems. Since 2005, Russia has been creating a system for monitoring of transients (a Russian analog of WAMS). The main measurement equipment in the system is SMART-WAMS recorders of voltage and current phasors. One of the applications, which will be significantly affected by the introduction of PMU, is the SE.

Until recently, SE in EPS was mainly based on the SCADA measurements: magnitudes of bus voltages, generation of active and reactive powers at buses, power flows in lines, and, less often, currents in lines and at buses.

SCADA systems are intended for receiving and processing information once per second, and the remote control systems themselves allow delays in information delivery up to several tens of seconds. Another serious flaw of existing remote control systems is the absence of a highly accurate synchronization of measurements with respect to astronomical time. The lack of simultaneity in receiving measurements is particularly noticeable in the calculation of simultaneously operating subsystems that have their own devices for collection and transfer of information. The state variables calculated on the basis of SCADA measurements are behind the current state variables of EPS and are only their approximation, which can cause errors in control.

The advent of WAMS that contains PMUs as the main measurement equipment makes it possible to synchronously and accurately control the EPS state and essentially improve the results of SE [11]. PMUs are employed in EPS both to solve the local problems and to obtain a general picture of the EPS state, which is further used for solution of control problems. As compared to a standard set of measurements received from SCADA, PMU installed in the node can measure voltage phasor in this node and current phasors in some or all lines adjacent to this node with high accuracy. The use of PMU measurements improves the observability of the calculated network, enhances the efficiency of methods for BDD, increases the accuracy and reliability of the obtained estimates, and offers new possibilities in decomposition of the SE problem.

While SCADA measurements provide observability of the network, the critical measurements and critical sets exist, that is, the redundancy of measurements is quite low. To increase the redundancy of measurements and eliminate critical measurements and critical sets, PMU shall be added to the network buses.

WAMS is a set of recorders of synchronized PMUs, phasor data concentrators (PDCs), channels for data transfer among the recorders, data concentrators, and dispatching centers of the JSC "SO UES" as well as systems for processing the obtained information. WAMS measurements are synchronized by the systems of GPS/GLONASS. The hierarchical architecture of WAMS [12] is presented in Fig. 4.

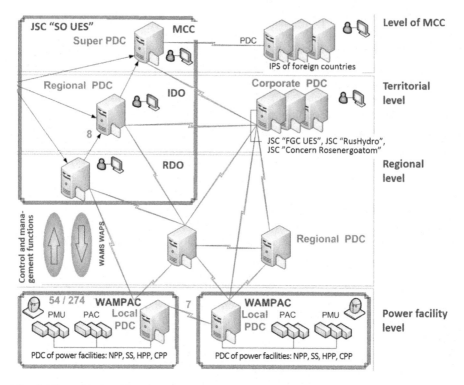

Fig. 4 Hierarchical architecture of WAMS.

PMUs measure the magnitudes and phases of nodal voltages and currents in lines, which are incidental to these nodes. The PDC collects, filters, processes, and retransmits the data. Besides, the PDC can register abrupt surges, distortions, parameters of switches, and parameters of loads and lines as well as identify generator parameters.

A super-PDC processes data and provides a dispatcher or an operator with a graphical interface and access to the data archive.

There are different approaches to the determination of sites for PMU placement in the system. In Russia's UES that consists of seven IPSs furnished to a different extent with SCADA measurements, the sites for priority placement of PMU for monitoring of transients were determined on the basis of practical engineering experience [13]. These are:

−at nodes limiting intersystem lines,
−at major generation nodes, and
−at the plants of secondary frequency control.

However, even the existing insignificant number of PMUs in combination with SCADA measurements make it possible to essentially improve the results obtained by solving the SE problem. Currently, there are algorithms for solving the problem of SE using only PMU or SCADA measurements, or a combination of both.

If there is a sufficient amount of PMUs to provide observability of the EPS scheme, SE can be performed using PMU data only.

Placement of PMU at a node of EPS provides a set of new measurements that are characterized by high accuracy. Inclusion of PMU measurements in the total set of measurements in the system increases the redundancy of measurements, which, in turn, helps in detecting bad data in the measurements and improves the quality of SE. The main types of measurements received from PMU are magnitudes U_i and phases δ_i of nodal voltages and currents I_{ij}, ϕ_{ij} in the outgoing lines.

Based on the vectors of nodal voltages and currents in the outgoing lines, it is possible to calculate the magnitudes and phases of nodal voltages at neighboring nodes, that is, to obtain the so-called calculated PMUs:

$$U_j^{\text{calc}} = \sqrt{U_i^2 - 2U_iI_{ij}\left(r_{ij}\cos\phi_i + x_{ij}\sin\phi_i\right) + I_{ij}^2\left(r_{ij}^2 + x_{ij}^2\right)} \tag{1}$$

$$\delta_j^{\text{calc}} = \delta_i - arctg\frac{I_{ij}\left(x_{ij}\cos\phi_{ij} - r_{ij}\sin\phi_{ij}\right)}{U_i - I_{ij}\left(r_{ij}\cos\phi_{ij} + x_{ij}\sin\phi_{ij}\right)} \tag{2}$$

where r_{ij}, x_{ij} is the resistance and reactance of line $i-j$.

Also, PMU measurements are used to calculate pseudomeasurements (PMs) of power flows. For example, PMs of power flows at the beginning of line P_{ij}^{PM} can be calculated as follows:

$$P_{ij}^{\text{PM}} = \sqrt{3}I_{ij_{\text{PMU}}}U_{i_{\text{PMU}}}\cos\phi_{ij_{\text{PMU}}} \tag{3}$$

$$Q_{ij}^{\text{PM}} = \sqrt{3}I_{ij_{\text{PMU}}}U_{i_{\text{PMU}}}\sin\phi_{ij_{\text{PMU}}} \tag{4}$$

Measurements from physical and calculated PMUs and flow PMs are the initial information for the SE problem.

Accuracy of flow PM owing to high accuracy of PMU measurements is considerably higher than accuracy of SCADA measurements. PMU measurements can be used to check the quality of SCADA measurements by the TE technique.

At the end of 2015, about 400 SMART-WAMS recorders were installed in the eastern synchronous zone of the UES of Russia. A priority set for creation of the smart grid in Russia is the intensive development of WAMS and WACS systems.

To obtain a general picture of the EPS state for further solving the control problems, the universal methods are necessary to place PMU to complement SCADA measurements. These methods should provide the best properties of the SE problem solution, such as observability of the calculated scheme, identifiability of bad data, and accuracy of obtained estimates. These methods should be based on the observability theory that was devised to place SCADA systems [14] and take into account the block character of PMU measurements. Different methods were offered to solve this problem. In the authors' opinion, the most promising are topological approaches based on different strategies of random search [15–17].

3.2 The electric power system state estimation problem. Specific features of state estimation for the control of IESAAN

SE is the most important procedure that provides EPS control with reliable and quality information. The result of SE is a model of the EPS steady state. The model is based on measured parameters of the system and data on the state of network components.

The SE problem consists of calculating the EPS steady-state conditions by measurements. The measurements applied in the EPS SE are mostly the measurements obtained from SCADA system and PMUs: U_i—magnitudes of nodal voltages; P_{g_i}, Q_{g_i}—generation of active and reactive powers at nodes; P_{n_i}, Q_{n_i}—loads of active and reactive powers at nodes; P_{ij}, Q_{ij}—power flows in transformers and lines; δ_i—voltage phases at the nodes of the scheme in which PMUs are placed; I_{ij}—magnitudes of currents in the lines incident to these nodes; and ϕ_{ij}—an angle between current and voltage.

The vector of measurements looks as follows:

$$\bar{y} = \left(P_i, Q_i, P_{ij}, Q_{ij}, U_i, \delta_i, I_{ij}, \phi_{ij} \right)$$

where $P_i = P_{n_i} + P_{g_i}$, $Q_i = Q_{n_i} + Q_{g_i}$.
The measurement model is as follows.

$$\bar{y}_i = y_{ss(i)} + \xi_{y(i)}$$

$y_{ss(i)}$—true value, $\xi_{y(i)}$—normally distributed random error $\xi_{i(y)} \in N(0, \sigma_{i(y)}^2)$, where $\sigma_{i(y)}^2$—a variance of the ith measurement.

The traditional mathematical statement of the SE problem consists in minimization of the objective function:

$$J(y) = (\bar{y} - \hat{y})^T R_y^{-1} (\bar{y} - \hat{y}) \tag{5}$$

subject to constraints in the form of steady state equations:

$$w(y, z) = 0 \tag{6}$$

where R_y is a diagonal matrix, the elements of which are equal to measurement variances.

The state vector x is the $(2n - 1)$-dimensional vector (where n is the number of nodes in the calculated scheme). The vector includes the magnitudes U and phase angles δ of voltages $x = (U, \delta)$. Such a state vector determines all the other state parameters.

By using dependences $y = y(x)$ as Eq. (6), the SE problem is reduced to a search for the values (estimates) of components of the state vector $\hat{x}$, such that the values $y(\hat{x})$ are maximum close to the measurements in a sense of the criterion:

$$J(x) = (\bar{y} - y(x))^T R^{-1} (\bar{y} - y(x)) \tag{7}$$

This method is presented in detail in Refs. [18–20].

The SE method based on the use of TEs was developed in Refs. [18,21]. TEs are steady-state equations that contain only measured state variables y

$$w_{te}(y) = 0 \tag{8}$$

which can be obtained from the system of steady-state equations (6) by elimination of unmeasured variables. By substituting the values of measurements in these equations, one can judge the presence of gross errors in measurements from the value of discrepancies obtained; therefore they are called TEs.

When TEs are used, the SE problem consists of the minimization of the criterion (5), that is, the calculation of estimates of measured variables $\hat{y}$ under the constraints represented by a system of TEs (8). After that, the estimates of basic measurements are used to calculate the state vector estimates, and these are used to calculate the estimates of unmeasured variables.

The TEs can be used to solve many problems related to the real-time SE: analysis of network topology, BDD, SE procedure, and identification of measurements variances.

The existing methods and approaches that are applied today in SE were developed in the conditions of insufficient telemetry data; low redundancy of measurements and incomplete visibility of the network; asynchronous measurements and considerable delays in their transfer; and low accuracy and reliability of analog measurements and remote signals. These conditions determined the main direction in the evolution of the theory and practice of EPS SE.

The creation of IESAAN suggests a wide use of

—new technologies and systems for measurement of state parameters (digital instrument transformers and transducers, PMUs, etc.),
—modern communication systems that radically change the processes of data collection, processing, and transfer, and
—digitalization of substations.

Change in the properties of an observed EPS itself due to saturation of the network with active components (FACTS, HVDC) forms new conditions for SE of EPS and puts forward new objectives to be accomplished in the modernization of the data computing system of the IESAAN.

—The authors outline the following main directions in the development of SE methods and technologies of their application to control of an intelligent energy system with an active-adaptive network [22]:
  (1) Form a mathematical model of EPS in terms of new systems for measurement of EPS state parameters (PMU, IED) and their control (FACTS, energy storage systems, etc.), and regular updating of the model parameters.
  (2) Develop methods for the decomposition of the SE problem for calculation of EPSs with a multilevel hierarchical structure, and their implementation on the basis of modern network technologies and multiagent approaches.
  (3) Provide a methodology for using the synchronized PMU data to increase the efficiency of algorithms for EPS SE and algorithms for the decomposition of the SE problem.

(4) Develop methods for checking the validity of measurements on the basis of a priori BDD methods to apply new high accuracy devices for measurement of EPS state variables (PMU, smart power meters).

(5) Develop methods for the DSE to forecast and monitor operating conditions on the basis of new systems and technologies (PMU, artificial intelligence methods, etc.).

(6) Develop methods ensuring observability in case of failure of individual lines, individual remote terminal units (RTUs), and/or PMU, or combinations thereof, in accordance with a given degree of reliability of observability.

(7) Apply the artificial intelligence methods (ANNs, genetic algorithms, multiagent systems, simulated annealing) in algorithms for EPS SE.

3.3 The main directions in the development of SE methods and technologies of their application to control of IESAAN

3.3.1 Phasor measurements in the state estimation problem

The issues of SCADA and PMU data integration are algorithmically resolved as follows: measurements of magnitudes and phases of currents are considered as additional components of the measurement vector $\bar{y}$, and measurements of voltage magnitudes and phases can be emphasized especially because they are simultaneously components of both measurement vector and state vector x:

$$\bar{x}_{\delta_i} = x_{\delta_i} + \xi_\delta, \quad \bar{x}_{U_i} = x_{U_i} + \xi_U \tag{9}$$

where ξ_δ, ξ_U—errors in measurements δ, U, that have normal distribution with zero mean, R_δ, R_U—diagonal covariance matrices with the measurement error variances of vector components x on their diagonal.

Problem statement: By using the given $\bar{y}, R_y, \bar{\delta}, R_\delta, \bar{U}, R_U$ find estimates $\hat{\delta}, \hat{U}$ by minimizing the objective function

$$\min_{\delta, U} \left[(\bar{y} - y(x))^T R_y^{-1} (\bar{y} - y(x)) + (\bar{\delta} - \delta)^T R_\delta^{-1} (\bar{\delta} - \delta) + (\bar{U} - U)^T R_U^{-1} (\bar{U} - U) \right] \tag{10}$$

Expressions for the calculation of estimates are obtained when derivatives with respect to δ and U equal zero:

$$\hat{\delta} = \left(\frac{\partial y}{\partial \delta} \right)^T R_y^{-1} (y(\delta) - \bar{y}) + R_\delta^{-1} (\bar{\delta} - \delta) \quad \text{and}$$

$$\hat{U} = \left(\frac{\partial y}{\partial u} \right)^T R_y^{-1} (y(U) - \bar{y}) + R_u^{-1} (\bar{U} - U) \tag{11}$$

Irrespective of the statement and method of solving the SE problem, the accuracy of the obtained estimates is determined by covariance matrices of erroneous estimates of the state vector $\widehat{\delta}, \hat{U}$:

$$P_\delta = \left[\left(\frac{\partial y}{\partial \delta} \right)^T R_y^{-1} \frac{\partial y}{\partial \delta} \right]^{-1} \tag{12}$$

$$P_u = \left[\left(\frac{\partial y}{\partial U} \right)^T R_y^{-1} \frac{\partial y}{\partial U} \right]^{-1} \tag{13}$$

When adding the measurements of magnitudes and phases of bus voltages obtained from PMU, the expressions (12), (13) are transformed into the form:

$$P_\delta = \left[\left(\frac{\partial y}{\partial \delta} \right)^T R_y^{-1} \frac{\partial y}{\partial \delta} + R_\delta^{-1} \right]^{-1} \tag{14}$$

$$P_u = \left[\left(\frac{\partial y}{\partial U} \right)^T R_y^{-1} \frac{\partial y}{\partial U} + R_u^{-1} \right]^{-1} \tag{15}$$

Criterion

$$\max \det [P_\delta, P_u] \tag{16}$$

where P_δ, P_U are determined by Eqs. (14), (15), can be used as a criterion for choosing the set of PMU placement from the viewpoint of maximum accuracy of estimates [17].

Results of the studies on the combined use of SCADA and PMU measurements, including those obtained by [11], indicate that in this case the main flaws characteristic of traditional SE still exist.

If the amount of PMUs is sufficient to provide the required observability of the EPS network, then SE can be made only on the basis of PMU data.

3.3.1.1 The use of TEs for validation of measurements

The TEs technique is used for a priori validation [23]. With substitution of measurements in the TE, the error in the measurements affects the value of the TE discrepancy. The presence of error can be detected by checking the condition

$$|w_k| < d \tag{17}$$

The value of threshold for the jth TE d_j is determined as

$$d_j = \gamma_\alpha \sqrt{\sum_{i \in \omega_j} a_{ij} \sigma_i^2} \tag{18}$$

where ω_j a set of measurements entering the jth TE, σ_i^2—variances of measurements, γ_α—inverse distribution $N(0,1)$, determined by the given probability of bad data penetration in the SE problem α; a_{ij}—coefficients of TE. If this (17) is not met, all the measurements in this TE should be analyzed. In this case, if all measurements except one are valid, the measurement with a gross error is replaced by a PM. If one TE contains several erroneous measurements, the additional information is required.

The algorithms of SE on the basis of the TE technique are less labor intensive and are very fast because the order of the TE set is, as a rule, considerably lower than the order of the initial set of steady-state equations that is used to obtain TEs.

The TE method was developed to detect bad data in SCADA measurements [23], then adapted to analyze the validity of the PMU measurements [18].

There are some reasons for the failures in PMU operation and the appearance of bad data in their measurements: low accuracy of instrument transformers at PMU inlets; difference in the error ranges in PMUs of different manufacturers, which manifests itself under large disturbances; failures in the system of data acquisition and transfer from PMU to the upper level; as well as the human factor (e.g., erroneous phase connection). Thus, the necessity arises to verify synchronized vector measurements coming from PMUs.

As is shown in Ref. [24], if calculations are made in rectangular coordinates of the state vector and the magnitudes and phase angles of currents are used as measurements, the measurement model becomes linear and the state vector estimates $\hat{x}$ can be obtained from the equation

$$\hat{x} = \left[H^T R_y^{-1} H\right]^{-1} H^T R_y^{-1} \bar{y} \tag{19}$$

at the constant Jacobian matrix $H = \frac{\partial y}{\partial x}$. State vector of the system x is defined by $x = \{\dot{U}_i\}$, where $\dot{U}_i = U_{ai} + jU_{ri}$—complex numbers of nodal voltages, vector of measurements $\bar{y}$ is also represented by the complex number vector $\bar{y} = \{\dot{U}_i, \dot{I}_{ij}, \dot{I}_i\}$.

The TEs are based on the exclusion of components of state vector x from the equations describing the relationships between measured variables y and state vector x in rectangular coordinates:

$$y - H(x) = 0 \tag{20}$$

Each measurement of the electrical current complex number in a line will correspond to two equations from (20):

$$\bar{I}_{ija} - (U_{ai} - U_{aj})y_{aij} + (U_{ri} - U_{rj})y_{rij} - U_{ai}y_{aiij} + U_{ri}y_{riij} = 0 \tag{21a}$$

$$\bar{I}_{ijr} - (U_{ai} - U_{aj})y_{rij} - (U_{ri} - U_{rj})y_{aij} - U_{ai}y_{riij} - U_{ri}y_{aiij} = 0 \tag{21b}$$

where y_{aij}, y_{rij}—series conductance and susceptance, and y_{aiij}, y_{riij}—real and imaginary part of shunt admittance of line $i - j$.

Each voltage phasor measurement at node i will correspond to two equations:

$$\overline{U}_{ai} - U_{ia} = 0 \tag{22a}$$

$$\overline{U}_{ri} - U_{ri} = 0 \tag{22b}$$

The problems of PMU-based linear SE and BDD are addressed in a great number of studies. An in-depth explanation of the linear SE procedure is presented in Ref. [25], where the authors suggest a three-phase SE. To speed up the computation process, the author divides all the complex numbers into real and imaginary components and transforms the matrices describing the state of the system. In Ref. [26] the authors show that before the procedure of linear SE starts, it is necessary to make sure that the considered object of estimation (substation, region, intersystem tie line) can be reliably monitored by PMU and the PMU data can be verified. Also, the authors of Ref. [26] note that (a) no traditional measurements including zero injections should be used for linear SE, and (b) no PMU can be critical.

The problem of SE using the TEs is solved in two stages [27].

The first stage suggests a search for the estimates (20) of measured state variables entering the TE through the minimization of criterion (9) under the constraints represented by the system of TEs (12).

$$\hat{y} = \overline{y} - R\left(\frac{\partial w_k}{\partial y}\right)^T \left[\left(\frac{\partial w_k}{\partial y}\right) R \left(\frac{\partial w_k}{\partial y}\right)^T\right]^{-1} w_k(\overline{y}) \tag{23}$$

Because the basic set of measurements y_b that enables us to determine $x = f(y_b)$ has already been chosen automatically by the software in the process of constructing TEs, it is only necessary at the second stage to determine the estimates of state vector $\hat{x}$ on the basis of the estimates of basic measurements $\hat{y}_b$ that were obtained from the equation:

$$H_b\hat{x} = -\hat{y}_b \tag{24}$$

and calculate the unmeasured variables.

3.3.1.2 Systematic errors in PMU measurements

The systematic errors caused by the errors of the instrument transformers that exceed the class of their accuracy are constantly present in the measurements and can be identified by considering some successive snapshots of measurements. The TE linearized at the point of a true measurement, taking into account random and systematic errors, can be written as:

$$w_k(\overline{y}) = \sum_{l \in \omega_k} \frac{\partial w}{\partial y_l}\left(\xi_{y_l} + c_{y_l}\right) = \sum a_{kl}\xi_{y_l} + \sum a_{kl}c_{y_l} \tag{25}$$

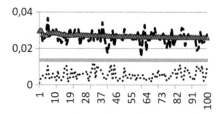

Fig. 5 Detection of a systematic error in the PMU measurements and identification of mathematical expectation of the test equation.

where $\sum a_{kl}\xi_{y_l}$—mathematical expectation of random errors of the TE, equal to zero; $\sum a_{kl}c_{y_l}$—mathematical expectation of systematic error of the TE, ω_k—a set of measurements contained in the kth TE.

The author of Ref. [28] suggests an algorithm for the identification of a systematic component of the measurement error on the basis of the current discrepancy of the TE. The algorithm rests on the fact that systematic errors of measurements do not change through a long time interval. In this case, condition (17) will not be met during such an interval of time. Based on the snapshots that arrive at time instants 0, 1, 2, …, $t-1$, t…, the sliding average method is used to calculate the mathematical expectation of the TE discrepancy:

$$\Delta_{w_k}(t) = (1-\alpha)\Delta_{w_k}(t-1) + \alpha w_k(t) \tag{26}$$

where $0 \leq \alpha \leq 1$.

Fig. 5 shows the curve of the TE discrepancy (a thin dotted line) calculated by (26) for 100 snapshots of measurements that do not have systematic errors.

It virtually does not exceed the threshold $d_k = 0.014$ (a light horizontal line). Above the threshold, there is a curve of the TE discrepancy (a bold dotted line) that contains a measurement with a systematic error and a curve of nonzero mathematical expectation $\Delta_{w_k}(t) \in [0.026; 0.03]$ (a black-blue thick line). However, the nonzero value of the calculated mathematical expectation of the TE discrepancy can only testify to the presence of a systematic error in the PMU measurements contained in this TE, but cannot be used to locate it.

3.3.1.3   Decomposition

Currently the decomposition problem of a large-dimension EPS is very important. The effective method of solution is the distributed processing of the information on the basis of decomposition. This approach is offered in articles of the Russian [29,30] and foreign scientists [31–34], etc. Distributed SE includes the following main procedures:

(1) decomposition of the entire calculated scheme into local areas,
(2) SE solving for each local area using available local area measurements,
(3) solution of the coordination problem consisting of calculation of the boundary variables and a check of the boundary conditions, and
(4) aggregation of calculation results for individual subsystems and transfer of aggregated data to the control center.

As boundary conditions at decomposition with boundary nodes, the equality of voltage modules U and angle phases δ of boundary nodes should be observed:

$$U_i = U_j = \cdots = U_k \tag{27}$$

$$\delta_i = \delta_j = \cdots = \delta_k \tag{28}$$

Also the relationships of boundary balance should be observed. For example, for a boundary node l, which is general for $i, j, \ldots, k$th subsystems

$$P_{\text{load}_l} - P_{\text{gen}_l} + \sum_{s=i,j,\ldots,k} \sum_{m \in \omega_s} P_{lm}(U_l, \delta_l, U_m, \delta_m) = 0 \tag{29}$$

$$Q_{\text{load}_l} - Q_{\text{gen}_l} + \sum_{s=i,j,\ldots,k} \sum_{m \in \omega_s} Q_{lm}(U_l, \delta_l, U_m, \delta_m) = 0 \tag{30}$$

where $P_{\text{load}_l}, Q_{\text{load}_l}, P_{\text{gen}_l}, Q_{\text{gen}_l}$—active and reactive powers of loads and generations in a node l, ω_s—set of nodes of sth subsystem, incident to lth node.

As boundary conditions at decomposition by boundary lines, the relationships for active P_{ij} and reactive Q_{ij} power flows from the ith subsystem to the jth subsystem should be observed:

$$P_{ij} = -P_{ji} + \Delta P_{ij} \tag{31}$$

$$Q_{ij} = -Q_{ji} + \Delta Q_{ij} \tag{32}$$

For boundary line

$$U_m^2 - \left(U_l - \frac{P_{ml}r_{ml} + Q_{ml}x_{ml}}{U_m}\right)^2 - \left(\frac{P_{ml}x_{ml} - Q_{ml}r_{ml}}{U_m}\right)^2 = 0 \tag{33}$$

$$\delta_m - \delta_l - \text{arctg}\frac{P_{ml}x_{ml} - Q_{ml}r_{ml}}{P_{ml}r_{ml} + Q_{ml}x_{ml}} = 0 \tag{34}$$

Installation of PMU at boundary nodes allows registering boundary variables U and δ on the values metered with high accuracy. In this case:

−boundary conditions Eqs. (27), (28) are fulfilled automatically and
−the solution of a coordination problem consists of the calculation of node injections in boundary nodes on Eqs. (29), (30), using estimations of line power flows received from results of separate subsystems.

For coordination of phase angles of the voltages received from local SE, only one PMU measurement of the phase angle is enough in each subsystem. Such a node is appointed for subsystems as a reference node. PMU measurements coordinate the results of the SE of separate subsystems. Therefore, it is necessary to install the device PMU in a reference node or to appoint the node with PMU as a reference. Thus SE of separate subsystems can be calculated in parallel, independently from each other; accomplishment of iterative calculations on subsystems is not required.

3.3.2 State estimation of electric power system involving FACTS models

Successful (reliable, quality, and economical) operation of IESAAN requires a wide range of advanced technical tools and technologies that afford the opportunity to endow the network with active-adaptive qualities.

There is a persistent trend in the international and Russian practices toward adopting the power electronics controlled devices in the electrical networks of EPSs, that is, flexible alternating current transmission systems (FACTS). Today, FACTS is among the most promising electrical network technologies [1,35,36], which makes it possible for the electrical network to turn from a passive facility for electricity transportation into the facility that takes an active part in the control of the network operation. The FACTS devices can be used to increase the transfer capability, improve static and dynamic stability, and provide better power quality. This technology provides control of interrelated parameters, including impedances, currents, voltages, phase angles, oscillation damping at different frequencies, etc., and opens up new opportunities to control EPSs.

Until recently, the models for the present-day FACTS devices based on power electronics were not involved in the design diagram when solving the SE problem. Developing mathematical EPS models for SE accounting for new devices is a topical problem nowadays. Such investigations and developments have been actively conducted in recent years [37–39]. A team of researchers at the Energy Systems Institute has been investigating modeling FACTS devices when solving the SE problem. By now, the models for the main FACTS devices and the algorithms to involve these models in the SE problem have been developed [40–43].

Because the parameters of equivalent circuits of FACTS devices change depending on EPS operating conditions, the problem of determining the parameters of equivalent circuits of these devices on the basis of PMU data is also topical. Therefore, developing the FACTS device models, identifying the parameters of these models, and involving them in the algorithms of the present-day EPS SE is a topical problem when creating the system of SG control.

Such an approach implementation when modeling TCSC and SVC at the EPS SE is considered below, together with a brief description of the models for these devices.

Thyristor-controlled series capacitors (TCSCs). Series capacitors (SCs) are the capacitor batteries connected in a series with a power line to compensate a part of a series inductive reactance. In different countries, SCs are widely applied in the regions where energy sources are far from customers, such as in Sweden. One of the first-ever SCs, the Tyret' SC, was installed on the lengthy 500 kV Bratsk-Irkutsk high-voltage line (~700 km) within the Irkutsk power system. The Tyret' SC is a group of capacitor batteries/banks, or bridges. Controlling the state by reactive power

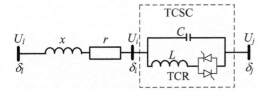

Fig. 6 Power transmission with TCSC.

occurs due to switching the bridges on and off. Switching the bridges on and off is quite problematic to perform in the process tempo. The alternative option is using TCSCs, in which a part of the capacitor bank is shunted by a thyristor controller that enables it to smoothly vary its equivalent capacitance depending on the line operation state.

TCSCs, in fact, is a standard SC but supplemented with a thyristor control block. Fig. 6 presents the power transmission scheme with a TCSC.

The TCSC impedance comprises the impedance of the parallel-connected bank and reactor [40], and depends on the thyristor firing angle α:

$$x_{\text{TCSC}}(\alpha) = \frac{x_{\text{TCR}}(\alpha) \cdot x_C}{x_{\text{TCR}}(\alpha) + x_C} \tag{35}$$

where

$$x_{\text{TCR}}(\alpha) = \omega L \cdot \frac{\pi}{\pi - 2\alpha - \sin 2\alpha} \tag{36}$$

At SE, TCSCs are set by the model with variable conductance, $x_{\text{TCSC}}(\alpha)$, or by the model with a variable firing angle α. An algorithm implementing the former model is developed. In this algorithm, TCSC is modeled by the $i-j$ line with a variable reactance ($y_r = 1/x_{\text{TCSC}}$ conductance) and by the active-power fixed overflow equal to the SCADA measurements. The nods restricting the $i-j$ line are transit (have zero injections). The set of equations, in the designations typical of SE [21] for the $i-j$ line shown in Fig. 6, looks like:

$$P_{i-j} = U_i U_j \sin\left(\delta_i - \delta_j\right) y_r$$

$$Q_{i-j} = U_i^2 y_r - U_i U_j \cos\left(\delta_i - \delta_j\right) y_r$$

The y_r susceptance is set as a state vector component, and is determined directly during the SE problem solution. The Jacobian matrix column corresponding to this state vector component contains derivatives of nodal injections at the nodes i and j, calculated through the formulas:

$$\frac{\partial P_i}{\partial y_r} = \frac{\partial P_{i-j}}{\partial y_r} = U_i U_j \sin\left(\delta_i - \delta_j\right)$$

$$\frac{\partial Q_i}{\partial y_r} = U_i^2 - U_i U_j \cos\left(\delta_i - \delta_j\right)$$

Similarly, one can calculate the derivatives of P_j, Q_j.

By using the Jacobian matrix formed in this way, the SE can be performed, during which the y_r estimation, and later the x_{TCSC} estimation are calculated. The x_{TCR} is determined from the x_{TCSC} using (35). The thyristor firing angle α is calculated iteratively from (36), represented in the form:

$f(\alpha_i) = \pi - 2\alpha_i - \sin(2\alpha_i) - \frac{\omega \pi L}{X_{TPr}}$, where i is the iteration number.

Static var. compensator (SVC) is a multipurpose static device providing constant voltage regulation and a smooth or stepwise variation in the consumed and (or) reactive power produced by it at its connection buses. The SVC basis is accumulative devices (capacities, inductances), reactor-thyristor, and capacitor-thyristor blocks. In most cases, an SVC device comprises thyristor-switched capacitors (TSC) and a thyristor-controlled reactor (TCR). There are other possible combinations of devices, for example, a separate TSC or a separate TCR.

The smooth control of the reactive power in SVC devices is performed through a variation in the reactor's thyristor firing angle α. To maintain the specified voltage at a node, one should determine angle α.

When solving the SE problem, SVC is modeled by a susceptance variable at node i of the SVC installation, and either conductance b_{SVC}, or angle α is involved in the state vector x instead of U_i that is fixed. In both cases, one should calculate derivatives by these variables from measuring the injection at this node, to calculate which nodal balance equation for a reactive power is used:

$$Q_i = Q_{gi} + Q_{li} + \sum_{j \in \omega_i} Q_{ij} + Q_i^{\text{sh}} \tag{37}$$

where: Q_{gi}, Q_{li} is the reactive power generated and consumed in node i, $\sum_{j \in \omega_i} Q_{ij}$ is the total of the reactive power overflows over the lines incident to node i, ω_i is a set of nodes incident to the ith, Q_i^{sh} is a shunt reactive power at node i determined for SVC by formula:

$$Q_i^{\text{SVC}} = U_i^2 b_{\text{SVC}}(\alpha) \tag{38}$$

In (37), only the last summand depends on $b_{\text{SVC}} \cdot (\alpha)$. Calculating the derivative of Q_i^{SVC} with respect $b_{\text{SVC}} \cdot (\alpha)$ does not cause difficulties, and the entire SE algorithm, in this case, practically coincides with the algorithm developed for TCSC. To calculate angle α after obtaining the $b_{\text{SVC}}(\alpha)$ estimation, the expression presented in [44,45] is used:

$$b_{\text{SVC}}(\alpha) = \frac{1}{X_C X_L} \left\{ X_L - \frac{X_C}{\pi} [2(\pi - \alpha) + \sin 2\alpha] \right\} \tag{39}$$

where X_C is the capacitive and X_L is the inductive SVC impedances.

The derivative of Q_i^{SVC} with respect to α is calculated as:

$$Q_i^{\text{SVC}} = \frac{U_i^2}{X_C X_L}\left\{ X_L - \frac{X_C}{\pi}[2(\pi - \alpha) + \sin 2\alpha]\right\}, \tag{40}$$

obtained upon substituting (39) into (38). The derivative of the injection measurement at the SVC installation node with respect to α is:

$$\frac{\partial Q_i}{\partial \alpha} = \frac{2U_i^2 b_L}{\pi}(1 - \cos 2\alpha) \tag{41}$$

Both algorithms were developed to test the method. The calculation results and their comparison are presented below.

A 19-node scheme of the Irkutsk power system is used for calculations (Fig. 7). The scheme contains 19 nodes, 28 lines, and 94 measurements (SCADA measurements and PMs of zero injections at transit nodes).

Modeling TCSC. The calculations were performed for a scheme with a TCSC connection instead of a discretely controlled Tyret' SC (line 3–4) for two states: when transmitting 1386 and 1852 MW over the Bratsk-Irkutsk 500 kV line. The Tyret' device has the full capacitance $X_{\text{SC}} = -26.3\,\Omega$; this value was accepted as an initial approximation when calculating with TCSC. Table 2 shows how the x_{TCSC} values varied during iterative calculations.

The calculations show that the series compensation rate varies with the change in the transmitted power value. For the 1386 MW transmission, it is necessary to compensate the Bratsk-Irkutsk transit reactance by $-18.36\,\Omega$. If the transmitted power is higher (1852 MW), the compensation will be also higher, by $-25.22\,\Omega$.

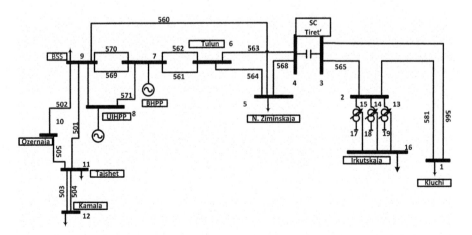

Fig. 7 Scheme of the Irkutsk electric power system.

Table 2 **Variation in x_{TCSC} (Ω) during iterative calculations**

Iteration no.	Transmitted active power	
	1386 MW	1852 MW
1	−26.30	−26.30
2	−18.51	−25.40
3	−18.46	−25.33
4	−18.39	−25.27
5	−18.36	−25.22

Modeling SVC. In the real scheme, to compensate the redundant reactive power at the 500 kV Irkutskaya substation, synchronous compensators (SCs) are installed at low-voltage nodes 17, 18, and 19. For calculations, an SVC installation (instead of SCs) was modeled at the same nodes. Table 3 presents the calculation results by the two algorithms.

As it follows from the table, practically conterminous values of SVC conductance and angles α are obtained in both calculations. These calculations show that SVCs operate in the inductive state, and compensate the redundant reactive power generated by the 500 kV high-voltage line of an extended Bratsk-Irkutsk transit, thereby maintaining the specified voltages at the 220 kV buses of the Irkutskaya substation (node 16).

3.3.3 Dynamic state estimation and its application

The quality of the results of the SE problem solved online can be improved by using retrospective information on the state variables. The algorithms are developed for this purpose which capable of considering interrelations between the time-varying state variables unlike the SE algorithms which use the only snapshot. The SE problem, for which a set of snapshots is the initial information, is called DSE [46–50].

The main DSE advantages are the ability to deliver good results in the cases of bad and incomplete data and to forecast state variables. In this chapter the DSE is based on the extended Kalman filter.

Based on a priori knowledge of the system behavior, the state space representation for discrete time-variant systems looks as follows [46].

$$x_{k+1} = F_k x_k + \xi_{F(k)} \tag{42}$$

$$y_k = H_k x_k + \xi_{y(k)} \tag{43}$$

The first type of equations (42) is a dynamics model of the process $x(t)$. The second type of equations (43) is the dependence of measurements $\bar{y}$ on state vector x ($\bar{y}(x)$).

The unmeasured state variables can be calculated at the availability of a certain set of variables that allows the values of all state variables to be obtained by the

Table 3 SE results with the SVC model

Node no.	kV	U_{SE} (kV)	δ (deg.)	P_{mes} (MW)	P_{SE} (MW)	Q_{mes} (MVAr)	Q_{SE} (MVAr)	b (Ω^{-1})	α
Algorithm I									
16	224	224	-8.1	-394	-393	-397	-397		
17	10.5	10.5	-8.2	0	0.0	–	-15	0.136	-1.086
18	10.5	10.5	-8.2	0	0.00	–	-15	0.136	-1.086
19	10.5	10.5	-8.2	0	0.00	–	-15	0.136	-1.086
Algorithm II									
16	224	224	-8.1	-394	-395	-397	-397		
17	10.5	10.51	-8.2	0	0.00	–	-15.1	0.137	-1.085
18	10.5	10.51	-8.2	0	0.00	–	-15.1	0.137	-1.085
19	10.5	10.51	-8.2	0	0.00	–	-15.1	0.137	-1.085

noniterative technique. Such a set of variables is the $2n$-dimensional voltage magnitude and phase $x = (U, \delta)$, where n—number of nodes in the calculated scheme and is called a state vector. In the SE problem, $2n - 1$ components of vector $x = (U, \delta)$ are calculated via the measured state variables by minimization of some objective function. One state vector component (voltage phase at the slack node) is considered as constant.

The objective function in the DSE looks as follows:

$$J(x) = (\bar{y} - y(x))^T R^{-1} (\bar{y} - y(x)) + (\tilde{x} - x)^T M^{-1} (\tilde{x} - x) \tag{44}$$

where $\tilde{x}$—forecast of the measured state vector components, R—covariance matrix of the measurement error, and M—covariance matrix of the forecasting errors, which is calculated in the process of the Kalman filter formation.

$$M_{k+1} = F_k P_k F_k^T + W_{F(k)} \tag{45}$$

$W_{F(k)}$—covariance matrix of dynamics model noise, P_k—covariance matrix of the state vector component estimates. The matrix of transition F is assumed to be a unit matrix because the time frames considered are small enough and the system changes extremely slowly. The covariance matrix $W_{F(k)}$ and P_k are calculated by the recurrence relations.

In the objective function (44), the measured and forecasting values are represented by different terms. The addend considers the process history.

The objective function (44) can be written in the form

$$J(x) = \left(\bar{y}_f - y(x) \right)^T R_f^{-1} \left(\bar{y}_f - y(x) \right) \tag{46}$$

where $R_f = \begin{bmatrix} R & 0 \\ 0 & M \end{bmatrix} - [(m + 2n - 1) \times (m + 2n - 1)]$ matrix, $\bar{y}_f = \begin{bmatrix} \bar{y} \\ \tilde{x} \end{bmatrix} - [(m + 2n - 1) \times 1]$, m—number of measurements.

The criterion (46) is minimized by the state vector to calculate estimates of the state vector components and the DSE problem is reduced to solving a system of nonlinear equations:

$$H_f^T R_f^{-1} \left(\bar{y}_f - y(x) \right) = 0 \tag{47}$$

where $H_f = \begin{bmatrix} H \\ H_m \end{bmatrix}$, which is a $[(m + 2n - 1) \times (2n - 1)]$ matrix.

The system of equations (47) is not underdetermined, that is,

$$rank \left(\frac{\partial y_f}{\partial x} \right) \geq 2n - 1 \tag{48}$$

Due to the nonlinear dependence $y(x)$, the problem is solved iteratively by Newton's method. The system is linearized at each iteration and a system of linear equations is solved where correction vector is calculated by the equation

$$\Delta x_k^i = P_{f(k)}^i H_{f(k)}^{T(i)} R_f^{-1} \Delta y \tag{49}$$

$$P_f^i = \left(H_f^{T(i)} R_f^{-1} H_f^i \right)^{-1} \tag{50}$$

where $\Delta y = \bar{y}_f - y(x_k^i)$, k—snapshot number, i—iteration number.

3.3.3.1 Criteria for the estimate accuracy
The estimate accuracy is determined according to the following equation:

$$\phi_{SE} = \sum_{i=1}^{m} \frac{(\bar{y}_i - \hat{y}_i)^2}{\sigma_{(y)i}^2} + \sum_{j=1}^{m1} \frac{(\bar{x}_j - \tilde{x}_j)^2}{\sigma_{(M)j}^2} \tag{51}$$

where $\sigma_{(y)i}^2$—the variance of ith measurement (component of matrix R), $\sigma_{(M)j}^2$—the variance of forecast error of jth state vector component (component of matrix M).

3.3.3.2 Detection of bad data in measurements by the methods of dynamic EPS state estimation
The calculation model of the current EPS state is constructed using the SE method. The SE result is the calculation of the EPS steady state (the current state) based on the measurements of state variables and the data on the scheme topology state. The power system state is estimated by one snapshot of telemetry through the static SE algorithms. Availability of several snapshots allows the application of the dynamic SE algorithms [28,50,51].

The SE result is correct if there are no gross errors in the initial data. Such errors in initial information are the source of distortion of the calculated state obtained by SE, which can lead to wrong decisions at the EPS control and development of severe emergencies. Therefore the detection of gross errors (hereafter—BDD) in measurements and suppression of their effect on the estimates of the EPS state variables is a most topical issue in the SE problem solution [11]. In real time, the preference is given to the algorithms of a priori BDD [11,52], which allows the bad data to be detected and eliminated before calculating the estimates of the state variables.

At present many BDD methods have been devised based on the analysis of:

−balance relations [11];
−innovations [46,53,54];
−estimation residues [55,56];
−retrospective information on one and the same operating parameter [57]; and
−correlation among measurements [11,57].

This work deals with the issues of applying the DSE methods to test the validity of measurements in the areas of low information redundancy, which contain critical measurements and critical sets [58].

3.3.3.3   Description of the devised method

The suggested method of bad data detection in measurements is based on the analysis of retrospective and forecasting information on state variables.

The measurements or estimates obtained at the previous snapshot are applied as retrospective information. The forecast value is calculated using DSE.

The retrospective information is tested in accordance with the conditions:

$$\left| \overline{y}_{i(k)} - \overline{y}_{i(k-1)} \right| \neq 0 \tag{52}$$

$$\left| \overline{y}_{i(k)} - \overline{y}_{i(k-1)} \right| < d_i \tag{53}$$

$$\left| \overline{y}_{i(k)} - \hat{y}_{i(k-1)} \right| < \hat{d}_i \tag{54}$$

Conditions (53), (54) are interchangeable. With the use of DSE, preference is given to (54). In the case of a change in the state, the results of the analysis of inequalities Eqs. (53), (54) will be interpreted as an error in the measurement. This disadvantage is eliminated by forming an additional condition consisting of innovation that takes into account forecasting information; it gives a correct answer in the case of the state change:

$$\left| \overline{y}_{i(k)} - \widetilde{y}_{i(k)} \right| < \widetilde{d}_i \tag{55}$$

In (52)–(55) $\overline{y}_{i(k)}$—ith measurement at snapshot k; $\widetilde{y}_{i(k)}$—forecast of the ith measurement at snapshot k; $\overline{y}_{i(k-1)}$—ith measurement at snapshot $k-1$; $\hat{y}_{i(k-1)}$—estimate of the ith measurement at snapshot $k-1$; d_i, $\widetilde{d}_i$, $\hat{d}_i$—thresholds calculated by the formulas:

$$d_i = \gamma \cdot \sqrt{\sigma_i^2}, \ \ \widetilde{d}_i = \gamma \cdot \sqrt{\sigma_i^2 + \sigma_{N_i}{}^2}, \ \ \hat{d}_i = \gamma \cdot \sqrt{\sigma_i^2 + Y_i} \tag{56}$$

where γ—quantile of distribution $N(0,1)$, is determined by the specified probability of error of the first kind α (at $\gamma = 3$ the measurement with the probability 0.997 is accepted as erroneous at the threshold violation); σ_N^2—diagonal element of matrix $\widetilde{N}$, Y_i—diagonal element of matrix Y, where $Y = HPH^T$, $\widetilde{N} = HMH^T$.

Conditions (52)–(55) can be processed in parallel. As a result of the analysis, the three-digit error code XYZ is formed, where the first element (X) indicates satisfaction—1 or dissatisfaction—0 of inequality Eq. (52), Y and Z are Eq. (53) or (54), (55), respectively. The errors of the first and second kinds are possible by virtue of the erroneous results when using the BDD methods [59]:

(1) The errors of the first kind appear in situations when, as a result of using the method, the true measurement was marked as erroneous and excluded from calculations. Such errors are not disastrous from the standpoint of the quality of SE results but deteriorate observability or even lead to its loss.
(2) The errors of the second kind occur as a result of incorrect work of the BDD methods when the distorted measurements were not detected. Hence, the calculated state turns out to be far from the real one, and the higher the error value is, the worse the result obtained.

The studies performed to minimize the errors of the first and second kinds are presented below.

3.3.4 Supporting cyber-physical security of the electric power system by the state estimation technique

The stability of the network toward physical and cyber intrusion is an attribute of the IESAAN [1]. While developing the conceptual smart grid models and projects, researchers nowadays pay great attention to the issue of cybersecurity.

Security is one of the most important properties of IESAAN, which ensures specified operation of the system in case of changes in conditions, failures of components, and unexpected disturbances. Malicious activities or cyberattacks should also be considered as such disturbances that may lead to a large-scale outage. Security of the IESAAN is necessary for the whole of its structure, which comprises:

−Physical (technological) infrastructure, including power generation, transmission, and consumption.
−Information and communication control infrastructure whose main components are SCADA/EMS and WAMS, WACS, and WAPS systems.

The systems of data collection and processing—SCADA and WAMS, which belong to the subsystems of the smart grid—are most vulnerable to the physical failures and information attacks that are dangerous in terms of their consequences [60–62].

The SE tool is considered as a link between physical and information-communication subsystems. It acts as a barrier to the corruption of data on current operating conditions of the EPS in the control problem, including the data corruption caused by cyberattacks against data collection and processing systems of the EPS.

This section presents the analysis of possible consequences of cyberattacks against these systems and the accuracy of solutions to the SE problem. Also, consideration is given to the methods applied to detect bad data in measurements as well as identifying and mitigating their consequences.

3.3.4.1 Cybersecurity of SCADA systems and WAMS

The SCADA/EMS systems aimed at supporting the dispatching personnel actions in the operation and emergency control of EPS include RTUs installed at EPS substations to record signals on the state of switching equipment and measurements of state variables; communication channels; databases; systems of online representation of state variables; a software package (a set of programs of EMS applications) for processing of measurement results; and formation of control commands for the dispatching control facilities.

There are several vulnerable points in the architecture of SCADA systems, which can be used for cyberattacks. These can be direct damage of RTU communication channels between the RTU and control center and damage of software located on the servers of the EPS control center, including the SE software and databases in the control center.

Cyberattacks against SCADA systems can result in a long-term failure of the data collection system, which can cause the emergence of bad data in measurements. Therefore, it is appropriate to apply the effective bad data detection methods capable of identifying a group of the most severe cyberattacks that can, at present, be missed by technical and physical protection systems of SCADA in most cases of modern intelligent "attacks" [63].

The WAMS is a set of recorders of synchronized PMUs, PDCs, channels connecting recorders, and data concentrators. The existing SCADA measurements and PMU measurements duplicating them make some energy companies think that WAMS is not a strategically important system and the technical means of WAMS are not cybercritical components of the energy infrastructure. However, WAMS is vulnerable to cyberattacks. Analysis of potential cyberattacks [64–66] has shown that the greatest harm to WAMS can be done in the following way:

(1) *The devices of PMU, PDC, and communication infrastructure:* reconnaissance attacks, communication links damage, false data injection, and denial-of-service attacks.
(2) *Reliability of GPS:* spoofing attacks, replay attacks, and jamming.

The specific feature of technological control in the modern EPS is imperative transmission of large data volumes to the upper levels of control. Normally, the data are transmitted through the corporate network, but recently points of interface with the Internet have appeared that make the network more accessible and thus vulnerable. In some cases, it is suggested to apply cloud technologies, particularly in WAMS, for collection and transmission of synchronized vector measurements from the PMU level to the PDC and super-PDC level. With a constantly growing number of cyberattacks against the information structure of energy facilities, the authors of [62] suggest enhancing cybersecurity, firstly, by a reduction in visible zones of potential attacks with the methods of traffic engineering, and, secondly, by the development of the next generation commercial anti-viruses and systems to detect the intrusion and block network attacks, and at the same time to detect the emergence of new network devices.

3.3.4.2 Technique analysis of the cybersecurity of SCADA and WAMS in a two-level state estimation

The PMU measurements coming at a high frequency make it possible to implement fast linear algorithms of SE for areas in the scheme of EPSs that are observed using PMUs.

The TE method makes it possible to perform a priori bad data detection (see Section 3.3.1.1).

The PMU devices are, first of all, placed at large power plants and high voltage substations. Historically, the same facilities were the first to be equipped with telemetry devices. Hence, they were monitored by SCADA measurements. And finally, the same facilities are currently the top priority targets for cyberattacks.

Let us see if the SE procedure copes with the detection of cyberattacks against WAMS system, using PMUs alone.

The modification of the Crout reduction method [27] for the linear estimation of such measurements gives both the estimates and the list of critical measurements at the same time, which is very important under the conditions of potential cyberattacks:

(1) The weaknesses in the system that make it vulnerable to external malicious activities are determined by the presence of critical measurements because the errors in these measurements cannot be detected.
(2) The estimates of the rest of the measurements can help in determining the nature of the cyberattack.

In a case where false information comes from individual recorders, the SE can independently surmount this problem by a priori detecting an erroneous measurement and replacing it with a PM. In the case where a large number of recorders are attacked, the number of erroneous measurements can lead to the situation that the computational process of SE will not converge. A simple check before the SE as to whether the current measurements lie within the technological limits can immediately show if an information failure has occurred in the system. According to the settings, the SE procedure is blocked in the case of a great amount of erroneous measurements to avoid startup of the SE, which can result in its failure.

A great number of rejected measurements can make the system unobservable. In this context a reminder is appropriate that, until now, the SCADA system and WAMS have been independent of one another. Hence, if a local region or an object is observable on the basis of PMU measurements, they are also observable by SCADA measurements. Therefore, it has been suggested, when needed, carrying out independent simultaneous (by one timestamp) bad data detection and SE by SCADA and WAMS measurements to find out if there is a malicious attack against one or another system.

3.3.4.3 Methodology of cyberattack identification
According to the above considerations, the following methodology to identify cyberattacks against SCADA and WAMS is suggested:

(1) Local areas that are totally observable by PMU measurements are singled out in the scheme of an EPS.
(2) For such local areas, the local SE is performed using linear algorithms with unconditional a priori verification of the initial data.
(3) A large number of bad data in measurements or the measurements that go beyond the technological limits—measurements diagnosed as doubtful and unchecked (belonging to one facility)—should mean a complete failure of PDC operation and initiate a search for the reasons for such a failure.

(4) If there are formal signs of a WAMS failure, an independent simultaneous (by one timestamp) bad data detection and SE by SCADA measurements are performed at this energy facility.

(5) The results of the independent procedures for the WAMS and SCADA measurement verifications are used to make a conclusion whether there is a malicious attack against one or another system.

3.3.4.4 Case study

To do calculations based on the suggested technique, a fragment of a real network is considered (Fig. 8). The fragment includes 750/330 kV lines. There are four PMUs (rectangles) at the ends of the lines 1–2 and 1–3 and a PDC installed at node 1 is common for them. Nodes 4–13 are observable by SCADA.

The calculations were done in the simulation experiment when the PMU measurements were simulated by distorting the parameters of the steady-state calculation.

The following types of cyberattacks were simulated:

(1) False data injection was simulated by the technique presented in [65]. A man-in-the-middle attack against the node was simulated to inject wrong measurements in the data. This node intercepted and modified the C37.118 block of data, which was transmitted from PMU-2 to PDC, by doubling the value of all captured vector measurements. The PDC receiving the data failed to distinguish changed packets from unchanged ones. Moreover, the process of changing the data occurred according to the requirements for PDC service life, therefore the changed packet was not marked as an old one and passed in the SE problem.

(2) A denial-of-service attack was simulated by the situation where the measurements did not arrive at PDC. This did not lead to the observability loss but considerably reduced redundancy of PMU measurements, which decreased the efficiency of the bad data detection algorithm.

(3) A desynchronization attack was simulated by changing the phase voltage angle at PMU-4 by 9°, which corresponds to missing one point when taking 40 readings in one 50-Hz cycle. Such a distortion of voltage angle led to a distortion of measurements in other angles: the angle of current in line 3–4 and load angle at the node of PMU placement.

Table 4 presents the SE results based on the data exposed to attacks, and measures undertaken by the SE methods to detect such data.

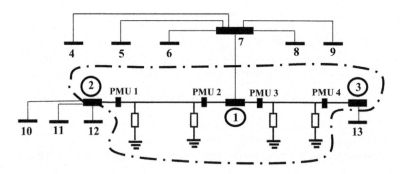

Fig. 8 Fragment of a real network.

Table 4 Detection of an attack by state estimation methods

Type of cyberattack	U (kV); P (MW), Q (MVAr)	Reference y_0	Measurement $\bar{y}$	Value of measurement at attack y_{att}	State estimation results		
					Detection of attack by the state estimation methods	Corrected measurement y_{corr}	Estimate $\hat{y}$
Attack 1, against phasor data concentrator	$U_1 \times 1.5$	757.1	758	1136	$\bar{U}_1$ exceeded the limits 2.7, true	Excluded	757.7
	δ_1,°	2.7°	2.7°	2.7°		2.7°	2.7°
	$P_{12} \times 2.25$	236	235.5	531	$\bar{P}_{12}$, rough error, is substituted:	236	235
	$Q_{12} \times 2.25$	−410	−411	−925	$P_{12} = P_{21} - \Delta P_{21}$ $\bar{Q}_{12}$—doubt measurement, its variance is increased $\sigma_{Q_{12}}$	−925, is not corrected	−402
	$P_{13} \times 2.25$	−1051	−1052	−2370	$P_{13} = P_{31} - \Delta P_{31}$ (see $\bar{P}_{12}$)	−1049.4	−1056
	$Q_{13} \times 2.25$	−126.5	−127	−295	$Q_{13} = Q_{31} - \Delta Q_{31}$ (see $\bar{P}_{12}$)	−81.2	−97
Attack 2, denial-of-service		757.1	—	—	$U_1 = f_1(U_2, I_{21}, \delta_2)$	750.0	757.1
	δ_1,°	2.7°	—	—	$\delta_1 = f_2(U_2, I_{21}, \delta_2)$	—	2.7°
	P_{12}	236	235.5	—	$P_{12} = f_3(U_1, I_{12}, \phi_{12})$	—	234
	Q_{12}	−410	−412	—	$Q_{12} = f_4(U_1, I_{12}, \phi_{12})$	—	−412
	P_{13}	−1051	−1052	—	$P_{13} = f_3(U_1, I_{13}, \phi_{13})$	—	−1050
	Q_{13}	−126.5	−127	—	$Q_{13} = f_4(U_1, I_{13}, \phi_{13})$	—	−128
Attack 3, de-synchronization	U_3	761.8	761.9	761.9	761.9—true	761.9	761.8
	δ_3,°	6.4°	6.4°	15.4°	Error is detected by the TE (6)	6.4°	6.4°
	P_{31}	1058	1056	1056	1056	1056	1057
	Q_{31}	−127	−126.5	−126.5	−126.5	−126.5	−127

Attack 1: Based on the PMU data alone, the SE procedure identified the measurements that arrived at PDC: 1 measurement goes beyond the limits, 3—contain bad data, 1—is doubtful. So many erroneous measurements at one PDC mean a complete failure in its operation and initiate the identification of reasons for such a failure. It is necessary to independently detect bad data by using SCADA measurements.

Attack 2: Accurate data of PMUs 1 and 4 provided accurate estimates at node 2.

Attack 3: As was said above, if an angle shift occurs in a certain PMU channel, the same shift will happen in all phase channels. Therefore, the same shift in the voltage angle δ_3 and current angle ψ_{31} will not affect the angle between voltage and current vectors $\phi_{31} = \delta_3 - \psi_{31}$; hence, this error will not change the values of active and reactive power. In this case, if the neighboring PMU gives a valid angle measurement, then δ_3 will be calculated correctly.

At present the SE procedure is becoming increasingly more important in the conditions of adjustment of the automated systems for acquisition of phasor measurements as well as in the conditions of various cyberthreats. The mathematical tools of SE allow straightening out a tangle of true and erroneous measurements and make certain conclusions on the operability of the devices for collection and primary processing of measurements. The results of bad data detection using two independent data collection and processing systems make it possible to conclude whether or not there is a malicious cyberattack against the considered energy facility.

4   Intelligent operation and smart emergency protection

4.1   Emergency control system in Russia

Emergency control, including emergency operation dispatching and automatic emergency control that provides reliability and survivability of the EPS, takes an important part in controlling the operating conditions of the UES of Russia. Emergency control is performed by the technological (dispatching and automatic) control systems that include the automatic systems of voltage, power, and frequency regulation, basic automatic systems of EPS elements, relay protection and automatic line control, and the ECS [67,68].

Integration of EPSs, liberalization, and modernization of the electric power industry increase the changeability and unpredictability of EPS operation and generate the need to improve and develop principles as well as systems of operation and emergency control. New conditions call for the development of a comprehensive system for monitoring, forecasting, and controlling EPS. Artificial intelligence application is an advanced way to carry out smart emergency control in EPS [4,69,70].

Emergency control philosophy in the UES of Russia is a hierarchical approach and is realized by the coordinated operation of many control devices, which maintain EPS stability and interrupt the expansion of an emergency situation in the case of stability violation and a threat of undesirable cascade emergency development. Coordinated emergency control is realized by joint participation of generators, networks, and consumers.

The electric power object called the UES of Russia is a power interconnection where seven IPSs are combined by weak ties. Under emergency conditions, the UES of Russia as a power interconnection is able to disintegrate into autonomously operating self-balanced IPSs without grave consequences.

At the same time, disintegration of any of the seven IPSs is far from smooth. Because of large power flows in the ties, practically any disintegration creates parts with power lack and power surplus. As a result, load and generator disconnections are quite possible and can acquire a cascade character, that is, turn into a system crash fault. For this reason, at the IPS level the generator and load disconnections by special system emergency devices (in order to unload cutsets and to maintain admissible values of operation parameters) are of course more preferable than cascade disconnection by protection devices (after these parameters go beyond the fixed limits). Just this principle is basic for arranging emergency control in Russian EPSs [67,68,71,72].

The ECS in Russia's IPSs includes the following components (Table 5):

(A) Dispatching emergency control: to provide admissible loading of the electric network ties according to the guidelines on power system stability [73].
(B) Automatic emergency control:
  ➤ *In preemergency conditions*—to maintain necessary transfer capabilities of ties by efficiently controlling them through the use of PSS, control systems of reactive power sources, DC links, etc.
  ➤ *In emergency conditions*
   • The above automatic controllers and control systems to maintain the voltage levels and damp the oscillations in EPS.
   • Automatic stability control of EPS.
   • Automatic frequency load shedding, automatic voltage deviation limiting, automatic isolation of power plants with maintenance of the auxiliary power supply, etc.

Table 5 Emergency operation control services

Conditions	Services	Aims
Preemergency	Emergency dispatching Transfer capability control	To maintain transfer capability margins of transmission lines To increase transfer capabilities of ties by voltage control
Emergency	Emergency control	(1) To provide EPS stability by increasing voltages and damping the oscillations (2) To provide EPS stability by automatic stability control (3) To interrupt emergency development by automatic devices
Postemergency	Restoration	To provide fast EPS restoration by observing the margins and excluding restoration disruption

(C) Restoration of EPS after large emergencies.

The emergency control system of Russia's UES has just now the elements of smart grid ideology, including:

—*adaptability* by coordination of control actions in order, depending on type, time, and severity of emergency; an echeloned control system is tried to interrupt the cascading emergency process sooner rather than later;
—*optimality* in the control actions, especially by a hierarchical two-level automatic centralized ECS that includes the top coordinating subsystem with fast algorithms for optimal choosing of control actions for control devices on the bottom level; and
—*intellectuality* of the decision-making process with the learning and optimal control actions in the hierarchical two-level automatic centralized ECS; such a system has the properties of multiagent systems.

4.2   Requirements for new emergency protection and operation systems

Modern electricity grids continue to be vulnerable to large-scale blackouts. During the past two decades, events in North America, Europe, and Russia have clearly demonstrated an increasing likelihood of large blackouts, which cause huge economic losses; disruptions in communication and transport, heating, and cooling; water supply; emergency services; and financial trading. Blackouts now typically spread across borders, escalating their impacts, and the trend toward national grid integration is likely to increase the risk of even larger power outages in the future.

In recent decades, power industries worldwide have experienced two major changes: liberalization of the electricity market and the expansion of renewable energy. Liberalization has resulted in the separation of power generation, transmission, and distribution businesses. This process has created additional boundaries that have adversely impacted communication and coordination activities between power system operators. The increasing penetration of renewable energy has led to an increasing dependency on the volatile nature of renewable energy sources. It is not only a shortage of power supply that can result from a blackout; oversupply can lead to frequency and voltage instability problems.

Because load centers are often remote from the sources of renewable generation, large amounts of energy often have to be transported over long distances or even through distribution grids that were not designed for this purpose. This leads to higher loading of the grid and the necessity for costly grid upgrades. However, due to several factors—including public resistance—planning and extending the electricity network can take many years. In the meantime, grids have to operate closer to their operational limits, and the probability of large-scale emergencies and blackouts is inevitably increasing.

The analysis showed that at present, the most relevant types of disturbances that lead to major system failures are the voltage collapse (violation of restrictions on voltage levels) as well as overloading equipment (violation of current restrictions). It has been found that currently the interconnections are well protected against these types of disturbances by means of stability control automation. In turn, the appearance and uncontrolled development of voltage instability and current overload in distribution power networks with a complex structure usually leads to serious consequences.

In the near future, the development of the networks in megalopolises and large industrial centers in Russia will result in the formation of systems with a complex multiloop structure. In these conditions, one should expect large-scale system emergencies that will occur according to the scenario in which the electrical current and voltage constraints become decisive in case of emergency operating conditions unfolding. The first such system blackout happened in the UES of Russia in the Moscow power system in May 2005.

It is necessary to note that the existing emergency control concept does not have full defense from voltage collapse because the running processes are much faster than the speed of control action implementation from the emergency control. The applied automatic devices intended for limitation of the voltage drop are extremely slow due to a time delay to consider short circuit duration. The existing automation for limiting the equipment overload does not regulate overload of active and reactive power sources.

Modern automatic stability control can theoretically cope with voltage collapse, but its principle of adjustment does not allow performing an adequate control in off-design conditions under nonstandard disturbances that occur at cascade emergencies in a complex interconnected network of megalopolises. The mentioned flaws of the existing automation as well as an increase in criticality of operating conditions and their control lead to the need to develop automatic intelligent systems for monitoring and control of security, which will operate online and be capable of harmoniously supplementing the modern automatic operation and ECSs without contradicting the concept of their operation. The complexity of the problem of the development of such systems lies in the fact that most of the dangerous preemergency states of EPSs that lead to large blackouts are unique and there is no single algorithm to effectively detect such conditions under the time shortage conditions. In addition, the problem is complicated by the fact that the security limits of EPS are permanently changing.

Large-scale blackouts that occurred in 1965–2014 in the power interconnections of different countries were analyzed in [74]. The analysis made it possible to identify the general regularities of their development, which are expressed in some typical phases (Fig. 9): preemergency state, initiating events, cascading development of emergency, final state, and restoration. It was established that the main types of emergency disturbances that occur in the quick phase of development were a voltage collapse and a considerable overload of equipment.

The analysis showed that in the phase of initiating events, the above-standard disturbances occur. The postemergency conditions that occur at the end of this phase are off-design for the existing emergency control devices and for the dispatching personnel. Therefore, the existing ECSs furnished with the up-to-date automation means and the actions of a transmission system operator may prove to be ineffective to prevent

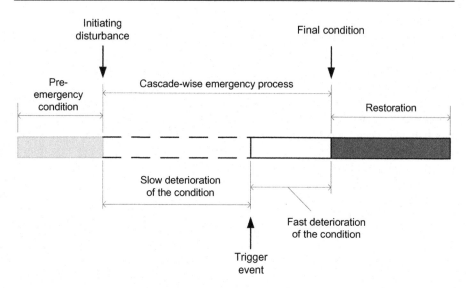

Fig. 9 Specific stages of the system blackout.

the subsequent catastrophic development of the emergency. Thus, the following drawbacks of the existing ECSs are:

(**1**) lack of ECSs for reliable protection against voltage collapse,
(**2**) low resilience of the ECSs,
(**3**) lack of adaptability and coordination of local devices, and
(**4**) critical redundancy of primary unprocessed data for the operator.

The results of the studies testify to the necessity of the development of next-generation intelligent systems to complement modern ECSs, taking into account its weak points [75,76]. Based on the above, the specific requirements for such intelligent ECSs have been developed. The systems should:

(**1**) include tools for the intelligent monitoring and assessment of the EPS operating conditions,
(**2**) be able to predict potentially dangerous states of the EPS,
(**3**) be highly resilient and able to coordinate local emergency control devices,
(**4**) include methods and models providing the protection of an EPS with a complex structure, and
(**5**) complement the existing ideology of ECSs but not contradict it.

4.3 The system of monitoring, forecasting, and control of power systems

4.3.1 General structure

Advanced smart devices and technologies such as PMU, artificial intelligence, and so on give new possibilities for solving the complex and comprehensive problems of emergency control using monitoring, forecasting, and control of EPS operating

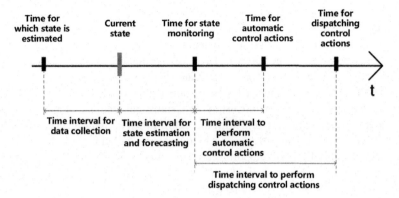

Fig. 10 Time diagram of events in the system for monitoring, forecasting, and control in electric power systems.

conditions. The time sequence of the individual stages of monitoring, forecasting, and control of operating conditions is shown in Fig. 10.

In essence, this combination of stages represents a comprehensive system providing stability, reliability, and controllability of modern EPSs. At the same time, from the viewpoint of the efficiency of this system, it is essential to increase the adaptability of control and improve coordination of the control stages, means, and systems. The blocks of monitoring and forecasting of the EPS normal, preemergency, and postemergency operating conditions solve the following problems:

−System SE.
−Forecasting the parameters of expected operating conditions. Forecasting is necessary because during the SE procedure, the current state is estimated with some delay while monitoring and control problems require some advance estimation of the system state ("to control is to foresee"). It is worth noting that, for these two blocks of problems, the advance time can vary.
−Detection of weak points in the system in the expected operating conditions.
−Determination of margins for transfer capabilities of ties in the expected conditions; this is necessary for efficient use of the margins in operating conditions and for automatic control through appropriate control actions.
−Visualization of expected conditions. Determination of indices and criteria for transition from normal to preemergency conditions and, vice versa, from postemergency to normal conditions.

Effective organization of the system of EPS operating condition monitoring and control is possible by an extensive involvement of new tools for the analysis and calculation of operating conditions and control actions, primarily technologies of artificial intelligence.

4.3.2 Forecasting

The balance between electricity consumption and generation must be kept in the electricity grid at any moment—otherwise disturbances in power quality or supply may occur. Forecasting EPSs parameters may be considered at different time scales, depending on the intended application. Forecasts for milliseconds up to a few minutes can be used, for instance, for turbine active control. This type of forecast is usually referred to as very short-term forecasts. Forecasts for the following 48–72 h are needed for power system management or energy trading.

The short-term forecasting of EPS parameters can be carried out both with the aid of classical approaches of dynamic estimation, statistical methods of analysis of time series, and regressive models, and with the aid of artificial intelligence. Many techniques have been employed for such purposes, including machine learning techniques—ANNs [77,78], support vector machines (SVMs) [79], random forest models [79,80], etc. Moreover, time series models, such as autoregressive integrated moving average (ARIMA), generalized autoregressive conditional heteroscedasticity (GARCH) models [81,82], and Kalman filter-based algorithms [83,84] have also been proven to be effective in EPS parameter forecasting. It is worth noting that EPSs parameters often feature sharply variable nonstationary behavior, which limits the efficiency of the stated technologies for the forecasting of time series. Studies have shown that hybrid approaches, that is a combination of different intelligent techniques, have great potential and are worth pursuing [85]. Previously, several hybrid forecasting approaches have been proposed, including ARIMA-ANN models [86], Fuzzy-ANN-based models [87], Wavelet-ANN-based models [88], and Fuzzy-expert system-based models [89], among others [90]. For a state-of-the-art review and bibliography of the methods for dynamical system identification and forecasting, readers may refer to the monograph [91].

It is well known that for effective forecasting of EPS parameters, a number of additional, correlated characteristics must be taken into account, which can be influenced by a predictable parameter. For instance, everything from salmon migration to forest fires can affect current and future electricity prices. But quite often, information about these characteristics is either unavailable or limited and contradictory. As a result, scientific researchers and power engineers have to work with only one retrospective predictable parameter time series without additional characteristics, which imposes limitations on forecasting performance. In such a case, forecasting performance can be improved by using a transition from a time series task to a regression task, when an initial time series is decomposed into some components. An investigation of properties of obtained components can help in improving the performance of an EPS parameter forecast.

4.3.3 Security monitoring and control

Analysis of the mechanisms of the development of large-scale blackouts an emphasizes the voltage instability as a main reason for blackouts in EPSs [92,93]. Enhancement of the relay protection systems, voltage controllers, and SCs as well as an increase in generator rotor speed led to a rise in the power flow dynamic stability limits of EPSs, which allowed the transmission of large amounts of power at considerable distances. Against the background of the rise in the transfer capability, it becomes necessary to meet the requirements for a necessary level of system security by providing sufficient backup sources of reactive power, which is not always fully implemented. This, in turn, facilitates voltage instability, which can lead to voltage collapse in a large-scale EPS.

At present, system operators worldwide widely use a "manual" approach to reactive power control. The control includes online adjustment of voltage settings as well as reactive power loading of generators according to the schedule, connection, and disconnection of static capacitor banks and shunt reactors, resetting of FACTS device settings, and automatic control of the transformation ratio at on-load tap changers. Moreover, the online control of EPSs deals with the notion of "prediction of emergency situation" (probability of its occurrence), which represents a problem of monitoring and security assessment. The prediction functions can be applied in various software packages, for example, dispatcher advisors. However, in the end, any control action is implemented directly by the operator that for some reason cannot make fast and effective decisions capable of preventing further emergency development.

In a general case, to assess voltage stability in EPS in real time, the problem of finding the distance of an operating point from the voltage stability is raised. The measure obtained may be qualitative or quantitative. The qualitative measure normally requires less computation time compared to the quantitative. It does not provide an estimate of the exact power reserve in megawatts but determines some dimensionless number (that normally varies from 0 to 1), which can be interpreted as a stability indicator. By definition, such indicators of security are scalar magnitudes and are determined by monitoring of change in certain parameters of EPSs. These stability indices seem quite convenient for dispatchers, first of all, to generate respective preventive control actions.

In the traditional statement, the application of such stability indices implies a purely algorithmic approach where the specified equations or partial derivatives are used to calculate numerical values of these indices for each current state of the EPS. However, as the practice of operation shows, such an approach has a number of significant downsides: low robustness to erroneous inputs, computational complexity erroneous identification of states, and some others [94]. One of the effective solutions to this problem is the use of a combination of traditional approaches on the basis of voltage stability indices and machine learning algorithms. The main idea here lies in an intelligent model learning to independently determine the current value of an assumed indicator on the basis of input data, thus identifying the current state of the EPS. As the studies [95,96] show, such a modified approach makes it possible to neutralize the drawbacks of traditional algorithmic approaches, owing to the original properties of the machine learning technologies.

To monitor that the system is within its limits, determined either online or offline, the primary measurement tools are SCADA systems and postprocessing by a state estimator [97]. The ENTSO-E network code on operational security requires each transmission system operator to classify its system according to the system operating states [98]. Significant advances have been made in the field of online security assessment technologies in [99], where 19 tools for dynamic security assessments in use, under testing, and under development were reviewed. A review of 15 of these state-of-the-art tools shows a wide variety of implementations. These systems include assessment of voltage security, transient security, and small signal security. The range of assessment capabilities includes determination of critical contingencies, transfer limits, and determination of remedial measures necessary to ensure security. The computational methods used for each type of security assessment are varied, and depend on the specific requirements, system characteristics, and, in some cases, the techniques available in the state-of-the-art tools used. All the tools reviewed in [97] were deterministic tools; however, commercially developed online risk tools do exist. In recent studies [100,101], state-of-the-art methodologies on risk assessment of cascading outages are presented. The methodologies are divided into two main categories: detailed modeling and simulation methods and the bulk analysis methods [102].

Computational intelligence techniques such as decision trees and ANNs have been applied to achieve the essential online intelligent security monitoring and assessment performance. A research on computational intelligence applications to EPS security [103] demonstrates that intelligent systems can play a major role in preventing major emergencies and blackouts by providing information that suggests optimal remedial action.

Beyond better monitoring, new control approaches are required. Six new control technologies and control schemes have been reported in the recent international European/Russian FP7 project ICOEUR [104]. The focus is on coordinated power flow control, transmission technologies, HVDC/VSC-HVDC technologies, optimization of interconnections and interarea oscillations, and different interconnection concepts.

Multiagent systems have been proposed for the last decade as a major component of intelligent control grids [105]. The research project ICOEUR also introduced a multilayer architecture for system monitoring and control. In this architecture, the physical layer was represented by the IPS, which consisted of several subsystems. Control actions that affected the subsystem were performed only if boundary conditions remained within the specified limits. If the boundary conditions were violated, coordinated control actions were applied using a multiagent control system [104]. The research of multiagent voltage and reactive power control system development is presented in [106]. The prototype of the system has been developed by the R&D Center at FGC UES (Russia). The control system architecture is based on the innovative multiagent system theory application that leads to the achievement of several significant advantages (in comparison to traditional control system implementation) such as control system efficiency enhancement, control system survivability, and cybersecurity.

4.4   Artificial intelligence applications

This section considers some specific possibilities of artificial intelligence applications to the development of principles and tools of automatic emergency control in EPS.

4.4.1   Forecast of state variables based on the dynamic state estimation method

DSE is used to forecast the state vector components. The DSE is based on the extended Kalman filter. The state vector is used to calculate all state variables. Forecasting all state vector components some time ahead makes it possible to forecast the values of all state variables. Forecasting of state vector components is considered as DSE if there are no new measurements.

The values of forecasts of all state vector components are calculated by the equation:

$$\widetilde{x}_{k+1} = \hat{x}_k + P_k \times M_{k+1}^{-1} \times \Delta\widetilde{x}_k \tag{57}$$

The forecast error $\Delta\widetilde{x}_k$ in the first forecasting cycle is determined by the equation:

$$\Delta\widetilde{x}_k = \hat{x}_k - \hat{x}_{k-1}$$

where $\hat{x}_k$—estimates of state vector components. In the subsequent snapshots the forecast error is calculated by the equation:

$$\Delta\widetilde{x}_k = \hat{x}_k - \widetilde{x}_k$$

where $\widetilde{x}_k$—the forecast of state vector components.

The forecast accuracy is determined according to the following equation

$$\phi_F = \frac{1}{k} \sum \frac{1}{m1} \sum_{j=1}^{m1} \frac{|\hat{x}_j - \widetilde{x}_j|}{\hat{x}_j} \times 100\% \tag{58}$$

where k—the number of snapshots and ϕ_F is the average error for forecasting the voltage magnitude and phase angle with regard to estimates of the total number of snapshots.

The extended Kalman filters were tested on the schemes with the 7 (Fig. 11), 33, and 100 nodes.

The archives of measurements were created for these schemes. The estimates of state variables for each scheme were obtained based on the formed snapshots through

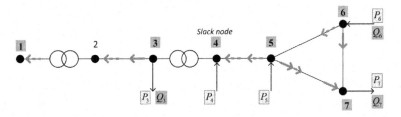

Fig. 11 A test scheme (solid color—reactive model, dashed shading—active model).

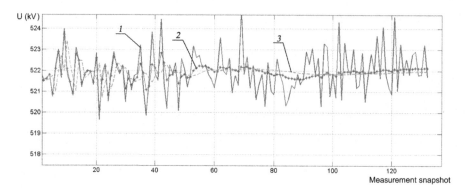

Fig. 12 The results of the error forecast and filtering in the voltage magnitude measurement at node 1 (1—measurements, 2—estimates, 3—forecasts).

the DSE on the basis of the extended Kalman filters. Fig. 12 shows the results of the error forecast and filtering in the voltage magnitude measurement at node 1.

DSE with application of the Kalman filter can be used to forecast all the EPS state variables for a short period of time. The regular filter adjustment improves the forecast quality. Measurements from PMU improve the forecast results.

4.4.2 Forecast of power system parameters based on a hybrid data-driven approach

A novel hybrid data-driven approach has been developed for forecasting EPS parameters (e.g., load, distributed embedded generation, wind energy, and distributed storage) to improve grid operation efficiency and reduce fossil fuel consumption and carbon emissions. The main goal of the hybrid forecasting approach development was to increase the efficiency of forecasting studies when dealing with limited retrospective data and a sharply variable, nonstationary time series of EPS parameters.

The proposed approach is based on mode decomposition and a feature analysis of initial retrospective data using the Hilbert-Huang transform and machine learning algorithms. The random forests and gradient boosting trees learning techniques were examined. The decision tree techniques were used to rank the importance of variables employed in the forecasting models. The Mean Decrease Gini index is employed as an impurity function. The resulting hybrid forecasting models employ the radial basis function neural network and support vector regression.

Fig. 13A shows a general forecasting scheme with five main blocks. Data collection is performed by the SCADA system, which merely identifies a set of EPS parameters. The second block preprocesses input data with HHT. HHT is a two-step algorithm; combining empirical mode decomposition (EMD) and the Hilbert transform (HT). After the HHT was employed, through EMD and HT, the sets of intrinsic mode functions (IMFs), instantaneous amplitudes, and frequencies are obtained. These sets are used as input values to the next block for feature selection and dimensionality reduction. In this block, learning methods are used—random forest and boosting trees to rank a variable's importance [107,108]. SVM and ANN forecasting

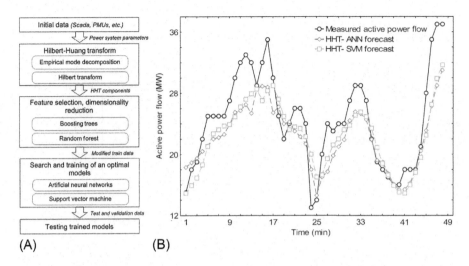

Fig. 13 Block diagram of the proposed hybrid approach scheme (A) and result of "1 min ahead" active power flow forecasting based on hybrid genetic models (B).

models with nonlinear optimization algorithms are employed in the next block for optimal selection of parameters. The last block is used for testing and a comparison of obtained forecasting models. As a result, an HHT-ANN-SVM-based forecasting system will be able to predict EPS parameters.

The novel hybrid approach is realized in STATISTICA 6.0 and MATLAB. The efficiency of the developed approach is demonstrated with a real-time series in the following electric power problems: active power flow forecasting for the western Baikal-Amur mainline (Russia) [109,110], electricity price forecasting for the Australian and Nord pool spot (EU) power markets [111], and wind speed forecasting for the Valentia region (Ireland) [84].

The Baikal-Amur mainline external power supply is carried on the 110 kV power network (from Taishet to Lena substations) and the 220 kV power network (from Lena to Taksimo substations). Continuous and random changes in time traction load and the presence of a two-way feed greatly complicates the forecasting of state variables.

The initial training data is represented by the telemetric data of the minute-by-minute active power flow for 3 days (4320 values). The initial time series were converted using the HHT algorithm. The obtained HHT components are used as input values of the trained machine learning-based forecasting models. The multilayer perceptron (MLP) neural network model is trained with the Broyden-Fletcher-Goldfarb-Shanno algorithm [112]. The radial basis kernel function is used in the epsilon-SVM regression model. Using a 10-fold cross-validation method, the following optimal training constants and kernel parameters were determined: the capacity constant, $C = 10$; epsilon, $\varepsilon = 0.4$; and degree, $\gamma = 0.001$. To compare the performance of forecasting approaches, the authors also calculated similar forecasts based on conventional approaches: ARIMA, single ANN, and multiplicative exponential smoothing.

Table 6 and Fig. 13B summaries the numerical results for a "1 min ahead" active power flow forecasting on based on the HHT-ANN and HHT-SVM models. As the

Table 6 **Comparison of a "1 min ahead" active power flow forecast on the basis of hybrid and conventional approaches**

No.	Forecasting models	Error		
		MAPE (%)	MAE (MW)	RMSE (MW^2)
1	HHT-ANN	12.5	3.2	3.9
2	HHT-SYM	11.6	2.9	3.5
3	ANN (MLP)	15.8	6.9	33.6
4	ARIMA	22.4	7.1	7.4
5	Exponential smoothing	34.2	10.8	10.9

table shows, the most accurate forecast was given by the hybrid HHT-SVM and HHT-ANN models (MAPE—11.6% and 12.5% respectively), compared with the conventional approaches: the ANN model—15.8%, the ARIMA model—22.4%, and exponential smoothing—32.2%.

4.4.3   Total transfer capability estimation method

TTC estimation means the calculation of the steady state with TTC in the controlled lines and with the parameters of current conditions in the remaining part of the EPS. The values of flows in the controlled lines change through the adjustment of EPS state variables determined in advance. The state of the resultant EPS operating conditions $\hat{y}_{res}$ is estimated using the measurements of the current operating conditions $\bar{y}$ and ideal power flow limits in terms of static stability or thermal stability of each controlled line. Further, these values will be referred to as the PMs of TTC ($P^{TTC\ PM}$). The vector of initial data for the TTC estimation looks as follows:

$$\bar{y} = \left(U_i, P_i, Q_i, P_{ij}, Q_{ij}, \delta_i, P_{ik}^{TTC\,PM}\right) \tag{59}$$

where U_i—magnitudes of nodal voltages; P_i, Q_i—injection of active and reactive powers at nodes; P_{ij}, Q_{ij}—power flows in transformers and lines, δ_i—voltage phases at the nodes of the scheme, in which PMUs are placed.

The main idea of the method is illustrated with a three-node scheme (Fig. 14A) in which five state variables are measured ($U_1, U_2, U_3, P_2, P_{2-1}$). The main idea of TTC estimation is showed in Fig. 14B.

The TTC estimation method makes the current values of flows in the controlled lines (P_{2-1}) as close to the ideal power flow limit ($P_{2-1}^{TTC\ PM}$) as possible in accordance with the topological and operational constraints.

$$U^{min} \leq U_2 \leq U^{max}, P^{min} \leq P_2 \leq P^{max} \tag{60}$$

and current values of uncontrollable state variables (U_1). Online generation of coordinated actions of dispatchers in order to obtain the resultant state variables optimal

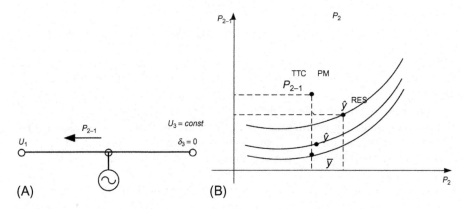

Fig. 14 Geometric interpretation of the main idea of the method (active model).

in terms of all stakeholders is provided by the values of weighting coefficients of $P_{2-1}^{\text{TTC PM}}$.

The TTC estimation parameters are adjusted offline [113] according to the operational constraints of the IPS. When all the requirements imposed by the dispatchers of individual systems are met, the modified SE program is considered to be adjusted for these operating conditions and the weighting coefficients are written in the database. A great number of various operating conditions, that is, constraints and snapshots of measurements, should be considered offline.

Weighting coefficients of $P^{\text{TTC PM}}$ that provide the optimal result of TTC estimation in the resultant condition in sense criterion (61).

$$\phi_P = \sqrt{\sum_1^{k0} \left(P_{lk}^{\text{TTC PM}} - P_{lk}^{\text{TTC}}(U, \delta)\right)^2} \to \min \tag{61}$$

where $k0$—the number of controlled lines.

There are determined by N cycles of SE of EPS operating condition by the TTC estimation method, taking into account constraints

$$U_i^{\min} < U_i < U_i^{\max} \tag{62}$$

where U_i—voltage at the control point; $U_i^{\min}$ and $U_i^{\max}$—lower and upper admissible limits for changes in the value.

$$P_l(U, \delta) \le P_l^{\lim} \tag{63}$$

where $P_l^{\lim} = P_l^{\text{av}} + P_l^{\text{res}}$; P_l^{av}—available power of power plant excluding equipment removed for maintenance; P^{res}—operating reserve.

$$Q_j^{\min} < Q_j(U, \delta) < Q_j^{\max} \tag{64}$$

where j—nodes at which reactive power is controlled.

The range of possible values for weighting coefficients for $P^{\text{TTC PM}}$ is divided into N points. In the first calculation the weighting coefficients of $P^{\text{TTC PM}}$ are taken equal to the least value. Each calculation consists of two stages. The first one is solving the SE problem disregarding inequality constraints (62)–(64), then checking if (62)–(64) are met. If constraints (62)–(64) are met, then the criterion (61) is checked.

If criterion (61) in the current calculation (cycle) is lower than in the previous calculation, the weighting coefficients for $P_{lk}^{\text{TTC PM}}$ are written in the database. Then weighting coefficients for $P_{lk}^{\text{TTC PM}}$ are changed and again a TTC estimation is made, constraints are checked, and criterion (61) is calculated. Thus, N cycles of TTC estimation are made offline by the TTC estimation method with different weighting coefficients of $P_{lk}^{\text{TTC PM}}$. The values from the database as weighting coefficients of $P_{lk}^{\text{TTC PM}}$ are used to perform the online TTC estimation method. ANN selects the weighting coefficients from the database.

The Kohonen ANNs [114] are applied to adjust the TTC estimation method in real time. In order to solve the problem using ANN, it is necessary to create a problem book and train as well as the ANN. The problem and testing books are created on the basis of an archive of realistic measurement snapshots. An approximation to the reality is achieved using typical or actual active and reactive load curves at the load nodes and reactive load curves at the nodes where compensators are installed for the calculation of steady states.

Based on the initial data, the algorithm calculates steady states for each point in the load curve. All the calculated steady states are written in the database of steady states. To create a standard measurement snapshot $\bar{y}$, the measured state variables y_{true} are chosen from this database. To form the measurement snapshot, random errors x_{rand} generated by the random number generator are applied to y_{true}:

$$\bar{y} = y_{\text{true}} + x_{\text{rand}} \sigma_y \tag{65}$$

where $x_{\text{rand}} \in N(0, 1)$, σ_y—measurement variance.

The problem book is compiled from the patterns represented by values of measurements. The problem book is used to train the ANN to identify the weighting coefficients that correspond to the given scenario and the current operating conditions. In the course of training all the patterns (input data) are classified into several classes through their comparison with typical elements of different classes and selection of the closest one. An answer of the trained ANN is the number of a class to which the considered pattern belongs. Each class corresponds to an individual set of weighting coefficients, which is determined by interpreting the answers of the ANN. Whether or not each set of weighting coefficients belongs to a certain class is determined offline. To this end the TTC estimation is performed N times for each snapshot, where N is the number of combinations of weighting coefficients.

The dispatcher of each system puts forward his or her requirements and suggests a set of control actions to use the TTC. The voltage deviation is allowed within $\pm 15\,\text{kV}$ of rated voltage at nodes. The range of reactive power control at nodes 4, 8, and 10 is possible at ± 100 MVAr of current reactive power. Active power generated at node 8 should not exceed 2000 MW.

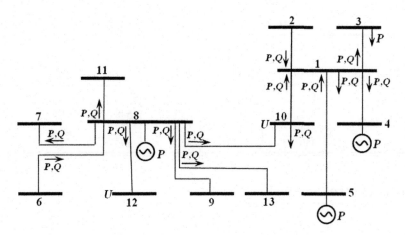

Fig. 15 Test scheme.

The calculations were done on a test scheme (Fig. 15) consisting of two IPSs according to the following scenario: it is necessary to calculate the TTC in lines 8–10, 1–10, and 1–4. Additional power will be provided by EPS_1 (node 8) and EPS_2 (node 4).

Two steps are required to solve this problem: offline and online.

(A) *The offline stage*

Creation of a problem book for ANN. Load curve at node 10 is used to create a problem book.

Training and testing the Kohonen network. The Kohonen network solves the problem of classifying the operating conditions into several classes. Relative testing error equals 0.83% (2 out of 240 are incorrect).

To interpret the ANN answers, the TTC estimation is performed 100 times for each of the 240 snapshots, where 100 is the number of combinations of weighting coefficients for $P_{lk}^{TTC\ PM}$ in lines 1–10, 8–10, and 1–4. Enumeration of weighting coefficients starts with the value 0.1 and changes by 0.3 at the next calculation. After the calculations there remains one set of weighting coefficients of each snapshot. The set provides optimal results of TTC estimation of the considered operating conditions in real time.

(B) *The online stage*

The obtained snapshot of measurements is used to perform SE of the current conditions under the weighting coefficients of measurements that correspond to the specified variances. New values of weighting coefficients for TTC estimation are identified using ANN. The input data for ANN are formed out of measurements. The ANN determines if this snapshot belongs to any of the classes, each of which is characterized by its set of weighting coefficients of TTC PMs in three controlled lines. The coincidence of weighting coefficients of the TTC PMs in controlled lines 8–10, 10–1, and 1–4 provides the maximum TTC value of all controlled lines in the current operating conditions of the EPS.

Table 7 **Results of state estimation and TTC estimation calculation**

Snapshot no.	Estimates			TTC		
	8–10	1–10	1–4	8–10	10–1	1–4
37	175	111	55.8	520	426	248
105	237	111	55.8	548	395	250
163	254	111	55.8	557	388	252
181	350	111	55.8	595	336	249

Table 8 **Control actions**

Snapshot no.	Control actions						
	U_8	U_{10}	P_8	Q_1	P_4	U_4	Q_{10}
37	13.8	5.15	330	−9.9	194	4.5	54.8
105	12.5	4.6	347	−10.5	194	4.5	53.1
163	11.7	4.3	365	−11.2	199	4.7	52.6
181	10	36	389	−12.2	196	4.7	49

The current snapshot of measurements, with the weighting coefficients selected by the ANN, is used to perform the TTC estimation. Table 7 presents the estimates of current power flows and TTC estimates in lines 8–10, 10–1, and 1–4 for four snapshots.

The values of control actions for four snapshots are presented in Table 8. The tables demonstrate that to achieve TTC in the controlled lines, it is necessary to increase active power generation at nodes 4 and 8, increase reactive power generation at node 10, and reduce reactive power generation at node 1.

4.4.4 Automatic decision tree-based system for online voltage security control of power systems

The authors tested the most popular machine learning models that have been proposed recently to solve the problems of security assessment of EPSs. These are: neural networks of Kohonen, SVMs, hybrid neural network models, and various decision-tree algorithms [96,115,116]. The capability of these algorithms to effectively identify operating conditions of EPSs was demonstrated on various test schemes of IEEE involving different stability indicators as target vectors for training the models.

The calculations, however, show that when security is assessed in real time, the most effective algorithm in terms of such criteria as accuracy, robustness, online adaptability, and versatility is the algorithm of proximity-driven streaming random forest (PDSRF). Therefore, it is the PDSRF algorithm implemented in the programming language C++, which was used as a basis for the developed automatic intelligent system for a real time security assessment of EPS. Moreover, the combination of the original properties of the PDSRF algorithm and capabilities of the L-index as a target vector for training a model made it possible to implement the following functions:

(1) the function of the online alarming and prediction,
(2) the function of control with localization of "critical nodes" in EPS, and
(3) the function of a direct interaction with the relay protection and automation systems.

As said above, it is more sensible to use system variables-based voltage stability indices of the EPS in real time. Therefore, the L-index is used in the study, being one of the effective indices from this group, as a target indicator of system stability when training the PDSRF model. The L-index is proposed by Kessel and Glavitsch [117] as an indicator of impeding voltage stability. Starting from the subsequent analysis of a power line equivalent model, the authors developed a voltage stability index based on the solution to power flow equations. The L-index is a quantitative measure for the estimation of the distance of the actual state of the system to the stability limit. The L-index describes the stability of the entire system with the expression:

$$L = \max_{j \in \alpha_L} (L_j) \tag{66}$$

where α_L is a set of load nodes and α_G is a set of generator nodes. L_j is a local indicator that determines the buses that can be sources of a collapse. The L-index varies in a range between 0 (no load) and 1 (voltage collapse) and is used to provide meaningful voltage instability information during dynamic disturbances in the system [118]. Kessel and Glavitsch reformulate the local indicator L_j in terms of the power as:

$$L_j = \left| 1 + \frac{\dot{U}_{0j}}{\dot{U}_j} \right| = \left| \frac{\dot{S}_j^+}{\dot{Y}_{jj}^{+*} \cdot U_j^2} \right| = \frac{S_j^+}{Y_{jj}^+ \cdot U_j^2} \tag{67}$$

where $\dot{S}_j^+$ is the complex conjugate injected power at bus j; Y_{jj}^+ is transformed admittance; and $\dot{U}_j$ is the complex voltage at bus j.

The voltage stability indicators are known to be not only an effective method to assess the system stability, but they also underlie the control of EPS security. In [118] the authors propose a variant of the L-index calculation, which was used to develop an L-Q sensitivity analysis method that eases the quantitative analysis of voltage influence between different nodes.

According to the basic differential property of the L-index, a common analytical algorithm is devised for reactive power optimization. The algorithm can be used to determine the required reactive power injection for the load node. Based on the applied methodology, a large-scale power system will operate in an optimal steady state under the minimum L_{sum}, which represents a sum of local indices L_j. In this case the function of the first partial derivative is defined as follows:

$$
\frac{\partial L_{\mathrm{sum}}}{\partial \Delta Q} = \begin{bmatrix} \dfrac{\partial L_{\mathrm{sum}}}{\partial \Delta Q_1} \\ \vdots \\ \dfrac{\partial L_{\mathrm{sum}}}{\partial \Delta Q_m} \end{bmatrix} = 0 \tag{68}
$$

The calculations carried out in Ref. [119] on different IEEE test systems show that the application of this approach makes it possible to improve the voltage stability by reactive power injections at load nodes. The injections are calculated from the L-index minimization conditions, and keep a system under heavy load conditions away from instability boundaries. Nevertheless, the authors state that despite the relative simplicity of the calculation, this method requires considerable computational efforts and its application in real-time problems can be complicated.

In this work supplementing and modifying this approach is suggested by using trainable models able "to learn" to calculate both the global L-index for the security assessment of an entire system, and the required injections ΔQ when determining the place and magnitude of corrective actions. This will allow applying this methodology in real time.

In many papers devoted to the studies of power system security on the basis of decision trees, the authors suggest using the model offline and, instead of adaptation, periodically updating the model [120,121]. However, even in the case of retraining, the complete training of the model is associated with additional time, which excludes the retraining in real time. This problem can be solved by online methods that update the existing model, using new data without its total restructuring [121]. A PDSRF algorithm proposed in this section can serve as one of such models.

The general structure of the developed system is presented in Fig. 16. The scheme shows that the proposed system consists of two main models on the basis of the PDSRF local and global algorithms.

Global model is trained to correctly identify the global L-index on the basis of system variables of an EPS such as voltage at nodes, loads, power flows, etc. Here an output value of the L-index is interpreted as a security signal (indicator) of the entire EPS. The local model on the basis of the PDSRF algorithm, in turn, is trained to determine the required reactive power injections ΔQ for load PQ-nodes. The inputs for the local model are represented by the same operating parameters and calculated values of the local L-index at a current time instant. Training of the local PDSRF model is based on the equation of partial derivative of (68). As a result, the trained model is able to determine the values ΔQ to perform corrective and/or preventive control actions online. Additionally, the security intelligent system can provide local signals in the form of local L-indices for each load at PQ-node.

In the end, such a structure makes it possible to implement the above functions of online alarming, localization, and interaction with automatic systems. For example,

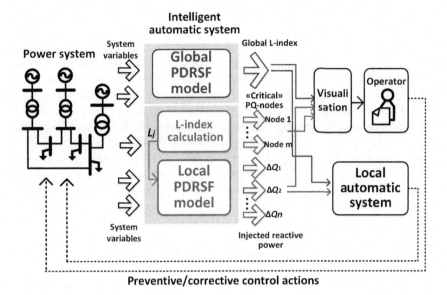

Fig. 16 Structural scheme of an automatic intelligent system for power system security control.

the system signal on the basis of the L-index, which is delivered to the operator through the visualization block, informs the dispatcher of the general level of security in the analyzed EPS ("high," "low," "emergency"), and allows the operator to predict (estimate) the extent to which the current state of the EPS is dangerous in terms of its closeness to voltage collapse. In the case of a dangerous state identification ("low security"), the local signals on the basis of the local L-index, formed by the local model, enable the localization of "critical load nodes" at which the system is at its closest to the stability loss. The corrective and/or preventive control actions can be implemented online on the basis of the injections ΔQ generated by the local model based on the PDSRF algorithm for PQ-nodes. Such control actions can both adjust the operating conditions in terms of their optimality according to some economic criteria or, in the case of a decrease in security, keep the conditions away from the instability boundaries.

It is important to note that the output signals of the alarming system are delivered both to the operator and directly to the operating automation. Interactions with the automation allow, where necessary, a correction of the agents actions because it is they that control the reactive power sources to regulate voltage in order to prevent the development of an emergency process. The operator, in turn, using the recommendations of the intelligent system (in the case of a security decrease), can adjust the protective relay settings by decreasing the settings with respect to time, increasing the sensitivity of startup signals of the emergency control functions through the selection of an appropriate group of settings, etc. Despite the fact that the proposed structure suggests a certain interaction with the dispatcher to control the EPS security, one can see the main mechanism of the developed intelligent system operation to be mostly automatic where many control actions are generated with the minimum involvement of the operator.

The efficiency of the proposed intelligent system was tested on various IEEE test power systems. The database of possible states of the test EPS for model training was generated by quasidynamic modeling in the MATLAB/PSAT environment. Global and local PDSRF models are implemented in C++. A set of the obtained system states was used to calculate the values of global L-index, and on the basis of local indices L_j, the reactive power injections ΔQ were found for each load node. These characteristics were applied as class marks for training and testing the models on the basis of the PDSRF algorithm. The testing was done in the developed modeling platform in the MATLAB/PSAT environment that involved the simulation models of the quasidynamic model. This model provides the dynamic modeling accuracy sufficient for research into voltage stability, overload of lines, algorithms for local automation devices, etc.

To demonstrate the effective adaptation of the PDSRF algorithm in real time, various significant disturbances were modeled in the IEEE 118 test system (Fig. 17A). The disturbances included losses of generation and connection of a large consumer at specified nodes (Fig. 17B). These operation changes in the system were specified randomly at quasidynamic modeling.

The testing results of the PDSRF algorithm compared to the other machine learning models are presented in Table 9. The table shows that in all the models, except the PDSRF model, the accuracy declines at modeling of significant disturbances in the system, that is, they could not adapt in real time and required additional training (updating). For example, the accuracy of state identification for the scheme IEEE 118 made up 97.24% (Table 9), which is approximately 10% higher than the accuracy of other known intelligent approaches presented in some international peer-reviewed journals in recent years. These approaches include neural networks of Kohonen, supporting vector machines, hybrid neural network models, and various algorithms of decision trees.

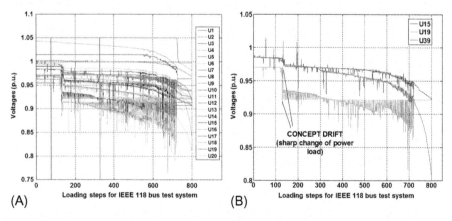

Fig. 17 Changes in voltage profiles of the IEEE 118 system (A) and voltage profiles at load buses with a concept-drift effect (B).

Table 9 Accuracy of various machine learning algorithms on repared dataset

Method	Accuracy (%)	Kappa (%)
Global PDSRF model	97.24	95.30
Support vector machine	81.54	64.92
Random forest	96.01	93.24
Gradient boosting trees	93.64	89.41
Elman neural network	80.80	65.08

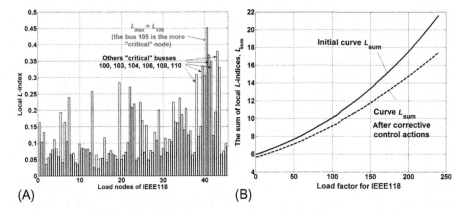

Fig. 18 The local L-index values for one of steady-state condition of IEEE118 (A) and the curves L_{sum} before and after corrective control actions (B).

Moreover, the calculations show that the proposed online approach on the basis of PDSRF provides lower errors (root-mean square error of order 11% for IEEE 118) and a high speed of solving process (about centiseconds for each steady state of IEEE 118 compared to 30–40 min in the traditional approach), when determining the additional reactive power injections.

This fact makes it possible to effectively apply the warning subsystem for monitoring to control security in electric EPSs of a large dimension in real time. The obtained values of additional injections were used for reactive power compensation by using reactive power sources, which decreased the sum of L-indices at load nodes. Their increase is indicative of an even greater proximity of voltage collapse, first of all for the heavy load and dangerous conditions of IEEE 118 system (Fig. 18).

In the former case, the gradual load increase led to equipment overload, the tripping of the generator, and, finally, to a voltage collapse. In the case where the proposed intelligent system was applied, in the same emergency scenarios the corrective control actions made it possible to maintain the stability of a test subsystem by the coordination of reactive power sources and without load shedding.

4.4.5 Multiagent coordination of emergency control devices

Analysis of the recent blackouts shows that the most severe contingencies that lead to voltage collapse occur in a highly loaded EPS without effective coordination of emergency control device operation.

A decentralized agent-based system for operation and emergency control, that is, a MAAS, has been developed in MATLAB and JADE. The system provides control of reactive power, prevents emergency tripping of generators, and maintains voltage at load buses within acceptable limits [122]. The MAAS is based on the decentralized algorithms of adaptive control of reactive power sources, where each source (agent) is an autonomous unit with a global goal to prevent voltage collapse not to the detriment of the local interests, that is, prevention of the local overload of the controlled reactive power source. The agents are interconnected by the common information environment that allows the exchange of messages. An agent's knowledge about a subsystem is formed as a base of sensitivity coefficients on the basis of the Jacobian matrix elements of steady-state equations.

The efficiency of the intelligent system was tested on the following schemes: IEEE RTS 96 (a modified 53-bus scheme) and IEEE 118 (a 118-bus scheme). The assumption is that using the two types of automatic control can be used in the IEEE RTS 96 scheme: conventional automatic control system (CACS) (includes turbine governor, automatic voltage regulator and over excitation limiters on each generator, and OLTCs on transformers connected to buses 204–210, excludes UVLS), and MAAS—in addition to the set of CACS local controllers, it includes OLTCs on transformers connected to buses 101, 102, and 103.

The multiagent automation testing performed for the IEEE RTS-96 system demonstrated that the coordination of local devices through the implementation of additional control actions from the proposed MAAS makes it possible to effectively cope with the equipment overload and, consequently, with voltage collapse (Fig. 19). In particular, where the local automation failed to tackle the situation, the multiagent automation prevented the voltage collapse without load shedding but only by coordinating the reactive power sources.

The testing carried out for the IEEE 118 system shows that the probability of a cascading failure in the case of applying the multiagent automation will be lower because, thanks to the reactive power redistribution, the generators will be loaded evenly compared to the situation where there is no multiagent automation at all (Fig. 20). Results of the test calculations testify to the fact that the proposed MAAS provides the opportunity to exactly determine the instant when it becomes necessary to transition from the secondary control to the emergency control, excluding human factors and ensuring a continuous process of operation and emergency control.

4.4.6 Intelligent system for preventing large-scale emergencies in power system

The research proposes supplementing the existing principles of EPS control with the method of "control with prediction," which suggests early detection and prevention of

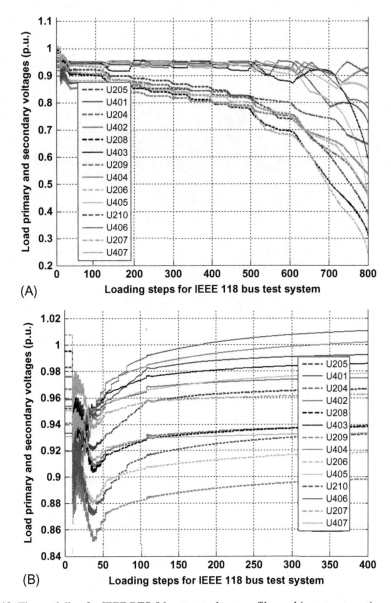

Fig. 19 The modeling for IEEE RTS-96 system voltage profile: multiagent automation system (A) and conventional automatic control system (B).

dangerous states and emergency situations before they lead to a voltage collapse [74,75]. Thus, the preemergency control represents a sort of "zero echelon (level)" in the whole system of emergency control and assists the system to avoid instability (Fig. 21). The control system based on this approach should adjust the operation of local emergency control devices both in steady state and postemergency conditions to reduce the probability of a severe emergency.

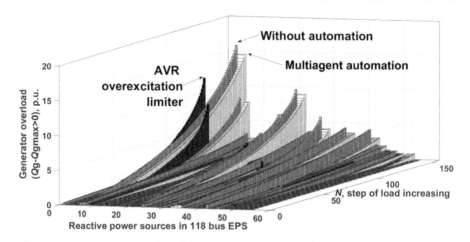

Fig. 20 The results of quasidynamic simulation on IEEE 118 network. Comparison of units overloading.

Fig. 21 The proposed approach to system monitoring and control and operating states and transitions.

The intelligent system developed in Ref. [85] makes it possible to compensate for the revealed downsides of some existing protection and emergency control devices by applying various artificial intelligence algorithms (system of agents, technology of decision trees) to early detection and prevention of dangerous conditions of complex power systems before they lead to a blackout. This fact makes it possible to improve the operation of domestic ECSs and considerably reduce the risk of massive blackouts by minimizing the economic, social, and technical losses.

An architecture of the intelligent system for prevention of blackouts in complex EPSs (Fig. 22) was developed based on the proposed intelligence algorithms. This architecture represents a software platform including the following components:

(1) a database of emergency disturbances,
(2) a block for simulation of the behavior of EPS (dynamic and quasidynamic models),
(3) a block of operation/emergency control on the basis of multiagent technologies,
(4) a block for monitoring and assessment of EPS security on the basis of machine learning algorithms,

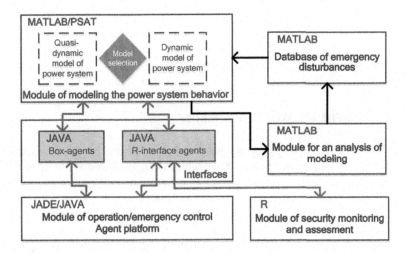

Fig. 22 Architecture of a proposed software platform for modeling decentralized intelligent monitoring and control systems of electric power system operation.

(5) interfaces providing interaction between the block of EPS behavior simulation with the blocks of monitoring and control, and

(6) a block for the analysis of results.

A developed structure of emergency disturbances contains the following elements:

(1) set of scenarios of load increase in the EPS,

(2) sequence of random and dependent discrete actions (tripping of lines, generators, transformers, load shedding, failure of automation, etc.),

(3) additional set defining the current state of automation devices in the event that the considered disturbance occurs, and

(4) results of test trials.

A database of emergency disturbances was created basing on the formulated principles. Its structure includes the following elements:

(1) a list of emergency disturbances and

(2) a set describing the state of the classical and agent-based automation systems.

The developed software allows an interaction with the described database by transmitting respective data to the block of EPS behavior simulation and receiving additional data from the block of analysis of results. Also a program for the formation of an emergency development scenario was developed. For a concrete test scheme of EPS based on the constructed database of emergency disturbances, the models of classical and agent-based automation systems were designed.

An alarm subsystem for monitoring and assessment of security to reveal dangerous states of EPS was developed. Within the alarm subsystem for monitoring, an innovative online method was devised for the assessment and control of the dynamic security of EPS using the technology of decision trees, that is, PDSRF, implemented in the language C++. The main qualitative distinction of this approach from the other modern approaches is the capability of PDSRF to independently and adaptively change in real time in case of serious changes in the received telemetry data without loss of accuracy while identifying the conditions of EPS.

5 Smart grid clusters in Russia

The process of IESAAN formation suggests implementation of pilot projects and creation of territorial smart grid clusters. These clusters are supposed to use information, technological, and control systems providing adaptive (depending on the operation situation in the system) control of network parameters, remote control of switching devices, and real-time estimation of the technical state of the network under normal, preemergency, and postemergency conditions in EPS [4,5].

Currently, along with implementation of the pilot projects and the creation of smart grid clusters, new equipment is being installed at the energy facilities of the UNEG of Russia as part of a modernization program. According to the investment program for 2011–14, the innovative equipment was installed at more than 50 facilities. First of all, there are new devices intended to compensate reactive power and maintain voltage (static thyristor compensators, STATCOMs, controlled shunt reactors).

5.1 Smart grid clusters in the east interconnected power system

5.1.1 Smart grid clusters

In 2011 the projects on formation of individual smart grid clusters and implementation of pilot projects aimed at creating the smart grid [4,5,123] were launched in the territory covered by the East IPS (Amur region, Sakha Republic, Primorye, and Khabarovsk Territories). The projects (Table 10 [4,5,9]) include:

−Smart grid cluster ElgaUgol.
−Smart grid cluster Nizhny Kuranakh-Maiya (Yakutian energy system).
−Smart grid cluster Vanino.
−Smart grid cluster of Primorye Territory.
−Pilot project on smart grid with distributed generation on Russky Island and Popov Island.
−Pilot project on reliable power supply from the Zeya HPP, and reliable and quality power supply to electric traction transit line Siberia-East.

Implementation of smart grid in the clusters Elgacoal, Vanino, and Primorye Territory suggests construction of compact digital substations furnished with innovative devices (new systems for reactive power compensation and voltage maintenance, active filters, equipment monitoring, and diagnosis systems, etc.) This addresses the enhanced reliability of the power supply to consumers from the substations located along the railway of the Khabarovsk Territory and provides a reliable electricity supply to the south of the Primorye Territory.

Table 10 Pilot projects on the creation of territorial smart grid clusters in the East interconnected power system

№	Pilot project	Project goal	Smart technologies
1	Smart grid cluster ElgaUgol	–To provide power quality and redundancy of tunneling and traction power supply (a two-circuit 220kV transit transmission line) –To ensure emergency and operation control, considering development of small-scale generation	–Digital 220kV substation ElgaUgol: equipment monitoring and diagnosis systems, 220kV SF6 switchgears, 50 MVAr STCs, active harmonic filters and voltage balancing systems, 50MW BESSs –Digital unattended 220kV substations A and B: equipment monitoring and diagnosis systems, 220 kVSF6 switchgears –WACS/WAPS technologies using PMU
2	Smart grid cluster Nizhny Kuranakh-Maiya (Yakutian electric power system)	To provide a high level of power supply reliability and power quality	–Construction of 220kV overhead transmission line (OTL) Neryungri CSS-Nizhny Kuranakh-Tommot-Maiya with the OTL monitoring system –Digital unattended 220kV substations Tommot and Maiya: equipment monitoring and diagnosis systems, 220kV SF6 switchgears, 50 MVAr STCs, active harmonic filters and voltage balancing systems –WACS/WAPS technologies using PMU
3	Smart grid cluster Vanino	To increase the reliability of the power supply to traction substations of the electrified railway in the Khabarovsk Territory	–220kV OTL Komsomolskaya-Selikhino-Uktur-Vysokogornaya-Vanino with the OTL monitoring system –100MVAr VAr-compensation system for 220kV, active harmonic filters and voltage balancing systems –WACS/WAPS technologies using PMU.

Continued

Table 10 Continued

№	Pilot project	Project goal	Smart technologies
4	Smart grid cluster of Primorye Territory	– To supply electricity to the southern part of the Primorye Territory – To increase transfer capabilities of 500 kV transit transmission lines by 350–400 MW	– 500 kV OTL PrimorskayaCPP-Dalnevostochnaya-Vladivostokskaya and Primorskaya CPP-Chuguevka-2-Lozovaya – Controlled series compensator, 500 MVAr VAr-compensation system for 500 kV, active harmonic filters and voltage balancing systems – WACS/WAPS technologies using PMU
5	Project of smart grid on Russky and Popov Islands with distributed generation	– To integrate wind generation and mini-CPP into the grid – To provide power quality and redundancy – To ensure emergency and operation control, considering small-scale generation expansion and involvement of energy storage devices	– 220 kV submarine cable line Zeleny ugol-Russkaya – 220 kV substation Russkaya: 20 MW BESS, system for monitoring of steady states and transients as well as conditions of energy plants and electric equipment, with integrated information communications intended to control generation and consumption facilities – Network of charging stations for electric cars – WACS/WAPS technologies using PMU
6	Project for reliable power supply from the Zeya HPP and to electric traction transit line Siberia-East	– To provide reliable power supply from the Zeya HPP and power supply to electric traction transit line Siberia-East – To ensure quality traction power supply (voltage levels, harmonics, voltage phase balance) – To install operation control devices	– 220 kV digital unattended substations Magdagachi and Birobidzhan: equipment monitoring and diagnosis systems, 220 kV SF6 switchgears (at the substation Birobidzhan), 50 MVAr STCs, active harmonic filters and voltage balancing systems – WACS/WAPS technologies using PMU

5.1.2 Pilot project for creation of territorial smart grid cluster in Russky and Popov Islands

The 220/35 kV substation Russkaya and 220 kV cable—overhead line Zeleny ugol—Russkaya with a submarine bridge through the Eastern Bosforus strait is under construction to provide centralized electricity supply to Russky Island [4,5,124].

The territorial smart grid cluster [4,5,125] was formed because of the summit of APEC countries in 2012 on Russky Island and involves the creation of a smart automated control system. The system aims to provide centralized monitoring, dispatching, and process control as well as solving the problems of creating the smart grid of the region and optimal control of electricity, heat, and gas supply facilities and consumers.

Alongside the network infrastructure, the integrated smart grid includes gas-fired power plants (mini-CHPPs), wind farms, boiler plants, electricity and heat storage systems, "smart houses," and a park of electric cars with charging stations. The control zones of the smart control system are listed in Table 11 [4,5,126].

The major components of the smart grid cluster are:

—Distributed software and hardware complex of the adaptive automatic system for optimal online control of the energy system.
—Software and hardware complex of the automated dispatching and process control system for facilities of the electric network, heat network, and gas supply system as an upper level of the control system of energy supply facilities (SCADA system).
—Automated control system of production process facilities as a lower level in the control loop of energy supply facilities that are directly connected with production equipment.
—System of smart metering of electricity consumption.

Table 11 Control zones of smart automated control systems

Control zone	Composition of objects
Generation	—Mini-CHPPs (Tsentralnaya, Okeanarium, Severnaya, Kommunalnaya) —Two wind farms, solar batteries, storage systems (in the future) —Generating units of industrial enterprises (in the future)
Electric network	—220 kV substation Russkaya (JSC FGC UES) —35 kV substations Tsentralnaya and Severnaya (specified) —10 kV transformer substation —Points of network partitioning (reclosers) —Transmission lines (if necessary)
Heat networks	—Heat supply stations —Boiler plants
Gas supply system	Gas distribution stations
Consumers	Summit facilities, afterwards—household and industrial consumers of Russky and Popov Islands

−Software and hardware complex of the smart grid cluster dynamic simulations for validation of engineering solutions, and for demonstration and education purposes.
−Information and computation infrastructure.
−Active elements of the energy system.

The total planned electric load of consumers on the island during the APEC summit is estimated at 320–380 MW, including 120 MW of residential communities. The first stage includes the construction of four mini-CHPPs: Severnaya, Tsentralnaya, Okeanarium, and Kommunalnaya on Russky Island. Besides, it is planned to construct the Dalnevostochnaya wind power plant consisting of a system of wind turbines on the islands. In terms of the estimated wind potential of the islands, the total capacity of the Dalnevostochnaya wind power plant is 36 MW. It is suggested to install 10 wind turbines in the southern part of Popov, and eight wind turbines (with the total capacity 16 MW) on the Kondratenko peninsula of Russky.

Wind power plants will supply electricity at a voltage of 35 kV with construction of a 35 kV cable line from Switchgear-2 on Popov to Switchgear-1 on Russky (crossing the strait of Stark) and from switchgear-1 to substation Russkaya. In this case the 35 kV cable line will be used to connect the centers of electricity supply to consumers on Popov that is to be switched to the centralized electricity supply from the EPS and residential consumers in the bay of Boyarin.

In the municipal zone of the Saperny peninsula of Russky, it is foreseen to create a residential district with low-rise buildings equipped with a substantial amount of automation devices ("smart houses"). Besides, it is envisioned to equip residential buildings with their own microgeneration sources (solar panels) and create infrastructure for electric cars.

5.2   Smart grid clusters in northwest interconnected power system

The following pilot projects of smart grid clusters will be implemented in the Northwestern region of Russia during the period through 2020 [4,5,127]: Karelskaya power system, power systems of Komi Republic and Arkhangelsk, and "Big Ring" and "Small Ring" of electric networks in St. Petersburg (Table 12 [4,5,9]).

The pilot projects are intended to furnish Northwest IPS with innovation technologies of smart grids that will effectively solve the regional problems (limitations on power output from power plants, insufficient reliability level of power supply to consumers, etc.).

5.3   Pilot project on electricity supply to the Skolkovo innovation center

Construction of the Skolkovo innovation center is a far-reaching project on the creation of the high-tech research and technological complex for design and commercialization of new technologies. The center is planned to be engaged in development of five priority directions of updating: energy, information technologies, telecommunication, biomedical technologies, and nuclear technologies. This "city of the future"

Table 12 Pilot projects on creation of territorial clusters of the smart grid for Northwest interconnected power system

No	Pilot project	Project goal	Smart technologies
1	Smart grid cluster Kola (Karelskaya power system)	To provide reliable power supply and power quality under conditions of parallel operation of 330 kV transit overhead transmission lines	–330 kV transit overhead transmission lines: construction of 2nd transmission line Loukhi-Putkinskaya-Ondskaya-Petrozavodsk-Tikhvin by using high temperature wires –330 kV substations (Loukhi, Putkinskaya, Ondskaya, Petrozavodsk, Tikhvin): • Elements of digital substation • Equipment monitoring and diagnosis systems, STATCOMs, STCs, BESSs • WACS/WAPS technologies using PMU
2	Smart grid cluster Komi (power system of Komi Republic and Arkhangelsk power system)	To provide high reliability level of power supply at the required power quality	–Construction of 2nd 220 kV transmission line Pechorskaya CPP-Ukhta-Mikun with application of high temperature wires –At 220 kV substation: • 220 kV outdoor switchgears of Pechorskaya CPP, Ukhta, Mikun • Elements of digital substation • Equipment monitoring and diagnosis systems, STCs, BESSs • WACS/WAPS technologies using PMU

Continued

Table 12 Continued

No	Pilot project	Project goal	Smart technologies

3	Smart grid cluster "Big Ring" of St. Petersburg	To ensure required reliability of power supply to urban consumers	−1000 MWDC transmission line LAES-2-Vyborg including 330kV submarine cable
			−Reconstruction of substation Vyborgskaya:
			  • 500 MW Complete inverter-converter system
			  • STCs, STATCOMs, asynchronized static compensators
			  • Equipment monitoring and diagnosis systems
			  • Digital substation
			  • Active filters
			  • WACS/WAPS technologies using PMU

4	Smart grid cluster "Small Ring" of St. Petersburg	To decrease current loading and redundancy of existing transmission lines	−Construction of 10kV DC HTSC cable line on a chosen facility (CHPP-5 to 110kV substation Oktyabrskaya, substation Volkhov-Severnaya to substation Sosnovskaya, substation Sinopskaya to 110kV substation Tsentralnaya) −Installation of current limiters based on mechanical releasers, HTSC cables and semiconductors −Construction of an ew underground substation

will host subsidiaries and laboratories of the leading Russian and international universities and companies.

The Skolkovo innovation center is planned to be built on the territory of the urban settlements Novoivanovskoye and Odintsovo near the village Skolkovo, in the eastern part of the Odintsovo district of the Moscow region at a distance of 2 km to the west of the Moscow ring road (MKAD) on the Skolkovo highway [128]. About 40,000 people will reside and work on the area of 370 ha.

For the electricity supply to objects of the Skolkovo innovation center, nine main electric network facilities will be built and reconstructed during 2011–14 [4,5,124]. They are [129]:

- Conversion of sections of seven 110–500 kV overhead transmission lines into cable lines from 500 kV substation Ochakovo to the Minsk highway with construction of 110–500 kV transfer points:
  - 500 kV transmission line Ochakovo-Zapadnaya;
  - Four 220 kV overhead transmission lines (Ochakovo-Krasnogorskaya, Ochakovo-Novovnukovo, Ochakovo-Choboty, and Ochakovo-Lykovo);
  - Two 110 kV cable lines Ochakovo-Odintsovo.

The cable routing is planned to be about 11 km long. It is planned to use cable made of cross-linked polyethylene by the leading world manufacturers.

- Two 220/20 kV indoor substations Skolkovo and Smirnovo with the total capacity 252 MVA on the territory of the Skolkovo innovation center. The variant of underground construction is discussed for substation Smirnovo with the use of gas-insulated power transformers for the first time in Russia. As a result, higher reliability and fire safety of substation operation will be achieved.

Transport links within the Skolkovo innovation center are planned to be implemented by innovation vehicles: five routes of electrobus that provide fast transit to the Moscow underground railway stations and electric cars (with a system of their hire and a network of charging stations).

6 Conclusion

(1) Essential sophistication of the operating conditions of current EPSs enhances the danger of heavy system emergencies and requires improvement and development of the principles and control systems of EPS operating conditions. It is necessary for these purposes to apply new methods and tools for measuring, transferring, and processing the EPS operating parameters. We have suggested in this chapter basic principles on the system of monitoring and forecasting the operating conditions and control of EPS, substantially enhancing the efficiency and adaptability of the coordinated operation and emergency control in EPS.

(2) Intelligent EPS is a customer-oriented EPS of a new generation that should provide high-quality, reliable, and efficient services for electricity supply through flexible interaction in all types of generation, electric networks, and consumers based on cutting-edge technologies and a common hierarchical system of control. A key role in the intelligent EPS is

assigned to the AAN that provides reliable and efficient connection between generation and consumers.

(3) The use of uniform principles of control and qualitatively new kinds of techniques and technologies is necessary for the effective solution of IESAAN control problems online in the conditions of incomplete information on parameters of EPS and disturbances. The structure of IESAAN control should be formed on the basis of the following principles:

- A control system realizing operating functions with a high degree of automation is a key link at the construction of IESAAN.
- IESAAN control systems should have the uniform (coordinated) principles of management regardless of the form of property and hierarchical level.
- When combining the local and system functions of management, the relations between controls of generation, networks, and load-controlled consumers should be traced explicitly.
- Information support of control systems should provide transmission of controlled parameters and signals into control loops in real time.
- The software of control systems should be realized in online and offline modes to trace and control both normal (optimization of conditions) and abnormal (prevention of overloaded and liquidation of preemergency conditions) EPS operation, and besides should be adaptive to changing situations.
- The communication systems included in the control loop should provide high speed, reliability of information, and control signals transmission.

(4) To obtain a general picture of the EPS state for further solving the control problems, the universal methods are necessary to place the PMU to complement SCADA measurements. These methods:

- should provide the best properties of the SE problem solution, such as the observability of the calculated scheme, the identifiability of bad data, and the accuracy of obtained estimates.
- should be based on the observability theory that was devised to place SCADA systems and take into account the block character of PMU measurements.

Among different ways of solving this problem, the topological approaches based on different strategies of random search seem to be the most promising.

(5) Quality of the results in the online SE can be improved by using retrospective information on the state variables. The algorithms were developed that consider interrelations between the time-varying state variables using a set of snapshots (DSE). The main advantage of the algorithms is the ability to deliver good results in the cases of bad and incomplete data and to forecast state variables.

(6) With a permanently growing number of cyberattacks against the information structure of energy facilities, the systems of data collection and processing (SCADA and WAMS) are most vulnerable to the physical failures and information attacks. If a local region or an object is observable on the basis of PMU measurements, they are also observable by SCADA measurements. Therefore, it is suggested, when needed, carrying out independent simultaneous (by one timestamp) bad data detection and SE by SCADA and WAMS measurements to additionally find out if there is a malicious attack against one or another system.

(7) The existing ECSs and the actions of the transmission system operator may prove to be ineffective to prevent the catastrophic development of the emergency. The development of next-generation intelligent systems is necessary to complement existing ECSs, taking

into account their "weak points." The new emergency protection and operation systems should

- include tools for the intelligent monitoring and assessment of the EPS operating conditions,
- be capable of predicting potentially dangerous states of the EPS,
- be highly resilient and able to coordinate local emergency control devices,
- include methods and models providing the protection of an EPS with a complex structure, and
- complement the existing ideology of ECSs but not contradict it.

(8) Computational intelligence techniques like decision trees, ANNs, and multiagent systems have been applied to achieve the essential online intelligent security monitoring and assessment performance. The following systems of monitoring, forecasting, and control of EPS are recommended for implementation:

- Novel forecasting systems such as *Forecast of EPS state variables based on the DSE method* and *forecast of EPS parameters based on the hybrid data-driven approach,* which have been developed to improve grid operation efficiency and to reduce fossil fuel consumption and carbon emissions. The main goal of the forecasting approach development was to increase the efficiency of forecasting studies when dealing with limited retrospective data and a sharply variable, nonstationary time series of EPS parameters.
- Automatic decision tree-based system for online voltage security control of EPSs, which provides the dynamics modeling accuracy sufficient for research into voltage stability, overload of lines, algorithms for local automation devices, etc. This fact makes it possible to effectively apply the system for monitoring to control security in EPSs of a large dimension in real time and to implement functions of online alarming, localization, and interaction with automatic systems. The operation of the developed intelligent system is mostly automatic, and the major part of the control actions is generated with the minimum involvement of the operator.
- A decentralized agent-based system for operation and emergency control (a MAAS), which provides reactive power control by coordinating the operation of different discrete and continuous control devices in a postdisturbance period. Such a coordinated control of local devices makes it possible to effectively cope with the equipment overload and voltage collapse, prevents generator tripping, and maintains load bus voltages within the normal range.
- Supplementing the existing principles of EPS control with the method of "control with prediction," which provides early detection and prevention of dangerous states and emergency situations before they lead to a voltage collapse. The proposed intelligent system for preventing large-scale emergencies in the EPS adjusts the operation of local emergency control devices both in steady state and in postemergency conditions to reduce the probability of a severe emergency.

(9) The areas for further developing IESAAN are:

- Design a subsequent application of new types of power equipment, giving active properties to an electric grid (on the basis of new kinds of materials such as power semiconductors, high-temperature superconductors, and so forth).
- Create a new means and systems of relay protection, automatic operation and emergency control, equipment diagnostics, and energy resources accounting.
- Design of control systems for substations of new generation.
- Improve existing as well as design and implementation of novel hierarchical systems for coordination and control of power flows, frequency, and generation.

- Provide a new quality of network monitoring and protection against external contingencies (lightning, icing, wind effects, wires sagging etc.).
- Monitoring the reliability and quality of electricity transmission services.

Finally, the creation of IESAAN should provide a qualitatively new level of efficiency of electric power industry development and functioning, raise system security, and increases the quality and reliability of the electricity supply to consumers.

References

[1] V.E. Fortov, A.A. Makarov, A Concept of Russia's Intelligent Power System With Active-Adaptive Network, JSC "NTC FSC EES", Moscow, 2012. 235p. (in Russian).

[2] Governmental Commission on High Technologies and Innovations, Procedure of Forming the List of Technology Platforms, Minutes No 4, 2010 August, 3 (in Russian). Available from: http://www.economy.gov.ru/wps/wcm/connect/debc870043f9c123b9 d9bbe7f74495eb/poradok_texplatf_030810.pdf?MOD=AJPERESandCACHEID=debc 870043f9c123b9d9bbe7f74495eb, 2010.

[3] Federal Grid Company of Unified Energy System, On Approval of Investment Program JSC "FGC UES" for 2010–2014, Resolution No 547, 2010 November, 12 (in Russian). Available from: http://www.fsk-ees.ru/media/File/stockholders/documents/docs/Prikaz_ 12.11.10.pdf, 2010.

[4] D.N. Efimov, Some developments, prospective ways and projects of smart grid technologies in Russia—an overview, Proc. of IEEE PES ISGT Europe 2011, Dec. 5–7, Manchester, UK, 2011. 6 pp.

[5] D.N. Efimov, in: Smart grid in Russia: today's and tomorrow's practices, Proc. of the 7th Conf. on Sustainable Development of Energy, Water and Environment Systems, Ohrid, Macedonia, July 01–06, 2012. 12 pp.

[6] N.I. Voropai, D.N. Efimov, V.G. Kurbatsky, N.V. Tomin, in: Smart control in the Russian electric power system, Proc. of Int. Conf., Smart Greens 2012 Porto, Portugal, April 19–20, 2012, pp. 133–136.

[7] N.I. Voropai, Intellectual electric power systems: the concept, a condition, prospects, Automat. IT Power Eng. 3 (2011) 11–15 (in Russian).

[8] O. Budargin, JSC "FGC UES": unified grid—United State, Trends Events Mark. Federal Business Mag. 10 (57) (2011) 14–17. (in Russian). Available from: http://www.tsr-media.ru/smi/archive/issue/article?cun=287936andissue=287934.

[9] JSC "R&D Center for Power Engineering", Conception for Development of Intelligent Energy Power System of Russia with Active-Adaptive Network, Working Materials for Scientific Report, 2011 (in Russian).

[10] G. Phadke, Synchronized phasor measurements. A historical overview, Proc. IEEE/PES Transmission and Distribution Conference, vol. 1, 2002, pp. 476–479.

[11] A. Gamm, Y. Grishin, A. Glazunova, I. Kolosok, E. Korkina, in: New EPS state estimation algorithms based on the technique of test equations and PMU measurements, Proc. of the International Conference "PowerTech'2007", Lausanne, 2007. CDROM, ID 256.

[12] A.V. Zhukov, Development of monitoring and control in UES of Russia on the base of Wide Area Monitoring System, in: Presentation for the 22th Conference on Relay Protection and Automation of Electric Power Systems, 2014 May, 27–29, 2014 (in Russian). Available from: http://www.bopd.ru/docs/04062014/2.pdf.

[13] A. Glazunova, I. Kolosok, E. Korkina, Monitoring of EPS Operation by the State Estimation Methods, ISGT, Manchester, 2011.

[14] R. Mogilko, in: SMART-WAMS recorder of transient variables. Experience in design and implementation experience. Prospects for development, Proc. of Conf. Monitoring of Power System Dynamics Performance, 28–30 April, Saint Petersburg, 2008.

[15] B. Milosevic, M. Begovic, Nondominated sorting genetic algorithm for optimal PMU placement, IEEE Trans. Power Syst. 18 (2003) 69–75.

[16] K.-S. Cho, J.-R. Shin, S.H. Hyun, in: Optimal placement of PMU with GPS receiver, PES Winter Meeting, IEEE, vol. 1, 2001, pp. 258–262.

[17] A.Z. Gamm, A. Glazunova, I.N. Kolosok, E.S. Korkina, PMU placement criteria for EPS state estimation, Proc. of Conf. DRPT 6–9 April, Nanjing, China, 2008.

[18] A.Z. Gamm, I.N. Kolosok, in: Test Equations and Their Use for State Estimation of Electrical Power System, Power and Electrical Engineering: Scientific Proc. of Riga Technical University, RTU, Riga, 2002, pp. 99–105.

[19] A. Monticelly, Electric power system state estimation, Proc. IEEE 88 (2) (2000) 262–282.

[20] A.E. Abur, Gymez, Power System State Estimation: Theory and Implementation, Marcel Dekker, New York, 2004.

[21] A.Z. Gamm, I.N. Kolosok, Detecting gross errors of SCADA-measurements in power systems, Nauka (Science), Novosibirsk, 2000. 152 pp. (in Russian).

[22] I. Kolosok, M. Khokhlov, in: Specific features of state estimation problem in control of electric power system with active-adaptive properties, Proc. of the 5th Intern. Conf. "Liberalization and Modernization of Power Systems: Smart Technologies for Joint Operation of Power Grid", Irkutsk, Russia, August 6–10, 2012, pp. 100–108.

[23] A.Z. Gamm, A.M. Glazunova, I.N. Kolosok, in: Test equations for validation of critical measurements and critical sets at power system state estimation, Proc. of the 2005 IEEE S. Petersburg, Power Tech, June 27–30, IEEE, Saint Petersburg, 2005 Paper 166.

[24] I. Kamwa, R. Grondin, PMU configuration for system dynamic performance measurement in large multiarea power systems, IEEE Trans. Power Syst. 17 (2) (2002) 385–394.

[25] K.D. Jones, Three-Phase Linear State Estimation With Phasor Measurements, Thesis of Master of Science in Electrical Engineering, http://scholar.lib.vt.edu/theses/available/etd-05022011-141649/unrestricted/Jones_KD_T_2011.pdf, 2011.

[26] B. Xu, Y. Yoon, A. Abur, in: Optimal placement and utilization of phasor measurements for state estimation, Proc. of the 15th Power Systems Computation Conf., Liege, Belgium, August, 2005.

[27] I. Kolosok, E. Korkina, E. Buchinsky, in: The test equation method for linear state estimation based on PMU data, Proc. of PSCC-2014, Poland, Wroclaw, 2014.

[28] A.Z. Gamm, Statistical methods for state estimation of electric power systems, Nauka, 1976. 220 pp. (in Russian).

[29] A. Gamm, Decomposition algorithms for solution of power system SE problem, Electronnoye Modelirovanie 3 (1983) 63–68 (in Russian).

[30] A. Gamm, Y. Grishin, Distributed information processing in automated power system control systems, in: Proc. V Int. Workshop "Distributed Information Processing", Novosibirsk, Russia, 1995, pp. 243–247 (in Russian).

[31] K. Clements, O. Denison, R. Ringle, in: A multy-area approach to SE in power system networks, Proc. IEEE Power Eng. Soc. Meeting, San Francisco, CA, 1972 Paper. C 72465-3.

[32] S. Iwamoto, M. Kusano, V. Quantana, Hierarchical SE using a fast rectangular-coordinate method, IEEE Trans. Power Syst. 4 (3) (1989) 870–879.

[33] A. El-Kleib, J. Nieplocha, H. Singh, D. Maratukulam, A decomposed SE technique suitable for parallel processor implementation, IEEE Trans. Power Syst. 7 (3) (1992) 1088–1097.

[34] D.M. Falcao, F.F. Wu, L. Murphy, Parallel and distributed state estimation, IEEE Trans. Power Syst. 10 (2) (1995) 724–730.

[35] E.V. Ametistov, Fundamentals of modern energy. Modern electric power engineering, Vol. 2, MEI, Moscow, 2010. 632 p.

[36] E. Acha, C.R. Fuerte-Esquivel, H. Ambriz-Perez, C. Angeles-Camacho, FACTS. Modelling and Simulation in Power Networks, John Willey & Sons. Ltd, England, 2004. 420 p.

[37] T. Okon, K. Wilkosz, WLS Stste Estimation in Polar and Rectangular Coordinate Systems for Power System with UPFC: Significance of Types of Measurements, Proceedings of the International Conference "PowerTech'2011", Trondheim, Norway, 19–23 June, 2011. USB #185.

[38] A. Zamora-Cárdenas, C.R. Fuerte-Esquivel, State estimation of power systems containing facts controllers, Electr. Power Syst. Res. 81 (2011) 995–1002.

[39] X. Bei, A. Abur, in: State estimation of systems with embedded FACTS devices, Proceedings of the International Conference "PowerTech'2003", Bologna, Italy, June 23–26, 2003. USB #207.

[40] I.N. Kolosok, A.V. Tikhonov, Algorithm of modeling static var compensators at state estimation of power systems, Promyshlennaya Energetika (Ind. Energ.) 10 (2015) 30–35 (in Russian).

[41] A.V. Tikhonov, in: Modeling FACTS devices at EPS state estimation, Proceedings of the ESI Young Scientists' Conference "System investigations in energetics." Issue 42, Irkutsk, 2012, pp. 103–108 (in Russian).

[42] I.N. Kolosok, A.V. Tihonov, E.S. Korkina, Studying the efficiency of SE algorithms when involving models of FACTS devices in the equivalent circuit, Coll. Papers "Methodical issues of studying the reliability of bulk energy systems", Issue 67 "Problems of power system reliability", "Komi Republican Printery" Ltd., Syktyvkar, 2016, pp. 206–214 (in Russian).

[43] I. Kolosok, A. Mahnitko, A. Tikhonov, State estimation of electric power systems including FACTS models (SVC and STATCOM), Power Electr. Eng. 33 (2016) 40–45.

[44] N.I. Voropai, A.B. Osak, Developing Equipment and Control Systems of Bulk Power Systems, Code 2008-0-2.7-31-01-007, Scientific and Technical Report, Irkutsk, 2009. 480 pp. (in Russian).

[45] V.I. Kochkin, New technologies of increasing the power-line throughput capacity. Controlled transmission of power, Novosti Elektrotekhniki (Electr. Technol. News) 4 (46) (2007) 2–6 (in Russian).

[46] M.B. Do Coutto Filho, J.C. Stacchini de Souza, R.S. Freund, Forecasting-aided state estimation—Part 2: implementation, IEEE Trans. Power Syst. 24 (4) (2009) 1678–1685.

[47] A.M. Leite da Silva, M.B. Do Couto Filho, J.M.C. Cantera, An efficient dynamic state estimation including bad data processing, IEEE Trans. Power Syst. 2 (4) (1987) 1050–1058.

[48] N. Zhou, Z. Huang, G. Welch, Dynamic state estimation of a synchronous machine using PMU data: a comparative study, IEEE Trans. Smart Grid 6 (1) (2015).

[49] M.A.M. Ariff, B.C. Pal, A.K. Singh, Estimating dynamic model parameters for adaptive protection and control in power system, IEEE Trans. Power Syst. 30 (2) (2015) 829–839.

[50] A.M. Glazunova, Dynamic state estimation, in: Monitoring, Control and Protection of Interconnected Power Systems, Springer-Verlag, Berlin Hiedelberg, 2014. pp. 107–123. Part 3.7.

[51] M.B. Do Coutto Filho, J.C. Stacchini de Souza, R.S. Freund, Forecasting-aided state estimation—Part 1: implementation, IEEE Trans. Power Syst. 24 (4) (2009) 1667.

[52] K. Wilkosz, Verification of the measurements of voltage magnitudes in electric power system, Second Int. Symp. on Security Power System Operation, Wroclaw, 1981, pp. 147–155. Paper E8.

[53] M. Pignati, L. Zanni, S. Sarri, R. Cherkaoui, J.-Y. Le Boudec, M. Paolone, in: A pre-estimation filtering process of bad data for linear power systems state estimation using

PMUs, Proceedings of the 18 th Power Systems Computation Conference, Wroclaw, Poland, August 18–22, 2014. #94.

[54] H. Karimipour, V. Dinavahi, Extended Kalman filer-based parallel dynamic state estimation, IEEE Trans. Smart Grid 6 (3) (2015) 1539–1549.

[55] Y. Yang, W. Hu, Y. Min, in: Projected unscented Kalman filter for dynamic state estimation and bad data detection in power system, 12th IET International Conference on "Developments in Power System Protection" (DPSP 2014), March 31–April 3, 2014, pp. 1–6.

[56] M. Gol, A. Abur, A modified Chi-squares test for improved bad data detection, Proceedings of the International Conference "PowerTech'2015", Eindhoven, Holland, 27 June–3 July, 2015. USB #448510.

[57] V.G. Neuymin, A.S. Alexandrov, D.M. Maksimenko, in: Verification of the model of scheme binding and identification of gross errors in telemetry data in the software "RastrWin3", Proceedings of the VI International Scientidic and Technical Conference "Power Energy Through the Eyes of Youth", 9–13 November, 2015, pp. 526–529.

[58] K.A. Clements, G.R. Krumpholz, P.W. Davis, Power system state estimation with measurement deficiency: an observability/measurement placement algorithm, IEEE Trans. Power Syst. PAS-102 (7) (1983) 2012–2020.

[59] H. Cramer, Mathematical Methods of Statistics, Publishing House MIR, Moscow, 1975.

[60] S. Zanouz, K.M. Rogers, R. Berthier, R.B. Bobba, W.H. Sanders, T.J. Overbye, SCPSE: security-oriented cyber-physical state estimation for power grid critical infrastructure, IEEE Trans. Smart Grid 3 (4) (2012) 1790–1799.

[61] C. Ten, C. Liu, M. Govindarasu, in: Vulnerability assessment of cybersecurity for SCADA systems, Proc. 24–28 June 2007 Power Engineering Society General Meeting, 2007, pp. 1–8.

[62] N. Voropai, D. Efimov, I. Kolosok, V. Kurbatsky, A. Glazunova, E. Korkina, A. Osak, N. Tomin, D. Panasetsky, Smart technologies in emergency control of Russia's unified energy system, IEEE Trans. Smart Grid 4 (3) (2013) 1732–1740.

[63] Y. Liu, M. Reiter, P. Ning, in: False data injection attacks against state estimation in electric power grids, Proc. of the 16th ACM Conf. on Computer and Communications Security, USA, 2009.

[64] M. Rihan, M. Ahmad, M.S. Beg, Vulnerability analysis of wide area measurement system in the smart grid, Smart Grid Renew. Energy (2013) 1–7. Available from: http://www.scirp.org/journal/sigre.

[65] T.H. Morris, P. Shengyi, U. Adhikari, in: Cyber security recommendations for wide area monitoring, protection and control systems, Proc. 22–26 July 2012 IEEE Power and Energy Society General Meeting, 2012, pp. 1–6.

[66] L. Heng, J.J. Makela, A.D. Dominguez-Garcia, R.B. Bobba, W.H. Sanders, G.X. Gao, in: Reliable GPS-based timing for power systems: a multi-layered multi-receiver architecture, Proc. 2014 Power and Energy Conference at Illinois (PECI), 2014, pp. 1–7.

[67] N.I. Voropai, V.V. Ershevich, Y.N. Luginsky, Management and Control of Bulk Power Grids, Energoatomizdat, Moscow, 1984, p. 256 (in Russian).

[68] S.A. Sovalov, V.A. Semenov, Emergency Control in Power Systems, Energoatomizdat, Moscow, 1988, p. 416 (in Russian).

[69] N.I. Voropai, D.N. Efimov, P.V. Etingov, D.A. Panasetsky, in: Smart emergency control in electic power system, 18th IFAC World Congress, Milano, Italy, Aug. 28–Sept. 2, 2011, pp. 1658–1664.

[70] N.I. Voropai, D.N. Efimov, P.V. Etingov, D.A. Panasetsky, in: Emergency control in electric power systems based on smart grid concept, APPEEC'2011 Int. Conf., Wuhan, China, March 25–28, 2011, 6 p.

[71] Y.V. Makarov, V.I. Reshetov, V.A. Stroev, N.I. Voropai, Blackout prevention in the United States, Europe and Russia, Proc. IEEE 93 (2005) 1942–1955.

[72] Y.V. Makarov, N.I. Voropai, D.N. Efimov, in: Complex emergency control system against blackouts in Russia, IEEE PES General Meeting, Pittsburgh, USA, July 21–24, 2008, p. 8.

[73] Russian National Standardization Authority, Methodological guidelines on power system stability, Approved by the order of Ministry of Energy of the RF of Jun. 30, 2003, No.277, Publ. House of RC ENAS, Moscow, 2004, 16 (in Russian).

[74] D. Panasetsky, Improving the Structure and Algorithms of Emergency Control in Power Systems to Prevent Voltage Collapse and Cascading Line Outages, Dissertation, Melentiev Energy Systems Institute, 2014.

[75] N. Voropai, M. Negnevitsky, N. Tomin, D. Panasetsky, C. Rehtanz, U. Haeger, V. Kurbatsky, Intelligent systems for preventing large-scale emergencies in power systems, Theor. Sci. J. "Electrichestvo" 8 (2014) (In Russian).

[76] N. Tomin, M. Negnevitsky, C. Rehtanz, Preventing large-scale emergencies in modern power systems: AI approach, J. Adv. Comput. Intell. Intell. Inform. 18 (5) (2014).

[77] H.H. Steinherz, P.C. Eduardo, S.R. Castro, Neural networks for short-term load forecasting: a review and evaluation, IEEE Trans. Power Syst. 16 (1) (2001).

[78] V. Gerikh, I. Kolosok, V. Kurbatsky, N. Tomin, Application of neural network technologies for price forecasting in the liberalized electricity market, Sci. J. Riga Tech. Univ. Power Electr. Eng. 5 (2009) 91–96.

[79] L. Fugon, J. Juban, G. Kariniotakis, in: Data mining for wind power forecasting, Proc. the European Wind Energy Conference, Brussels, Belgium, April, 2008.

[80] E.J. Natenberg, D.J. Gagne II, J.W. Zack, J. Manobianco, G.E. Van Knowe, T. Melino, in: Application of a random forest approach to model output statistics for use in day ahead wind power forecasts, Proc. the Symposium on the Role of Statistical Methods in Weather and Climate Prediction, Austin, USA, 2013.

[81] R.G. Kavasseri, K. Seetharaman, Day-ahead wind speed forecasting using f-ARIMA models, Renew. Energy 34 (5) (2009) 1388–1393.

[82] R.C. Garcia, J. Contreras, M. van Akkeren, J. Batista, C. Garcia, A GARCH forecasting model to predict day-ahead electricity prices, IEEE Trans. Power Syst. 20 (2) (2005).

[83] A.M. Glazunova, in: Forecasting power system state variables on the basis of dynamic state estimation and artificial neural networks, Proc. the IEEE Region 8 SIBIRCON-2010, Listvyanka, Russia, 2010.

[84] C.S. Watters, P. Leahy, in: Comparison of linear, Kalman filter and neural network downscaling of wind speeds from numerical weather prediction, Proc. 2011 10th International Conference on Environment and Electrical Engineering, Rome, Italia, 2009.

[85] N.I. Voropai, N.V. Tomin, V.G. Kurbatsky, et al., A Set of Intelligent Tools for Prevention of Large-Scale Emergencies in Power Systems, Nauka, Novosibirsk, 2016. 332 p.

[86] P. Areekul, T. Senjyu, H. Toyama, A. Yona, A hybrid ARIMA and neural network model for short-term price forecasting in deregulated market, IEEE Trans. Power Syst. 25 (1) (2010) 524–530.

[87] A.Z. Khotanzad, H. Elragal, A neuro-fuzzy approach to short-term load forecasting in a price sensitive environment, IEEE Trans. Power Syst. 17 (4) (2002) 1273–1282.

[88] N. Sinha, L.L. Lai, P. Kumar Ghosh, Y. Ma, in: Wavelet-GA-ANN based hybrid model for accurate prediction of short-term load forecast, Proc. the IEEE Inter. Conf. on ISAP, 2007, Toki Messe, Niigata, 2007, pp. 1–8.

[89] I.G. Damousis, P. Dokopoulos, A fuzzy model expert system for the forecasting of wind speed and power generation in wind farms, Proc. of the IEEE International Conference on Power Industry Computer Applications PICA 01, 2001, pp. 63–69.

[90] A.M. Foley, P.G. Leahy, A. Marvuglia, E.J. McKeogh, Current methods and advances in forecasting of wind power generation, Renew. Energy 37 (1) (2012) 1–8.

[91] D.N. Sidorov, Methods of Analysis of Integral Dynamical Models: Theory and Applications, ISU Publ, Irkutsk, 2013. 293 p. (Russian).

[92] I. Dobson, P. Zhang, et al., Initial review of methods for cascading failure analysis in electric power transmission systems IEEE PES CAMS task force on understanding, prediction, mitigation and restoration of cascading failures, in: In Proc. IEEE PES General Meeting, Pittsburgh, PA, USA, July, 2008, p. 1.

[93] W.R. Lachs, Voltage instability in interconnected power systems: a simulation approach, IEEE Trans. Power Syst. 7 (2) (1992) 753–761.

[94] M.S.S. Danish, Voltage Stability in Electric Power System: A Practical Introduction, Logos Verlag Berlin, Berlin, 2015.

[95] M.J. Karki, Methods for Online Voltage Stability Monitoring, PhD Dissertation, Iowa State University, 2008.

[96] M. Negnevitsky, N. Tomin, V. Kurbatsky, D. Panasetsky, A. Zhukov, C. Rehtanz, in: A random forest-based approach for voltage security monitoring in a power system, Proc. 2015 IEEE Eindhoven Powertech, 2015.

[97] K. Morison, L. Wang, P. Kundur, Power system security assessment, IEEE Power Energy Mag. 2 (5) (2004) 30–39.

[98] ENTSO-E, Network Code on Operational Security, 24 September 2013; Available from: http://networkcodes.entsoe.eu/operational-codes/operational-security/, 2013.

[99] CIGRE, Working Group C4.601, Review of On-Line Dynamic Security Assessment Tools and Techniques, 2007.

[100] M. Vaiman, et al., in: Risk assessment of cascading outages. Part I. Overview of methodologies, Power and Energy Society General Meeting, 2011 IEEE, 2011.

[101] N. Ming, et al., Software implementation of online risk-based security assessment, IEEE Trans. Power Syst. 18 (3) (2003) 1165–1172.

[102] GARPUR Project, State of the Art on Reliability Assessment in Power Systems, Deliverable D1.1, 2014.

[103] M. Negnevitsky, Computational intelligence approach to crisis management in power systems, Int. J. Autom. Control. 2 (2/3) (2008) 247–273.

[104] U. Häger, C. Rehtanz, N. Voropai, in: ICOEUR project results on improving observability and flexibility of large scale transmission systems, Presented at the IEEE PES General Meeting, 2012, San Diego, USA, 2012.

[105] S.D. Ramchurn, et al., Putting the 'smarts' into the smart grid: a grand challenge for artificial intelligence, Commun. ACM 55 (4) (2012) 86–97.

[106] I. Arkhipov, A. Molskij, A. Ivanov, D. Novickij, D. Sorokin, D. Holkin, Multiagent voltage and reactive power control system, EAI Endorsed Trans. Energy Web 14 (3) (2014) e3.

[107] J.H. Friedman, Stochastic gradient boosting, Comput. Stat. Data Anal. 38 (2002) 367–378.

[108] L. Breiman, Random forests, Mach. Learn. 45 (2001) 5–32.

[109] V. Kurbatsky, N. Tomin, D. Sidorov, V. Spiryaev, Hybrid model for short-term forecasting in electric power system, Int. J. Mach. Learn. Comput. 1 (2) (2011) 138–147.

[110] V. Kurbatsky, D. Sidorov, V. Spiryaev, N. Tomin, On the neural network approach for forecasting of nonstationary time series on the basis of the Hilbert-Huang transform, Autom. Remote. Control. 72 (7) (2011) 1405–1414.

[111] V. Kurbatsky, D. Sidorov, V. Spiryaev, N. Tomin, Forecasting nonstationary time series on the basis of Hilbert-Huang transform and machine learning, Autom. Remote. Control. 75 (4) (2014) 12–16.

[112] C. Zhu, R.H. Byrd, P. Lu, J. Nocedal, Algorithm 778: L-BFGS-B, Fortran subroutines for large scale bound constrained optimization, ACM Trans. Math. Softw. 23 (4) (1997) 550–560.

[113] A.M. Glazunova, E.S. Aksaeva, Actions generation for online power system control, Proceedings of the International Conference "PowerTech'2015", Eindhovene, The Netherlands, 29 June–3 July, 2015. USB #459689.

[114] F. Wasserman, Neurocomputer hardware: Theory and Practice, MIR, 1992. 186 p.

[115] M. Negnevitsky, N. Voropai, V. Kurbatsky, N. Tomin, D. Panasetsky, in: Development of an intelligent system for preventing large-scale emergencies in power systems, IEEE/PES General Meeting, Vancouver, BC, Canada, 2013.

[116] A. Zhukov, N. Tomin, D. Sidorov, D. Panasetsky, V. Spirayev, in: A hybrid artificial neural network for voltage security evaluation in a power system, Proc. 2015 5th International Youth Conference on Energy (IYCE), Pisa, Italy, 2015.

[117] P. Kessel, H. Glavitsch, Estimating the voltage stability of a power system, IEEE Trans. Power Deliv. PWRD-1 (3) (1986) 346–353.

[118] Y.N. Nguegan Tchokonte, Real-time Identification and Monitoring of the Voltage Stability Margin in Electric Power Transmission Systems Using Synchronized Phasor Measurements, Kassel University Press GmbH, 2009, p. 344.

[119] X. Gong, B. Zhang, B. Kong, A. Zhang, H. Li, W. Fang, Research on the method of calculating node injected reactive power based on L indicator, J. Power Energy Eng. 2 (2014) 361–367.

[120] R. Diao, A.Z. Tempe, K. Sun, V. Vittal, R.J. O'Keefe, M.R. Richardson, N. Bhatt, D. Stradford, S.K. Sarawgi, Decision tree-based online voltage security assessment using PMU measurements, IEEE Trans. Power Syst. 24 (2) (2009) 832–839.

[121] M. Beiraghi, A.M. Ranjbar, Online voltage security assessment based on wide-area measurements, IEEE Trans. Power Deliv. 28 (2) (2013) 989–997.

[122] N.I. Voropai, D.N. Efimov, P.V. Etingov, D.A. Panasetsky, in: Smart emergency control in electric power systems, Proc. 2011 IFAC World Congress, 2011, pp. 1658–1664.

[123] O. Budargin, Modernization via innovation development—creation of smart grid, Presentation for Commission on Modernization and Technological Development of Russian Economy, 2010 October, 26, 2010 (in Russian). Available from: http://www.fsk-ees.ru/media/File/press_centre/speeches/Presentation_Budargin_26.10.10.pdf.

[124] Federal Grid Company of Unified Energy System, Moving Forward With New Energy, FGC UES, 2010.

[125] JSC "DALENERGOSETPROJECT", Main Directions for Development of Primorye Region Energy Till 2012 and for Perspective to 2020, 2009. R&D Report, Vladivostok (in Russian).

[126] JSC "Far-East Energy Management Company", Creation of Smart Grid Cluster at Russky Island, 2011. Customer Requirements (in Russian).

[127] O. Budargin, in: Smart Grid—First Outcomes, Presentation for Round Table "Smart Grid—Projects of Future", 2011 June, 16–18, 2011 (in Russian). Available from: http://www.fsk-ees.ru/upload/docs/01_ayladwlg.pdf.

[128] The official web-site of the Skolkovo Foundation, http://www.i-gorod.com/en/.

[129] Federal Grid Company of Unified Energy System, The Electricity Supply to the Skolkovo Innovation Center, (in Russian). Available from: http://www.fsk-ees.ru/investments/electricity_innovation_centre_skolkovo/.

Part Two

North America

Demand response: An enabling technology to achieve energy efficiency in a smart grid

Chen Chen
Argonne National Laboratory, Argonne, United States

1 Introduction

The US electricity grids face increasing pressure to match supply and demand. This pressure comes from two sources. On the one hand, electricity demand is increasing in quantity and changing in quality. For example, the Energy Information Administration estimates that overall electricity demand will increase 30% from 3.9 trillion kWh in 2009 to 5.0 trillion kWh in 2035 [1]. A portion of this increase will come from wide adoption of electric vehicles, whose demand profile is significantly different from current loads. On the other hand, driven by increasing prices for fossil fuels and concerns about greenhouse gas emissions, renewable energy resources are rapidly being introduced into the existing electricity supply portfolio. For example, the United States Department of Energy estimates that wind power will meet 20% of the US electricity demand by 2030, which means that the wind power capacity of the United States may reach more than 300 GW [2]. With emerging smart grid development, a key solution for the power grids to alleviate the stresses of increasing demand and intermittent renewable generation is to encourage demand response (DR) as demand-side participation in the electricity market [3]. According to the Federal Energy Regulatory Commission (FERC), the DR is defined as:

> *Changes in electric usage by end-use customers from their normal consumption patterns in response to changes in the price of electricity over time, or to incentive payments designed to induce lower electricity use at times of high wholesale market prices or when system reliability is jeopardized.*

The capabilities of "changes in electric usage" in the above definition are the resources available for DR. It is observed that certain types of appliances with operational flexibility may serve as candidate resources for DR. For example, some appliances offer flexibility in the operation time (e.g., dish washers or clothes washers/dryers) so that their start time can be delayed under the constraint of a hard deadline. The plug-in electric vehicle (PEV) is another important emerging load type that offers flexibility in the power consumed in charging PEV batteries. This type of load is a great DR asset because individual PEV battery charge can be modified (and coordinated across multiple PEV batteries) as long as its total energy requirement is fulfilled

Application of Smart Grid Technologies. https://doi.org/10.1016/B978-0-12-803128-5.00004-0

by a given deadline. Thus, DR schemes that exploit loads with flexible operation time and power draw can shift electricity usage in time and benefit both consumers and the grid [4]. From the grid's perspective, shifting some load away from peak periods: (1) lowers the cost of employing less efficient generation during peak demand; (2) alleviates transmission congestion; and (3) helps maintain grid stability during these critical peak hours. In addition, the DR could have the potential to facilitate the increasing penetration of renewable energy, which has intermittent and stochastic features. From the customer's perspective, shifting load to off-peak periods results in reduced electricity bills and occurrences of blackouts and brownouts. In this sense, the DR is an enabling technology to achieve the energy efficiency in the future smart grid.

Due to the underlying two-way communication infrastructures enabled by future smart grid renovation, the interactions between customers, utilities, and the market can be largely enhanced. A central challenge in implementing DR in future smart grids is how to manage DR resources, that is, how to exploit the underlying information infrastructure and operational flexibilities of loads to achieve the promised benefits of DR. In this chapter, we will describe the DR programs and implementation in the United States, and then a distributed direct load control (DLC) architecture and associated algorithms for large-scale residential DR is discussed. Numerical results are also provided to validate the effectiveness of the proposed DR scheme.

2   Demand response development in the United States

Depending on the different levels of aggregation, DR implementation and participation in the layered structure of the electricity market can be divided into three categories, as shown in Fig. 1:

- *Consumer-premise level*: The operation of appliances can be scheduled, either manually or by automated devices (e.g., the energy management controller [EMC]) on behalf of customers, within the home to save money or fulfill the control task. This level involves the homes or buildings that are the load end of the electric power distribution grids.

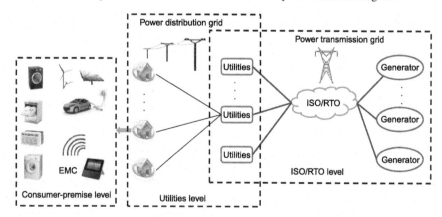

Fig. 1 Three levels of DR implementation in the electricity market.

- *Utilities level*: With operational flexibilities of appliances, the EMCs interact with utilities in the retail electricity market. This level involves the electric power distribution grids.
- *ISO/RTO level*: Large-scale aggregation of DR resources can participate in and affect the wholesale electricity market. This level involves the electric power transmission grids.

2.1   Demand response at consumer-premise level

The DR implemented in this level usually aims to schedule operation of appliances within the home to save money or fulfill the control task. It can be done by customers manually, but a more sophisticated EMC could ease the task in a more effective and optimal way [5].

Besides the operational flexibilities provided by some appliances such as clothes washers/dryers or dishwashers, thermal comfort flexibilities of customers provide additional freedom in scheduling thermostatically controlled appliances such as HVAC (heating, ventilation, and air conditioning) systems. In addition, several utilities already provide time-varying electricity tariffs to subscribed customers, which will be described in Section 2.2. Such customers can utilize their load flexibilities to save on electricity bills. However, current residential load-control activities are mainly operated manually, which poses great challenges to customers in optimally scheduling the operations of their appliances. Customers may not have time to make such scheduling decisions and if prices vary fast and frequently, scheduling may be too complex. Hence, an automated EMC is necessary to optimize the appliances' operation on behalf of customers.

Several appliance vendors have smart appliances in the market that have the communication and control capability regarding their operations; this could facilitate the consumer-premise DR implementation. For example, GE Appliances provides Wi-Fi Connect appliances including refrigerators, ranges and ovens, dishwashers, clothes washers and dryers, air conditioners (ACs), and water heaters [6]. These appliances can be connected to smart phones via Wi-Fi and can be monitored and controlled by the GE Appliance applications (Apps) so that adjusting the operation schedule is enabled. Whirlpool also provides smart washers and dryers with built-in Wi-Fi communications to smart phones to enable control capabilities [7]. In addition, the washer and dryer can work with the Nest Learning Thermostat to make better choices for fabric care and energy efficiency, for example, auto-delay laundry cycles during energy rush hours [7].

2.2   Demand response at utilities level

The interactions between customers and utilities are usually involved when DR is implemented in this level. By these interactions, utilities can make customers change their power usage profile, either directly or indirectly, to reduce peak demand and alleviate the mismatch between supply and demand. Direct approaches allow utilities to remotely control appliances during high-demand periods or power supply emergencies while indirect approaches control appliances' operation by sending price signals. Several utilities in the United States have different types of DR programs for residential customers. Below we describe DR programs provided by some utilities in geographic regions (West Coast, Mid-west, and East Coast) in the United States.

2.2.1   PG&E

Pacific Gas and Electric Company (PG&E), which is an investor-owned electric utility (IOU) serving most of the northern two-thirds of California with 5.2 million households, provides two types of DR programs for residential customers [8]:

- *SmartRate Program*: In this program, the typical electric rate is reduced from June 1 to September 30, except on SmartDays, which are days when electric rates are higher from 2 p.m. to 7 p.m. SmartDays are usually called or especially hot days when electricity demand peaks, and the program limits a maximum of 15 SmartDays called each year. PG&E will notify customers by 2 p.m. on the day before a SmartDay (by text, email, or phone call), so that customers can plan to shift or reduce the electricity usage the following day. This program is voluntary and customers can opt out at any time.
- *SmartAC Program*: In this program, PG&E will install a free SmartAC device on customer's AC and give the customer $50 for participating. In case of an energy shortage from May 1 to October 31, the utility sends a signal to the SmartAC device directing the AC to run at a lower capacity. If a SmartAC event is called at a time that is inconvenient for a customer, he/she can easily return the AC to its normal settings, and customers can opt out at any time. This program will help prevent outages when the electrical grid is typically under the most strain.

2.2.2   SCE

Southern California Edison (SCE), which is the primary electricity supply company for much of southern California with 14 million people, provides a DR program called "Save Power Days" for residential customers, and this program can be divided into two plans [9]:

- *Smart Thermostat Plans*: In this program, customers will install a qualified smart Wi-Fi connected thermostat and enroll through smart thermostat service provider. When a Save Power Days event is called, the smart thermostat will be adjusted automatically (e.g., precool the home and/or change the temperature setting) to reduce energy usage by the service provider. Customers will receive a $75 bill credit on enrollment, and can earn up to $60 in bill credits every year, with $1.25 in bill credits for every kilowatt hour saved during the event. Customers also have the flexibility to adjust the temperature setting based on their comfort and choice.
- *Time-of-Use (TOU) Rate Plans*: In this program, there are several TOU rate options with different peak periods and associated pricing. Customers have the potential to save on their electricity bills by reducing electricity usage during on-peak hours and taking advantage of lower rates during other times of the day/year.

2.2.3   ComEd

Commonwealth Edison (commonly known as ComEd), which is the largest electric utility in Illinois and serves Chicago and Northern Illinois with more than 3.8 million customers, provides three DR programs for residential customers [10]:

- *Hourly Pricing Program*: In this program, customers pay a rate based on hourly wholesale market prices (ComEd Zone PJM wholesale market prices) for electricity. Customers get access to electricity price updates online or by phone. In this way, customers can manage the electricity costs by running appliances when the price of electricity is lower. Customers in this program also receive services including real-time high price alerts, day-ahead high price alerts, and an online bill comparison tool to guide the energy usage decision. Typical program participants have saved an average of more than 15% on their electricity costs compared to what they would have paid on the standard ComEd fixed-price rate. Hourly Pricing participants must also remain in the program for at least 12 consecutive monthly billing periods. Thereafter, participants are free to leave the program if they wish.
- *Peak Time Savings Program*: The customer with a smart meter installed at home can enroll in this program. Upon enrollment, customers can receive credit on their electric bill when they reduce their electricity use during Peak Time Savings Hours. Peak Time Savings Hours will typically occur for a few hours between 11 a.m. and 7 p.m. during the summer—when most ACs are on, stores are open, and factories are running. The notification of the Peak Time Savings Hours can be through phone call, text, or email. There is no penalty for not reducing electricity usage.
- *Smart Ideas Central AC Cycling Program*: This program can help customers save on summer electricity bills by reducing electricity demand on the hottest days of the summer. The program offers two device options. The first option is that a DLC switch will be installed by ComEd at customer's home so that during a cycling event, the compressor can be turned off and on to use less power safely. The fan still stays on to circulate already cooled air. The second option is through the authorized smart thermostat installed by customers, and ComEd can control the customers' thermostat through home Wi-Fi. Currently, Nest is the Smart Thermostat authorized for this program by ComEd.

2.2.4   WPS

Wisconsin Public Service Corporation (WPS), which is a utility company serving more than 440,000 electric customers in Wisconsin and a small portion of Michigan's Upper Peninsula, provides two DR programs for residential customers [11]:

- *Cool Credit Direct Load Control Program*: In this program, customers can earn credits on their electric bills for allowing their air conditioning and/or electric water heater to be occasionally shut down for short periods. Customers will earn $8 a month in credits from June through September for their air conditioning, and all year for their water heater at $2 a month. A small, remote-controlled switch is installed free of charge on the customer's central AC, electric water heater, or both. If a power crunch occurs, the AC and/or water heater will be automatically shut off or cycled on and off without interfering with the rest of the electrical appliances.
- *Three-Level Pricing Program*: In this program, customers can lower their electricity bills by reducing electric use on short notice or shift electric use to lower-rate periods. The program offers two plan options. The first option is called Response Rewards, in which customers are notified of on-peak and critical-peak hours in advance so that they have time to temporarily shut down or stop using high-use electrical items, such as the AC, clothes dryer and dishwasher, to avoid those higher rates. The notification will be through pager, text message, email, or fax at least 1 h in advance. The second option is Three-Tier Time-of-Use in which the hourly electricity prices in a day are divided into on-peak hours, mid-peak hours, and off-peak hours. Customers can save money when they can shift the greatest share of their

electricity use to off-peak saving hours, for example, doing a load of laundry later at night or setting the AC to a higher/lower temperature for heating/cooling in winter/summer while sleeping. On average, customers on the Three-Tier Time-of-Use rate have been saving $50 per year by shifting their electricity use.

2.2.5   Con Edison

Consolidated Edison (commonly known as Con Edison), which is a large investor-owned energy company serving the New York metropolitan area with more than 10 million customers, provides two DR programs for residential customers [12]:

- *Time-of-Use Rates Program*: In this program, customers will pay less during off-peak periods and more during peak and super-peak times—particularly during the summer. By reducing electricity use during peak times or shifting electricity use to off-peak times, customers should be able to save money on their electricity bills. Customers enrolled in this program must remain for 1 year.
- *Smart Thermostat Program*: In this program, customers allow Con Edison to make brief, limited adjustments to their central AC setting during the summer to reduce demand and high energy use. This includes a maximum of up to 10 events each summer, unless there is a power emergency. A Wi-Fi network and a Con Edison-approved thermostat operating the central cooling system is needed for the enrollment. Customers are always in control and can override the settings at any time. Upon enrollment, customers can get a check for $85 to reward participation for the 2-year program.

2.2.6   Gulf Power

Gulf Power Company (GPC), which is an IOU serving northwestern Florida with 428,000 customers, provides a home energy management program called Energy Select [13]. This program combines a unique variable price, an online programming portal, and a smart thermostat so that energy consumption could have a lower electricity price 87% of the time. The program has the following components:

- *Residential Service Variable Price (RSVP) Rate*: The RSVP rate features four variable prices that are based upon the time of day and time of week, with time periods that vary by season (summer and winter). It is designed so that residential customers can save money by using the bulk of the energy during the low and medium periods so that a lower amount than the standard residential rate will be paid.
- *Wi-Fi-Enabled Thermostat*: The Energy Select programmable thermostat, along with timers for the customer's appliances, enables automatic control over the home's central cooling and heating system, electric water heater, and pool pump. Customers can program much of their home's energy usage to fit their schedule and comfort level while managing their level of energy conservation and monthly bill savings. It has a price-response feature that allows customers to program it to automatically respond to the price of electricity in effect at any given time.
- *Wi-Fi Access for Home Area Network*: The customer engagement portal, smart thermostat, and the home's Wi-Fi are all key components of the energy management system. They work together to put the customers in control of their energy (e.g., how much they use, when they purchase it, and how much they purchase).

2.3   Demand response at ISO/RTO level

When DR resources participate in the wholesale electricity market, aggregation is usually needed. Depending on the independent system operator's (ISO's) or regional transmission organization's (RTO's) market design, DR may provide energy, reserve ancillary service (A/S), or capacity in the wholesale electricity market [14]. Currently, several wholesale market level DR programs have been implemented by some ISOs/RTOs in the United States (e.g., New York ISO [15], Electric Reliability Council of Texas (ERCOT) [16–18], PJM Interconnection [19], California ISO [20], and ISO New England [21]). In this section, we will briefly describe current DR programs in these ISOs/RTOs wholesale markets. At the same time, stimulus of DR in the wholesale electricity market is also provided by regulatory institutions such as the FERC. For example, FERC Order No. 719 [22] and No. 745 [23] specify how ISOs can permit DR providers to bid DR on behalf of retail customers directly into the ISO's organized markets, and get compensated for the service they provide at the locational marginal price (LMP). These efforts help integrate DR resources into the wholesale electricity market to cope with future supply-demand matching challenges.

2.3.1   NYISO

The New York Independent System Operator (NYISO), which operates competitive wholesale markets to manage the flow of electricity across New York, has developed DR programs that pay program participants that can meet the appropriate qualifications to reduce their electricity consumption for discrete periods of time. The NYISO DR program includes two categories [15]:

- *Reliability-based DR program*: This program includes two subprograms—(a) Installed Capacity-Special Case Resource (ICAP-SCR) Program and (b) Emergency Demand Response Program. During periods of increased demand or when the grid is affected by unplanned events such as inclement weather, the NYISO's market pays participants in Reliability-based DR programs for load reductions that lessen stress on the electric grid. Program rules unique to the ICAP-SCR program also enable participants to receive monthly payments (called "capacity payments") based on the obligated level of load reduction (i.e., the committed level of load reduction at the facility when the NYISO requests that participants reduce load).
- *Economic-based DR program*: This program also includes two subprograms—(a) Day-Ahead Demand Response Program (DADRP) and (b) Demand-Side Ancillary Services Program (DSASP). Economic-based DR programs provide participants the opportunity to offer load reduction into New York's electricity markets in response to high electricity prices. DADRP participants submit to the NYISO an "energy offer" to reduce consumption at the price the participants determine. Similarly, DSASP participants submit "reserves" and/or "regulation" service offers to the NYISO. If the offer is accepted and scheduled by the NYISO, DSASP participants are eligible to receive market payments based upon actual performance.

2.3.2   ERCOT

ERCOT, which manages the flow of electric power to 90% of the Texas electric load, provides three DR programs:

- *Emergency Response Service (ERS)*: ERCOT procures ERS by selecting qualified loads and generators (including aggregations of loads and generators) to make themselves available for deployment in an electric grid emergency. ERS is a valuable emergency service designed to decrease the likelihood of the need for firm load shedding (a.k.a, rolling blackouts) [16].
- *Load Resources (LRs)*: Customers who are capable of changing their load in response to an instruction and can meet certain performance requirements may qualify to become LRs. Qualified LRs may participate in ERCOT's real-time energy market (Security-Constrained Economic Dispatch or SCED) and/or may provide operating reserves in the ERCOT ancillary service (A/S) markets. In the ERCOT markets, the value of an LR's load reduction is equal to that of an increase in generation by a generating plant. LRs in SCED submit bids to buy power "up to" their specified level, and are instructed by ERCOT to reduce load if wholesale market prices equal or exceed that level. LRs that are scheduled or selected in the ERCOT Day-Ahead A/S Markets are eligible to receive a capacity payment regardless of whether they are actually curtailed. An LR needs to register with ERCOT as a resource entity and must be represented in the ERCOT markets by a qualified scheduling entity [17].
- *Voluntary Load Response*: A customer may decide independently to reduce consumption from its scheduled or anticipated level in response to price signals or high demand on the ERCOT system. This is known as voluntary load response. Depending on how the retail contract with their load serving entity is structured, these customers may have the opportunity to benefit financially during periods when wholesale market prices are high [18].

2.3.3 PJM interconnection

PJM is a regional transmission organization that coordinates the movement of wholesale electricity in all or parts of Delaware, Illinois, Indiana, Kentucky, Maryland, Michigan, New Jersey, North Carolina, Ohio, Pennsylvania, Tennessee, Virginia, West Virginia, and the District of Columbia. Its DR program is a voluntary PJM program that compensates end-use (retail) customers for reducing their electricity use (load) when requested by PJM during periods of high power prices, or when the reliability of the grid is threatened. End-use retail customers have access to PJM's DR program through the agents that are PJM members, known as curtailment service providers (CSPs). These customers receive payments from CSPs [19].

2.3.4 California ISO

California ISO manages the flow of electricity across up to 80% of California and a small part of Nevada's grid. It provides the reliability DR product, which is a wholesale DR product that enables compatibility with, and integration of, existing retail emergency-triggered DR programs into the California ISO market and operations. This includes newly configured DR resources that have a reliability trigger and desire to be dispatched only under particular system conditions. On May 1, 2014, the ISO implemented a new resource model that reduces barriers to participation for DR providers by providing them the ability to submit day-ahead economic and real-time emergency energy bids on behalf of retail emergency-triggered DR programs [20].

2.3.5 ISO New England

In ISO New England, which ensures the reliable flow of competitively priced wholesale electricity to New England areas, the DR is generalized as demand resources. By reducing consumption, demand resources can help ensure enough electricity is available to maintain grid reliability. Demand resources can take many forms. They can be a capacity product, a type of equipment, system, service, practice, or strategy almost anything that verifiably reduces end-use demand for electricity from the power system (reductions must be verified using an ISO-accepted measurement and verification protocol) [21]. These resources fall into two general types:

- Active demand resources (also known as DR resources) are activated only when needed. An example would be the practice of powering down machines or using electricity from an onsite generator rather than from the grid.
- Passive demand resources are principally designed to save electricity use at all times. Examples include energy-efficiency measures, such as the use of energy-efficient appliances and lighting, advanced cooling and heating technologies, electronic devices to cycle ACs on and off, and equipment to shift electricity use to off-peak hours.

The ISO also currently administers the transitional price-responsive demand program for active demand resources, in which demand resources must respond to ISO dispatch instructions. This program replaced the day-ahead load-response program, which ended on May 31, 2012. Market participants can offer load reductions in response to day-ahead LMPs. Market participants are paid the day-ahead LMP for their cleared offers and are obligated to reduce load by the amount cleared day ahead. The participant is then charged or credited at the real-time LMP for any deviations in curtailment in real time compared with the amount cleared day ahead. Participants may also be eligible for payments if the real-time LMP exceeds their offer and they did not clear day ahead [21].

2.4 Incentive-based approaches vs. pricing-based approaches for residential DR

Although using different names, current residential DR programs provided by utilities can be divided into two categories depending on the schemes of how the demand is altered: incentive-based approach and pricing-based approach [24], as shown in Fig. 2.

(1) *Incentive-based approach*: The incentive-based approach, also called DLC, is usually contract-based provided by utilities, that is, by signing up for the contract, customers give utilities the option to remotely shut down appliances during high-demand periods or power supply emergencies, or customers will get notice of the peak event and can voluntarily shut down appliances. As a result, customers can receive credit on electricity bills for this participation. As discussed in Section 2.2, SmartAC Program by PG&E, Smart Thermostat Plans by SCE, Peak Time Savings Programs and Smart Ideas Central AC Cycling Program by ComEd, Cool Credit Direct Load Control Program by WPS, and Smart Thermostat Program by Con Edison belong to this type of approach.

(2) *Pricing-based approaches*: In this approach, the electric utility controls customers' appliances indirectly by sending price signals. These prices reflect supply-demand condition of

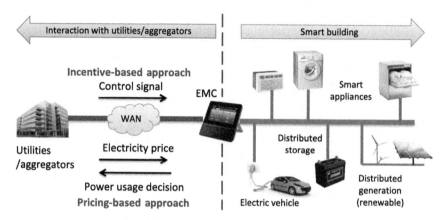

Fig. 2 Two types of approaches for demand response.
(Reproduced with permission from C. Chen, J. Wang, S. Kishore, A distributed direct load control approach for large-scale residential demand response, IEEE Trans. Power Syst. 29 (5) (2014) 2219.)

the grids by setting different time-of-use prices, critical-time prices, or more complex time-varying prices considering the variation in wholesale electricity prices, thus eventually affecting customers' electricity bills. Because of time-varying prices, customers may be induced to make smart decisions, for example, shifting some power usage from high-price periods to low-price periods. In such cases, customers save money while the utility benefits from peak demand reduction. Examples of implementations of this type of approach as discussed in Section 2.2, include SmartRate Program by PG&E, Time-of-Use Rate Plans by SCE, Hourly Pricing Program by ComEd, Three-Level Pricing Program by WPS, Time-of-Use Rates Program by Con Edison, and Energy Select Program with Residential Service Variable Price Rate by Gulf Power.

There are both advantages and disadvantages to the two types of DR programs. Current pricing-based approaches have resulted in notable savings in customers' electricity bills when there is a small percentage of customer participation [25]. Several studies formulated the optimization problems for the individual appliance control within the building by exploiting the operational and thermal comfort flexibilities to minimize the electricity costs for customers [5, 26]. However, pricing-based approaches also have potential issues that may emerge with a higher penetration of DR. One major concern is stability, for example, a "rebound" peak of demand may occur during an originally low demand period due to a significant amount of shifted load. With a large customer base, prices will become highly dynamic so that customers may worry about price risks, which in turn might undermine the customer base. The reason for these issues is the lack of coordination among customers, that is, each customer controls appliance operations to minimize his/her own cost without considering other customers' behaviors. In this sense, the control of a large amount of residential flexible appliances is needed to achieve system-level benefits of DR.

On the other hand, using the DLC approaches, utilities can coordinate the control of all appliances to avoid the stability issue in price-based approaches. However, several limitations also exist for incentive-based approaches: First, they are generally only for emergency cases so that they do not fully exploit operational flexibilities of appliances to help balance supply and demand. Second, when the customer base is large, the design of a centralized control scheme used in current DLC programs is complicated both in terms of computation and communication requirements. Third, another shortcoming of existing DLC programs lies in customers' privacy concerns because their power usage profile is exposed whenever each individual appliance is remotely controlled by a central controller.

3   A distributed direct load-control mechanism for residential DR

As discussed in Section 2.4, if the amount of customer participation in DR programs becomes large, both incentive-based approaches and pricing-based approaches will face technical challenges, that is, stability issues due to lack of coordination in price-based approaches and complexity challenges due to large-scale features in incentive-based approaches. In this section, we describe a new DLC approach [24] that has the advantage of coordination while tackling the challenges for current DLC approaches in the scenarios of large-scale DR participation.

Scheduling the responsive demand at the aggregated level has been investigated extensively in the literature (e.g., [27–32]). Some papers employ the utility function, a concave, and an increasing function to model how customers favor the usage of demand, together with generation costs in the system social welfare maximization problem to calculate the optimal dispatch of responsive demand (e.g., [27–29]). Another method is to apply the price elasticity of demand in the market-clearing scheme to get the optimal demand schedules (e.g., [31, 32]). However, the aggregated optimal demand used in the market analysis cannot specify how to control each individual appliance to achieve the scheduled demand. In [30], although authors compute the optimal schedules of individual appliances in the energy consumption game model, the assumption that the start/stop time of an appliance is known is not suitable for real-time control because the control can only be conducted once the customer requests it. In a word, existing research on DR at the aggregated level cannot provide a real-time control scheme of individual appliances to achieve the benefits of DR.

To bridge this gap, we propose a two-layer distributed DLC approach for the large-scale residential DR [24]. The approach utilizes the underlying two-layer communication networks. Specifically, the lower-layer network is within each building where the EMC schedules operation of appliances upon request according to a local power consumption target via wireless links. The upper-layer network consists of EMCs in a region where demand is served by a load aggregator. The load aggregator wants the actual aggregated demand over this region to match a desired aggregated demand profile determined day-ahead, which is viewed as supply from customers' perspective. Our approach utilizes the average consensus algorithm to distribute portions of the desired aggregated demand to each EMC in a decentralized fashion. The allocated portion corresponds to each building's local power consumption target, which its EMC then uses to schedule the in-building appliances. The result will be an aggregated demand over this region that more closely reaches the desired demand.

Compared to the existing methods and programs for DLC-based DR, our method has the following advantages that can tackle the challenges of large-scale participation of residential customers in DR [24]:

- *Low complexity for communication and computation*: The distributed algorithm in the upper-layer network that sets local power consumption targets involves only local communications between neighboring EMCs. This approach is therefore of low complexity compared to the centralized method that requires links between each EMC and the centralized controller. Scheduling is conducted in each building by its EMC, which incurs lower computation complexity than if the centralized controller made such decisions for each appliance in each building over the larger region.
- *Powerful load-shaping tools to alleviate the mismatch between demand and supply*: By dividing the desired demand among each building's local power consumption target and designing EMCs that exploit load flexibilities to closely meet these targets, the mismatch between the actual aggregated demand and the desired demand can be alleviated.
- *Privacy protection for customers*: In our approach, only the aggregated power consumption information of customers is visible in the upper-layer network. Thus, the individual appliance usage profile of customers is kept private.
- *Fairness*: In the proposed scheme, the local power consumption targets for each building are set according to DR resources each customer can provide, with no bias to any individual customer.
- *Nonintrusiveness for appliance operation*: Our approach integrates nonintrusiveness features for appliance scheduling. This includes options for overriding the control of the EMC as well as a scheme for preventing frequent ON/OFF switch and for guaranteeing operation deadlines.
- *Reliable and robust*: Because local power consumption targets are set in a distributed manner, our approach does not rely on a central controller for the load aggregator, which implies our scheme is more reliable and robust.

3.1 Two-layer communication-based direct load-control architecture

We consider a residential region with B buildings whose power consumptions are served by a load aggregator. The set of buildings is denoted by $\mathcal{B}$. We propose a two-layer communication-based distributed DLC architecture, shown in Fig. 3, as the DR of B buildings over this region. Time is segmented into frames denoted by τ as the time slot for control.

The lower-layer network is within each building $i \in \mathcal{B}$ (as shown in Fig. 3), where an EMC is installed to locally schedule operation of appliances within the building via two-way wireless links between the EMC and appliances. We assume that each appliance is equipped with a wireless communication module that can send/receive messages containing the load information to/from the EMC. Because the EMC is part of the infrastructure of the DR program provided by the load aggregator, we assume that the security of the load information of individual appliances within the building can be protected by the load aggregator (i.e., the EMC is assumed to be a trusted device).

Two types of appliances with operational flexibilities are considered as DR resources in our scheme: (1) Some appliances are flexible in operation such that they

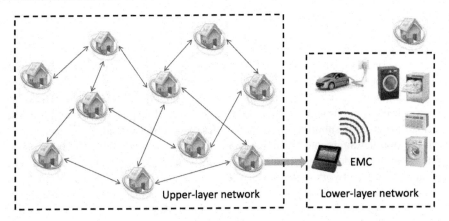

Fig. 3 Two-layer communication-based distributed direct load-control architecture. (Reproduced with permission from C. Chen, J. Wang, S. Kishore, A distributed direct load control approach for large-scale residential demand response, IEEE Trans. Power Syst. 29 (5) (2014) 2221.)

can be switched on and/or off either when requested or possibly some time later (e.g., clothes washer/dryer and dishwasher). Some appliances provide operational flexibilities such that their executions can be interrupted and resumed as long as it is completed before the deadline specified by customers. We denote this type of appliance as the *Type I flexible appliance* and by the set of $\mathcal{K}_i$ for building i; (2) another type of appliance provides flexibility in that its power consumption can be altered, for example, the PEV whose battery charging rate can be adjusted as long as the charging is completed before the departure [33], or the electric heater whose power consumption can be changed if the resulting thermal comfort is within a range specified by the customer [34]. We denote this type of appliance as the *Type II flexible appliance* and by the set of $\mathcal{Z}_i$ for building i. The illustration of lower-layer network control with two types of flexible appliances is shown in Fig. 4

The upper-layer network consists of EMCs of the B buildings in the region. This EMC network is designed without a central controller (i.e., each EMC only

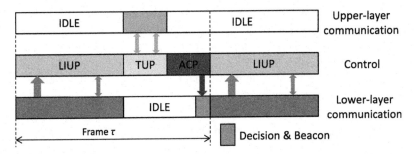

Fig. 4 Lower-layer network control of two types of appliances with flexibilities within buildings.

communicates to its neighboring EMCs). We assume that each pair of EMCs with a direct link has two-way communication capability, and the EMC network is strongly connected (i.e., no island exists in the network). This assumption is consistent with the smart grid communication network implemented in practice, for example, the Smart Grid Infrastructure provided by Silver Spring Networks [35].

The load aggregator serving this region has a desired demand Z_τ for each frame τ, which can be the optimal day-ahead (real-time) cleared responsive demand bid that the load aggregator submitted into the wholesale electricity market for each frame τ, or the optimal aggregated responsive demand in the social welfare maximization problem in the work of Samadi et al. [27] and Li et al. [28]. In real-time operations, the upper-layer network coordinates the EMCs such that the actual aggregated demand of the B buildings over this region to be lower than Z_τ, that is, the actual demand always fulfills the demand commitment in advance (Z_τ), and the actual aggregated demand should be as close to this Z_τ as possible to minimize the deviation cost. Note that the load-shaping control according to Z_τ discussed in this chapter is a general case of peak load reduction or load shifting for DR, because the desired demand Z_τ can be viewed as the target of peak reduced load or shifted load. In other words, by defining the shape of the target demand Z_τ over a time horizon, our objective function includes the cases of load shifting or peak reduction.

The main idea underlying the proposed approach is to allocate the desired demand value Z_τ to each EMC as a local power consumption target, denoted by θ_i^τ, via coordinations among EMCs utilizing the upper-layer network. This target is used by the EMC to control appliance operations within the building, exploiting the lower-layer network. Thus, by local appliance control according to θ_i^τ, the actual aggregated demand can approach the target Z_τ.

To do this, we design the frame-based protocol to integrate the two-layer modules of communication and control in the EMC, as shown in Fig. 5. Each frame τ has three phases: load information update phase (LIUP), target update phase (TUP), and admission control phase (ACP). The coordination of these three phases in the proposed protocol is illustrated in Fig. 5 and explained in the following sections.

3.1.1   Load information update phase

During the LIUP, the EMC lower-layer communication interface module receives the power request information from appliances in the building. Specifically, for Type I flexible appliances, the power request information refers to the power-on or power-off request of $P_{i,j}^\tau$ for appliance j in building i during frame τ. For Type II flexible appliances, the power request information refers to the bounds of power consumption, denoted by $v_{i,j}^{\tau,\min}$ and $v_{i,j}^{\tau,\max}$, that specify the range within which the appliance may consume.

With this data collected, the control module updates the aggregated fixed load in the building, denoted by $\hat{q}_i^\tau$, as the demand that must be satisfied, and the aggregated flexible load in the building, denoted by $\widetilde{q}_i^\tau$, as the demand that can be altered, respectively. The details of the scheme during this phase are described in Section 3.3.1.

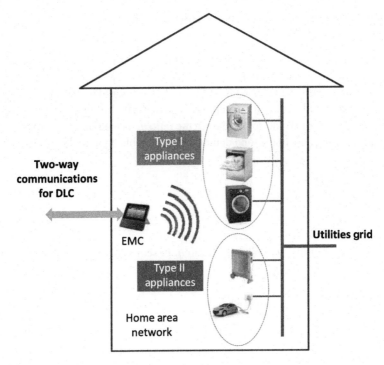

Fig. 5 Two-layer communication and control protocol.
(Reproduced with permission from C. Chen, J. Wang, S. Kishore, A distributed direct load control approach for large-scale residential demand response, IEEE Trans. Power Syst. 29 (5) (2014) 2221.)

3.1.2 Target update phase

During the TUP, with the $\hat{q}_i^\tau$ and $\widetilde{q}_i^\tau$ obtained in LIUP, the upper-layer communication interface module at each EMC communicates with its neighboring EMCs to compute the local demand target value θ_i^τ for the frame τ. We employ the average consensus algorithm as the coordination among the EMCs to distributively allocate the desired demand Z_τ to each local target θ_i^τ. The details of the scheme during this phase are described in Section 3.2.

3.1.3 Admission control phase

During the ACP, the EMC's control module makes the admission decision, that is, decides which Type I flexible appliances can be turned on and power consumptions of Type II flexible appliances, given the local demand target value θ_i^τ. At the end of the ACP, the lower-layer communication interface module broadcasts the admission decision and synchronizes to the next frame. Details of the admission control scheme are described in Section 3.3.2.

3.2 Distributed demand target allocation in upper-layer EMC network

As discussed in Section 3.1, the fixed load $\hat{q}_i^\tau$ and the flexible load $\widetilde{q}_i^\tau$ within the building i obtained in the phase of LIUP will be used in the upper-layer network to compute the local demand target θ_i^τ in the phase of TUP. The local demand target θ_i^τ should reflect the aggregated demand and supply status specified by $\sum_{i\in\mathcal{B}}\hat{q}_i^\tau$, $\sum_{i\in\mathcal{B}}\widetilde{q}_i^\tau$, and Z_τ (as shown in Fig. 6). To do so, we design the ratio η_τ as the ratio of the total demand target gap, $Z_\tau - \sum_{i\in\mathcal{B}}\hat{q}_i^\tau$, to the total demand from flexible appliances, $\sum_{i\in\mathcal{B}}\widetilde{q}_i^\tau$, that is,

$$\eta_\tau = \frac{Z_\tau - \sum_{i\in\mathcal{B}}\hat{q}_i^\tau}{\sum_{i\in\mathcal{B}}\widetilde{q}_i^\tau} \tag{1}$$

and the EMC for building i applies this ratio η_τ to set its local target power consumption, θ_i^τ, as

$$\theta_i^\tau = \hat{q}_i^\tau + \eta_\tau \cdot \widetilde{q}_i^\tau \tag{2}$$

If each EMC controls power requests of appliances at frame τ to follow this target value θ_i^τ, then we can verify that $\sum_{i\in\mathcal{B}}\theta_i^\tau = \sum_{i\in\mathcal{B}}\hat{q}_i^\tau + \eta_\tau\sum_{i\in\mathcal{B}}\widetilde{q}_i^\tau = Z_\tau$, that is, the aggregated target demand matches the desired demand Z_τ for time τ. In this approach, the status of aggregated demand reflected by the ratio η_τ is applied in the local target demand computation for each EMC. Besides guaranteeing the fixed demand $\hat{q}_i^\tau$, the additional part of the local target demand θ_i^τ in building i is proportional to the demand of flexible appliances requested in time τ. In other words, this demand target allocation only depends on the size of flexible demand that each customer has, without bias to any individual customer. This scheme is fair in the sense that it conforms to the rules of general DR programs that the incentives received by the customer are dependent on the size of DR resources the customer can provide (e.g., the Distribution Load Relief Program by Con Edison [36]).

As mentioned earlier, a centralized approach in this scenario will suffer from high communication infrastructure costs when a large number of EMCs send $\hat{q}_i^\tau$ and $\widetilde{q}_i^\tau$ to the central controller, and the single-point failure issue. An alternative approach is to employ a distributed method in which the local demand target θ_i^τ is computed at each

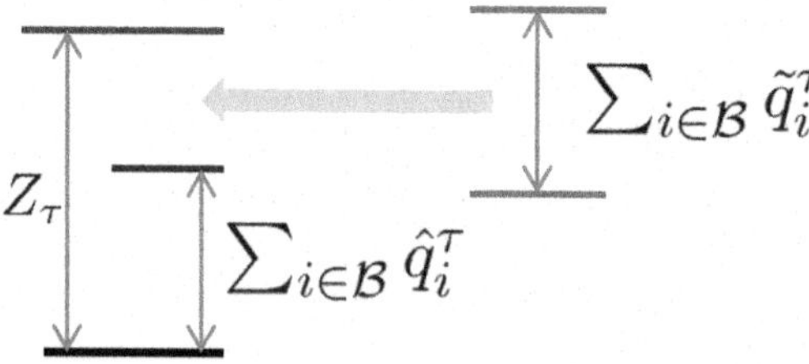

Fig. 6 Illustration of relation among aggregated demand and supply status specified by $\sum_{i\in\mathcal{B}}\hat{q}_i^\tau$, $\sum_{i\in\mathcal{B}}\widetilde{q}_i^\tau$, and Z_τ to set the target θ_i^τ.

EMC level via information exchange with its neighboring EMCs using the upper-layer EMC network.

Dominguez-Garcia et al., in their work [37–39], proposed innovative distributed algorithms for control and coordination of loads and distributed energy resources (DERs) in distribution networks via information exchange among neighboring DER nodes in the network. Here we employ the ideas of the distributed algorithms proposed in [37–39] to our distributed coordination mechanism for the upper-layer EMC network. The topology of the upper-layer network of EMCs can be represented by an undirected graph. We see that the terms $\sum_{i\in B}\hat{q}_i^\tau$ and $\sum_{i\in B}\widetilde{q}_i^\tau$ are the same for all EMCs, so they can be viewed as a consensus agreement that all EMCs reach. In this way, the average consensus algorithm [40] can be applied to compute the η_τ iteratively.

Let $\hat{Q}_i^\tau(k)$ and $\widetilde{Q}_i^\tau(k)$ denote the iteration variables at the step k, and their initial values are set as $\hat{Q}_i^\tau(0)=\hat{q}_i^\tau$ and $\widetilde{Q}_i^\tau(0)=\widetilde{q}_i^\tau$. By linear updates iteratively as:

$$\hat{Q}_i^\tau(k+1)=\hat{Q}_i^\tau(k)+\varepsilon\cdot\sum_{j\in\mathcal{N}_i}\left[\hat{Q}_j^\tau(k)-\hat{Q}_i^\tau(k)\right] \tag{3}$$

$$\widetilde{Q}_i^\tau(k+1)=\widetilde{Q}_i^\tau(k)+\varepsilon\cdot\sum_{j\in\mathcal{N}_i}\left[\widetilde{Q}_j^\tau(k)-\widetilde{Q}_i^\tau(k)\right] \tag{4}$$

the values of $\hat{Q}_i^\tau(k)$ and $\widetilde{Q}_i^\tau(k)$ will converge to the average value $\hat{q}^{\tau^*}=\frac{1}{B}\sum_{i\in B}\hat{q}_i^\tau$ and $\widetilde{q}^{\tau^*}=\frac{1}{B}\sum_{i\in B}\widetilde{q}_i^\tau$, respectively, with a proper step size ε. The $\mathcal{N}_i$ denotes the set of neighbors of the EMC i in the upper-layer network. After the updates converge, the global ratio η_τ can be computed locally at each EMC level as $\eta_\tau=\frac{Z_\tau-B\cdot\hat{q}^{\tau^*}}{B\cdot\widetilde{q}_i^\tau}$ where the overall desired target demand (Z_τ) and total number of EMCs (B) are assumed to be known to all EMCs beforehand.

The convergence of the average consensus algorithm in Eqs. (3), (4) over a time-invariant, connected, undirected network is related to the eigenvalues of the Laplacian matrix representing the graph, which has been studied extensively in the literature (e.g., [40–42] and references therein). Specifically, as shown in [41], average consensus can be asymptotically achieved when $\varepsilon\in(0, 2/\lambda_B(L))$, where $\lambda_i(L)$ is the ith largest eigenvalue of the graph Laplacian matrix L, which is a real number due to the symmetric feature of the Laplacian matrix for an undirected graph. In addition, the minimum convergence time is obtained when $\varepsilon=2/[\lambda_2(L)+\lambda_B(L)]$ [42]. With regard to convergence speed, Olfati-Saber et al. [40] show that a discrete-time consensus is globally exponentially reached with a speed that is faster or equal to $\kappa_2=1-\varepsilon\lambda_2(L)$. These convergence features help in designing the network topology to achieve a better trade-off between the infrastructure cost and the convergence rate.

Given the ratio η_τ from the average consensus algorithm, the local power consumption target θ_i^τ for building i can be computed locally according to Eq. (2). Depending on the values of η_τ, three scenarios need to be considered:

If $\eta_\tau \leq 0$, that is, the desired demand Z_τ is less than the aggregated fixed demand, the local power consumption target could be set as $\theta_i^\tau = \hat{q}_i^\tau$, by letting $\eta_\tau = 0$. In this case, all flexible appliances will not be turned on during that time slot.

If $0 < \eta_\tau \leq 1$, the local power consumption target, θ_i^τ, can be computed directly using Eq. (2).

If $\eta_\tau > 1$, then the local power consumption target can be set as $\theta_i^\tau = \hat{q}_i^\tau + \widetilde{q}_i^\tau$, that is, by letting $\eta_\tau = 1$. This represents the best effort that the EMC can do to schedule appliances at time τ to reach the target.

Besides the advantage of only local communications between neighboring EMCs and not relying on a central controller, our distributed algorithm protects customers' privacy because only aggregated information of each building ($\hat{q}_i^\tau$ and $\widetilde{q}_i^\tau$) is transmitted instead of individual appliance information.

3.3 Lower-layer communication and admission control scheme

3.3.1 Load information update

To obtain the fixed demand $\hat{q}_i^\tau$ and flexible demand requests $\widetilde{q}_i^\tau$ during the phase of LIUP in each frame τ, each EMC i maintains the following sets of appliances in building $i \in \mathcal{B}$:

- *Active Appliance Set* ($\mathcal{A}_i^\tau \subseteq \mathcal{K}_i^\tau$) contains already powered-on appliances during the frame τ in building i.
- *Request Appliance Set* ($\mathcal{R}_i^\tau \subseteq \mathcal{K}_i^\tau$) tracks appliances that are requested to power on by the customer and their power-on request messages (PRQMs) are successfully received by the EMC in building i during frame τ. If appliance $j \in \mathcal{K}_i \backslash (\mathcal{A}_i^\tau \cup \mathcal{R}_i^\tau)$ is requested to power on, it sends a PRQM; upon receiving this message, the EMC updates $\mathcal{R}_i^\tau = \mathcal{R}_i^\tau \cup \{j\}$.
- *Release Appliance Set* ($\mathcal{X}_i^\tau \subseteq \mathcal{K}_i^\tau$) tracks appliances that wish to power off and their power release messages (PRLMs) have been successfully received by the EMC in building i during frame τ. Upon receiving the PRLM from an appliance $j \in \mathcal{A}_i^\tau$, the EMC updates $\mathcal{X}_i^\tau$ to $\mathcal{X}_i^\tau = \mathcal{X}_i^\tau \cup \{j\}$.

Note that for a Type I flexible appliance whose PRQM (PRLM) is received by the EMC, the appliance is not added to (removed from) *Active Appliance Set* $\mathcal{A}_i^\tau$ immediately; instead it is added to $\mathcal{R}_i^\tau$ ($\mathcal{X}_i^\tau$) and waits until the ACP phase to be scheduled on (off) by the EMC. Thus, $\mathcal{R}_i^\tau$ and $\mathcal{X}_i^\tau$ only contain Type I flexible appliances. Inflexible appliances, which will power on/off as requested, will be discussed in Section 3.4.

The EMC updates these appliance sets according to the received packets (PRQMs/PRLMs) transmitted by appliances. Table 1 shows the essential fields in these packets required for our proposed approach.

The EMC transmits an acknowledge (ACK) packet to acknowledge successful receipt of the PRQM/PRLM packets sent by an appliance. The ACK is integrated into the general EMC transmission packet, the essential fields of which are shown in Table 2.

Carrier sense multiple access with collision avoidance (CSMA/CA), which is widely used in wireless local area networks such as IEEE 802.11 [43], is applied

Table 1 **Fields description for the appliance's packet**

Field name	Description
ID	Appliance/EMC ID
RRI	PRQM/PRLM indicator (0/1 binary)
ORI	Override indicator (0/1 binary)
PCV	Power consumption
PDU	Duration of power consumption
MON	Minimum on time
MSUS	Minimum suspension time
DT	Delay tolerance
STAT	ON/OFF status (appliances in $\mathcal{Z}_i$)

Table 2 **Fields description for the EMC's packet**

Field name	Description
ID	Appliance/EMC ID
BCI	Beacon packet indicator
ACKID	ACK destination ID, ACKID = 0 for broadcast
OPD	Admission control decision vector, valid for the beacon and broadcast (ACKID = 0) packet
PCD	Power consumption decision (appliances in $\mathcal{Z}_i$)
RXD	Duration of LIUP, valid for beacon packet

for packet transmissions in our scheme. At the end of the LIUP, if an appliance's request is not received by the EMC due to collision, it will wait until the next frame to send its request; however, if the duration of LIUP is long and the number of requests is not large, the impact of collisions can be neglected.

For a Type II flexible appliance $j \in \mathcal{Z}_i$, the EMC obtains the power consumption bounds ($v_{i,j}^{\tau,\min}$ and $v_{i,j}^{\tau,\max}$) from the request of the appliance or some optimization problems conducted by the EMC, as discussed in Section 3.1.

At the end of the LIUP phase, with updated sets $\mathcal{R}_i^{\tau}$ and $\mathcal{X}_i^{\tau}$ as well as $\mathcal{A}_i^{\tau}$ for the frame τ, the fixed demand ($\hat{q}_i^{\tau}$) and requested demand from flexible appliances ($\widetilde{q}_i^{\tau}$) of building i can be computed as $\hat{q}_i^{\tau} = \sum_{j \in \mathcal{A}_i^{\tau} \setminus \mathcal{X}_i^{\tau}} P_{i,j}^{\tau} + \sum_{j \in \mathcal{Z}_i} v_{i,j}^{\tau,\min}$, $\widetilde{q}_i^{\tau} = \sum_{j \in \mathcal{R}_i^{\tau}} P_{i,j}^{\tau} + \sum_{j \in \mathcal{Z}_i} (v_{i,j}^{\tau,\max} - v_{i,j}^{\tau,\min})$, respectively.

3.3.2 Admission control mechanism

In the phase of ACP, the EMC makes admission decisions on which appliances in $\mathcal{R}_i^{\tau}$ to turn on and adjusts the power consumptions of appliances in $\mathcal{Z}_i$, according to the local demand target value θ_i^τ. The aim of the controller is to make the aggregated load in building i approach θ_i^τ as close as possible, without being over θ_i^τ. We employ a binary decision variable x_j associated with each appliance $j \in \mathcal{R}_i^\tau$ indicating whether it can be turned on, and a continuous variable $v_{i,j}^\tau$ associated with each appliance $j \in \mathcal{Z}_i$ representing the power consumption value. The EMC's optimization can be formulated as:

$$\max_{x_j, v_{i,j}^\tau} \sum_{j \in \mathcal{R}_i^\tau} x_j \cdot P_{i,j}^\tau + \sum_{j \in \mathcal{Z}_i} v_{i,j}^\tau \tag{5}$$

$$\text{s.t.} \quad \hat{q}_i^\tau + \sum_{j \in \mathcal{R}_i^\tau} x_j \cdot P_{i,j}^\tau + \sum_{j \in \mathcal{Z}_i} v_{i,j}^\tau \leq \theta_i^\tau \tag{6}$$

$$\xi_{i,j}^\tau \cdot v_{i,j}^{\tau,\min} \leq v_{i,j}^\tau \leq \xi_{i,j}^\tau \cdot v_{i,j}^{\tau,\max}, \quad j \in \mathcal{Z}_i \tag{7}$$

$$x_j \in \{0,1\}, \quad j \in \mathcal{R}_i^\tau, \tag{8}$$

where $x_j = 1$ decides that appliance j is scheduled to turn on. The binary parameter $\xi_{i,j}^\tau \in \{0,1\}$ indicates the ON/OFF status of Type II appliances in $\mathcal{Z}_i$. We see that Eqs. (5)–(8) are a mixed-integer linear programming (MILP) problem that can be solved efficiently using off-the-shelf solvers due to the small size of decision variables and constraints.

At the end of the ACP phase, the EMC broadcasts a packet containing the admission results (set the OPD and PCD fields in Table 2). This packet also serves as a beacon signal (set BCI = 1 in Table 2) to specify the transmission/receiving duration (length of LIUP) for the next frame by setting the RXD field in Table 2. With the decision result $\{x_j\}$, the *Admitted Appliance Set* $\mathcal{D}_i^\tau$ can be updated as $\mathcal{D}_i^\tau = \{j | j \in \mathcal{R}_i^\tau, x_j = 1$ in Eqs. (5)–(7)$\}$. The *Active Appliance Set* for the next frame $\tau + 1$ is updated as $\mathcal{A}_i^{\tau+1} = \mathcal{A}_i^\tau \cup \mathcal{D}_i^\tau \backslash \mathcal{X}_i^\tau$.

3.4 Nonintrusive operation for appliances

Unlike flexible appliances discussed earlier, some appliances with critical time requirements, such as TVs, desktops, and emergency devices, are required to turn on immediately once the customer wants to use them. Some appliances have flexibilities on their turn-on times, but have strict turn-off times due to their specifications (e.g., a clothes washer). In some cases, even for flexible appliances, customers may not prefer to delay usage for a variety of reasons. Our approach fulfills these requirements by providing override options to these appliances for customers. Besides, our approach avoids frequent appliance ON/OFF switches in the scheduling and guarantees the appliance's task is completed before a deadline set by the customer.

3.4.1 Customer override option

Our scheme provides customers with override options, for example, to specify time-critical appliances or change request preferences. In such cases, the appliance should be turned on/off as soon as the customer requests; meanwhile the appliance sends a PRQM/PRLM packet with the ORI $= 1$, as shown in Table 1, to indicate the request is due to an override option.

If the PRQM packet with an override option from appliance j is received by the EMC, the *Active Appliance Set* is updated as $\mathcal{A}_i^\tau = \mathcal{A}_i^\tau \cup \{j\}$. However, because during this frame τ, the demand of the building needs to be kept below the demand target value computed at the end of the last frame, that is, $\theta_i^{\tau-1}$, this newly added appliance may cause the actual demand to exceed this target. On the other hand, if the PRLM packet with an override option from appliance j is received by the EMC, the *Active Appliance Set* is updated as $\mathcal{A}_i^\tau = \mathcal{A}_i^\tau \setminus \{j\}$. However, this newly released appliance will increase the gap between the actual demand and the target value $\theta_i^{\tau-1}$. In both cases, ON/OFF switches of Type I appliances and/or power consumption adjustments of Type II appliances are needed to be conducted for better matching the demand target $\theta_i^{\tau-1}$.

Define the *Curtailable Appliance Set* $\mathcal{C}_i^\tau$, where $\mathcal{C}_i^\tau \subseteq \mathcal{A}_i^\tau$, as the set of appliances that can be curtailed in frame τ. Also define the *Admissible Appliance Set* $\mathcal{S}_i^\tau$, where $\mathcal{S}_i^\tau \subseteq \mathcal{R}_i^\tau$, as the set of appliances that can be admitted to power on at this moment. An adjustment optimization is formulated by the EMC as:

$$\max_{y_j, z_j, v_{i,j}^\tau} \sum_{j \in \mathcal{C}_i^\tau} (1 - y_j) \cdot P_{i,j}^\tau + \sum_{j \in \mathcal{S}_i^\tau} z_j \cdot P_{i,j}^\tau + \sum_{j \in \mathcal{Z}_i} v_{i,j}^\tau \tag{9}$$

$$\text{s.t.} \quad \sum_{j \in \mathcal{C}_i^\tau} (1 - y_j) \cdot P_{i,j}^\tau + \sum_{j \in \mathcal{S}_i^\tau} z_j \cdot P_{i,j}^\tau + \sum_{j \in \mathcal{Z}_i} v_{i,j}^\tau$$

$$\leq \left(\theta_i^{\tau-1} - \sum_{k \in \mathcal{A}_i^\tau \setminus \mathcal{C}_i^\tau} P_{i,k}^\tau \right), \tag{10}$$

$$\xi_{i,j}^\tau \cdot v_{i,j}^{\tau,\min} \leq v_{i,j}^\tau \leq \xi_{i,j}^\tau \cdot v_{i,j}^{\tau,\max}, \quad j \in \mathcal{Z}_i \tag{11}$$

$$y_j \in \{0,1\}, \quad j \in \mathcal{C}_i^\tau \tag{12}$$

$$z_j \in \{0,1\}, \quad j \in \mathcal{S}_i^\tau \tag{13}$$

where the binary variable $y_j = 1$ indicates appliance j in $\mathcal{C}_i^\tau$ is scheduled to be curtailed. The curtailed appliances will be added to $\mathcal{R}_i^\tau$ for future scheduling, that is, $\mathcal{R}_i^\tau = \mathcal{R}_i^\tau \cup \{j | j \in \mathcal{C}_j, \ y_j = 1 \text{ in Eqs. (9)—(13)}\}$. The binary variable $z_j = 1$ indicates appliance j in $\mathcal{S}_i^\tau$ is scheduled to turn on and be added to $\mathcal{A}_i^\tau$, that is, $\mathcal{A}_i^\tau = \mathcal{A}_i^\tau \cup \{j | j \in \mathcal{S}_j, \ z_j = 1 \text{ in Eqs. (9)–(13)}\}$.

For both cases, after the decision is made, the EMC (via lower-layer communication interface module) broadcasts an ACK packet in which the BCI and ACKID fields, as shown in Table 2, are set to 0, indicating a nonbeacon and broadcast packet, and the OPD and PCD fields are set as the admission decisions to inform all appliances. Upon receiving this broadcast ACK, the corresponding Type I appliances will be suspended or turned on for power consumption, and Type II appliances will adjust their power consumption.

Note that the override action may happen during the TUP period, so the local target demand θ_i^τ computation by the average consensus algorithm may not be accurate in terms of actual demand status. However, this problem is trivial because the duration of TUP can be designed to be very short (e.g., several seconds). In such a case, the probability that the override operations of many appliances all happen during this phase is negligible; the override option during this short time can also be disabled, which is feasible for customers.

3.4.2 Preventing frequent ON/OFF switching

With the override option, some flexible appliances may be shut down temporarily (suspended) or resumed. However, too frequent ON/OFF switching is harmful to appliances. To accommodate this nonintrusive constraint, min-ON and min-OFF parameters are specified for each appliance to indicate the minimum power-on time after it is turned on and minimum off time after it is suspended. These two parameters are passed to the EMC by the MON and MSUS fields, respectively, of the PRQM packet shown in Table 1.

Accordingly, there are ON and OFF timers for each appliance in the EMC's control module to record the ON/OFF times. When the appliance is turned on, either for the first time or when it resumes from a suspended state, the ON timer counts down with an initial value min-ON. The appliance cannot be added to the *Curtailable Appliance Set*, C_i^τ until the timer reaches 0; this prevents the appliance from being curtailed and suspended again. Similarly, when the appliance is curtailed, the OFF timer counts down with an initial value of min-OFF. The appliance will not be added to the *Admissible Appliance Set*, S_i^τ until the timer reaches 0. This mechanism helps avoid frequent ON/OFF switching for appliances due to curtailment and admission resulting from the override operation.

3.4.3 Operation deadline constraint

Although some flexible appliances allow delayed power-on or suspension, an operation deadline is usually needed to guarantee that the appliance's task is completed by the customer's desired deadline. Our scheme integrates this requirement by specifying a DT field in the PRQM packet (shown in Table 1), which represents the delay tolerance. This delay tolerance can be directly specified by the customer, or be computed if the customer indicates the deadline for the operation completion.

After the EMC receives the PRQM packet containing the DT, a deadline timer is set to the value of DT. When the appliance waits to get access or suspends, the timer counts down. If the appliance is in power-on status, the timer pauses. If the value of the timer is less than the min-OFF parameter (specified by the MSUS field shown in Table 1), it cannot be scheduled to be suspended. If the timer reaches 0 before the task completion, the appliance powers on as an override operation and continues to consume power without interruption until the task is completed.

4 Numerical results

We provide simulation results to validate our approach in this section. Specially, we consider a region with B buildings over which the corresponding EMCs form a mesh network. Each building has $|\mathcal{K}_i|$ Type I flexible appliances and $|\mathcal{Z}_i|$ Type II flexible appliances to schedule. As the input to the proposed control scheme, the power request information of appliances needs to be generated. In reality, the parameters of appliance operations have large diversities due to a huge variety of appliance specifications, customer behaviors, flexibility requirements, etc. Hence, for the simulations of the proposed large-scale DR scheme, picking a small set of certain appliances may not cover all these varieties. Instead, we randomly generate the appliance power request information to reflect the statistics of the residential appliances usage characteristics. The following assumptions are made for the power request information: (a) Because typical residential appliances' power consumption range from 50 to 2000 W [44], the power consumption of appliances is uniformly distributed as U(50W, 2000W); (b) as shown in [45], the residential appliances have operation durations ranging from 15 to 135 min, so we assume the cycle durations of appliances follow a uniform distribution of U(15, 135) min in the simulation; (c) the request interval (i.e., the duration from the time at which an appliance becomes idle to the time when the appliance is requested to power on by the customer) is assumed to be exponentially distributed with a mean of $\alpha_{i,k}^{\tau}$; and (d) type II flexible appliances have power consumption ranges of $\pm 5\%$ of the rated power. Note that although these assumptions do not make the generated power request information exactly the same as the appliance operations in reality, they are valid in the sense that under the same appliance power request profile settings, if the proposed scheme can control the individual appliance operations such that the deviation between the actual aggregated demand and the desired demand target is improved compared to the deviation between the original (uncontrolled) aggregated demand and the desired demand target, the effectiveness of the proposed scheme can be demonstrated.

In addition, we tune the appliance power-on request interval parameter $\alpha_{i,k}^{\tau}$ for frame τ according to the actual real-time demand data of New York City from New York ISO [46] to make aggregated demand follow the trend of real power demand. Specifically, we denote the base value of this parameter by $\alpha_{i,k}^{0}$, and the tuned value is computed as $\alpha_{i,k}^{\tau} = \gamma^{\tau} \cdot \alpha_{i,k}^{0}$, where the ratio γ^{τ} is computed in terms of actual real-time demand data D^{τ} for a length of L frames as $\gamma^{\tau} = \left[\left(\sum_{\tau=0}^{L-1} D^{\tau} \right) / (L \cdot D^{\tau}) \right]^{\omega}$, $\omega \geq 1$, where ω is the tuning parameter, and we choose $\omega = 1$ in our simulation. We can see the higher actual demand results in a smaller power-on request interval parameter $\alpha_{i,k}^{\tau}$; thus the higher demand in the simulation is due to more frequent power-on requests. Fig. 7 shows the aggregate demand of our test system using the tuned parameters (solid curve), and the actual average demand of N.Y.C used for tuning parameter $\alpha_{i,k}^{\tau}$ (dash curve) is also shown for comparison. We can see that the simulated demand follows the trend of the actual demand data.

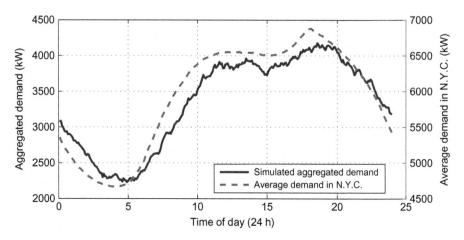

Fig. 7 Aggregated demand profile by the tuned power requests for the simulation. (From C. Chen, J. Wang, S. Kishore, A distributed direct load control approach for large-scale residential demand response, IEEE Trans. Power Syst. 29 (5) (2014) 2226.)

4.1 Scheduling results

We consider a case where $B = 400$ forming a 20×20 mesh network, and each building i has 40 flexible appliances, of which $|\mathcal{K}_i| = 36$ are Type I flexible appliances and $|\mathcal{Z}_i| = 4$ are Type II flexible appliances. In this scenario, the step size of the average consensus algorithm in Eqs. (3), (4) can be computed as $\varepsilon = 2/[\lambda_2(L) + \lambda_B(L)] = 0.2508$, where the second smallest eigenvalue of graph Laplacian matrix $\lambda_2(L) = 0.0246$, which gives the convergence speed parameter $\kappa_2 = 1 - \varepsilon\lambda_2(L) = 0.9938$. Assuming the initial standard deviation of $\{\hat{q}_i^\tau\}$ and $\{\widetilde{q}_i^\tau\}$, $i \in \mathcal{B}$ to be φ_0, with an exponential convergence speed faster than $\kappa_2 = 1 - \varepsilon\lambda_2(L)$, the converged deviation to be φ_c, the upper bound iteration time I can be computed as

$$I = \frac{\ln(\varphi_c/\varphi_0)}{\ln(\kappa_2)} \tag{14}$$

Consider the maximum value of φ_0 in which half of EMCs have the largest value $(|\mathcal{K}_i| \cdot P_i^{\max})$ and half of EMCs have zero values, so $\varphi_0 = |\mathcal{K}_i| \cdot P_i^{\max}/2$. We set the converged deviation as $\varphi_c = 0.1$, so from Eq. (14) the upper bound of convergence iteration is computed as $I = 2083$. Assuming the time of each iteration in the upper-layer network (transmission and processing time) as 0.1 ms, we can set the duration of TUP as 0.3 s to guarantee convergence.

Fig. 8 shows the scheduling results for 1 day under our proposed scheme. We can see the scheduled aggregated demand (thick solid curve) is closer to the desired demand (thin solid curve) than the original demand (dash curve), where the given desired demand is obtained by averaging actual aggregated power consumption over 100 days. Note that the desired demand can also be obtained by the economic dispatch

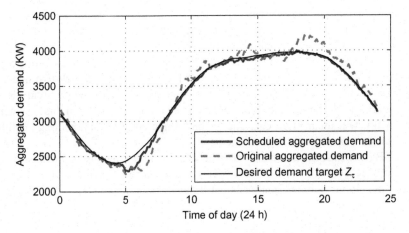

Fig. 8 Scheduling result for 1 day.
(From C. Chen, J. Wang, S. Kishore, A distributed direct load control approach for large-scale residential demand response, IEEE Trans. Power Syst. 29 (5) (2014) 2226.)

of the elastic demand discussed in Samadi et al. [27] and Li et al. [28]. With the proposed scheme, the average deviation between the desired and actual demand levels over the day decreases by 54.6% compared to that of the original demand. The decrease of deviation actually validates the effectiveness of the proposed method that achieves the system-level benefit of DR.

4.2   Effects of EMC network size

When the size of the upper-layer EMC network varies, we investigate its effects in two aspects in Fig. 9: convergence of the distributed algorithm and the deviation between the desired and actual demand levels. Fig. 9A shows the number of iterations needed

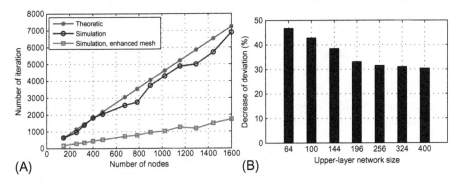

Fig. 9 Effects of the EMC network size. (A) Effects on convergence; (B) effects on deviation.
(From C. Chen, J. Wang, S. Kishore, A distributed direct load control approach for large-scale residential demand response, IEEE Trans. Power Syst. 29 (5) (2014) 2227.)

for convergence under different network sizes. The theoretic result (curve with dot marker) is computed as Eq. (14), which upper bounds the simulation result (curve with circle marker). We observe that the number of iterations for convergence increases linearly with the number of EMCs (i.e., our scheme is scalable to network size). Fig. 9A also shows the result for an enhanced mesh network in which two nodes at far ends of each row and each column are connected to form a cyclic ring (curve with square marker). We can see the convergence speed increases largely in this case while this network will incur more cost due to long-range communications (e.g., cellular networks incur more communication costs).

Fig. 9B shows the deviation decrease (in %) compared to a nonscheduled case with different EMC network sizes. We can see as the network size increases, the improvement of deviation becomes smaller but still has notable effects.

4.3 Effects of DR resources

Fig. 10 shows how the delay tolerance of Type I flexible appliances affects the deviation between the actual demand and the desired demand Z_r. The delay tolerances of appliances are assumed to be uniformly distributed with different parameters shown in the horizon axis. We can see that with larger delay tolerance, the deviation decreases; however, this decrease saturates as more delay tolerance is allowed.

Table 3 shows how the portion of flexible appliances affects the deviation between the desired and the actual demand by varying the number of flexible appliances as 40, 30, 20, and 10 in the simulation (i.e., the percentage of flexible appliances to be 100, 75, 50, and 25, respectively), where the ratio of the number of Type I appliances to the

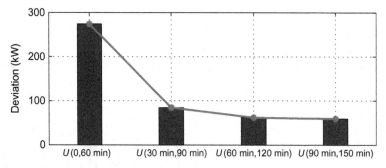

Fig. 10 Effect of delay tolerance.
(From C. Chen, J. Wang, S. Kishore, A distributed direct load control approach for large-scale residential demand response, IEEE Trans. Power Syst. 29 (5) (2014) 2227.)

Table 3 Effect of portion of flexible appliances

Flexible appliances (%)	100	75	50	25
Deviation (kW)	69.4	73.1	84.9	89.6

number of Type II appliances remains the same. We can see that more flexible appliances (i.e., more DR resources) incur the lower deviation.

5 Summary

In this chapter, we briefly reviewed the DR implementation and programs in the United States via some examples from consumer-premise, utilities, and ISOs/RTOs perspectives. When the amount of DR participation from residential customers becomes large, the existing DR methods will face several challenges. Based on this observation, we describe an innovative two-layer communication-based distributed DLC approach [24] for large-scale residential DR. The idea is to allocate the overall control task into each building's EMC in a distributed way. We employ average consensus algorithms in the upper layer network, which consists of connections between neighboring EMCs, to allocate target power consumption levels for each building. The EMC in each building then schedules appliance operations according to the local power consumption target. The protocol integrates these two layers seamlessly and also enables nonintrusive operation of appliances. Numerical results show notable improvement in the system's ability to match a prescribed load consumption profile, with actual demand levels. The proposed DR scheme provides a viable load-shaping tool to facilitate the energy efficiency benefits of DR in the future smart grids.

References

[1] U.S. Energy Information Administration (EIA), Annual Energy Outlook 2011 with Projection to 2035, 2010. Available from: https://www.eia.gov/outlooks/archive/aeo11/ (accessed March 6, 2018).

[2] U.S. Department of Energy (DOE), 20% Wind Energy by 2030: Increasing Wind Energy's Contribution to U.S. Electricity Supply, DOE Office of Energy Efficiency and Renewable Energy Report, in: DOE Office of Energy Efficiency and Renewable Energy Report, 2008. Available from: http://www1.eere.energy.gov/wind/pdfs/41869.pdf (accessed March 28, 2017).

[3] U.S. Department of Energy, The Smart Grids: An Introduction, 2009. Available from: http://energy.gov/oe/downloads/smart-grid-introduction-0 (accessed March 28, 2017).

[4] U.S. Department of Energy, Benefits of Demand Response in Electricity Markets and Recommendations for Achieving Them, 2006. Available from: https://eetd.lbl.gov/sites/all/files/publications/report-lbnl-1252d.pdf (accessed March 6, 2018).

[5] C. Chen, J. Wang, Y. Heo, S. Kishore, MPC-based appliance scheduling for residential building energy management controller, IEEE Trans. Smart Grid (3) (2013) 1401–1410.

[6] General Electric Appliances, GE WiFi Connect, 2017. Available from: http://www.geappliances.com/ge/connected-appliances/ (accessed March 28, 2017).

[7] Whirlpool, Whirlpool Smart Appliances, 2017. Available from: https://www.whirlpool.com/smart-appliances/) (accessed March 28, 2017).

[8] Pacific Gas and Electric Company, Explore PG&E Energy-Saving Programs, 2017. Available from: https://www.pge.com/en_US/residential/save-energy-money/savings-programs/savings-programs-overview/savings-programs-overview.page? (accessed March 28, 2017).

[9] Southern California Edison, Save Power Days—Program Update, 2017. Available from: https://www.sce.com/wps/portal/home/residential/rebates-savings/save-power-day/ (accessed March 28, 2017).

[10] Commonwealth Edison, Manage My Energy, 2017. Available from: https://www.comed.com/WaysToSave/ForYourHome/Pages/ManageMyEnergy.aspx (accessed March 28, 2017).

[11] Wisconsin Public Service, Wisconsin Electric Rate & Programs, 2017. Available from: http://www.wisconsinpublicservice.com/home/wi_electric.aspx (accessed March 28, 2017).

[12] Con Edison, Energy Saving Programs, 2017. Available from: https://www.coned.com/en/save-money/energy-saving-programs (accessed March 28, 2017).

[13] Gulf Power, Mange your energy use to control your savings, 2017. Available from: https://www.gulfpower.com/residential/savings-and-energy/rebates-and-programs/energy-select (accessed March 28, 2017).

[14] F. Rahimi, A. Ipakchi, Demand response as a market resource under the smart grid paradigm, IEEE Trans. Smart Grid 1 (1) (2010) 82–88.

[15] New York ISO, Demand Response Programs, 2017. Available from: http://www.nyiso.com/public/markets_operations/market_data/demand_response/index.jsp (accessed March 28, 2017).

[16] ERCOT, Emergency Response Service, 2017. Available from: http://www.ercot.com/services/programs/load/eils (accessed March 28, 2017).

[17] ERCOT, Load Resource Participation in the ERCOT Markets, 2017. Available from: http://www.ercot.com/services/programs/load/laar (accessed March 28, 2017).

[18] ERCOT, Voluntary Load Response, 2017. Available from: http://www.ercot.com/services/programs/load/vlrp (accessed March 28, 2017).

[19] PJM Interconnection, Demand Response, 2017. Available from: http://www.pjm.com/markets-and-operations/demand-response.aspx (accessed March 28, 2017).

[20] California ISO, Reliability Demand Response Product, 2017. Available from: http://www.caiso.com/informed/Pages/StakeholderProcesses/CompletedStakeholderProcesses/ReliabilityDemandResponseProduct.aspx (accessed March 28, 2017).

[21] ISO New England, About Demand Resources (DR) in the ISO's Wholesale Electricity Markets, 2017. Available from: https://www.iso-ne.com/markets-operations/markets/demand-resources/about (accessed March 28, 2017).

[22] Federal Energy Regulatory Commission (FERC), Order No. 719, Wholesale Competition in Regions with Organized Electric Markets, 2008. Available from: http://www.ferc.gov/whats-new/comm-meet/2008/101608/E-1.pdf (accessed February 13, 2014).

[23] Federal Energy Regulatory Commission (FERC), Order No. 745, Demand Response Compensation in Organized Wholesale Energy Markets, 2011. Available from: http://www.ferc.gov/EventCalendar/Files/20110315105757-RM10-17-000.pdf (accessed February 13, 2014).

[24] C. Chen, J. Wang, S. Kishore, A distributed direct load control approach for large-scale residential demand response, IEEE Trans. Power Syst. 29 (5) (2014) 2219–2228.

[25] Ameren Illinois Utilities, Ameren Power Smart Pricing, 2012. Available from: http://www.powersmartpricing.org/how-it-works/ (accessed February 13, 2014).

[26] F. Oldewurtel, A. Ulbig, A. Parisio, G. Andersson, M. Morari, Reducing peak electricity demand in building climate control using real-time pricing and model predictive control, Proceedings of IEEE Conference on Decision and Control, 2010.

[27] P. Samadi, A.H. Mohsenian-Rad, R. Schober, V.W. Wong, J. Jatskevich, Optimal real-time pricing algorithm based on utility maximization for smart grid, Proceedings of IEEE SmartGridComm, 2010, pp. 415–420.

[28] N. Li, L. Chen, S.H. Low, Optimal demand response based on utility maximization in power networks, Proceedings of PES General Meeting, 2011.

[29] H. Zhong, L. Xie, Q. Xia, Coupon incentive-based demand response: theory and case study, IEEE Trans. Power Syst. 28 (2) (2013) 1266–1276.

[30] A.H. Mohsenian-Rad, V.W. Wong, J. Jatskevich, R. Schober, A. Leon-Garcia, Autonomous demand-side management based on game-theoretic energy consumption scheduling for the future smart grid, IEEE Trans. Smart Grid 1 (3) (2010) 320–331.

[31] C.L. Su, D.S. Kirschen, Quantifying the effect of demand response on electricity markets, IEEE Trans. Power Syst. 24 (3) (2009) 1199–1207.

[32] A. Botterud, Z. Zhou, J. Wang, J. Sumaili, H. Keko, J. Mendes, R.J. Bessa, V. Miranda, Demand dispatch and probabilistic wind power forecasting in unit commitment and economic dispatch: a case study of Illinois, IEEE Trans. Sustain. Energy 4 (1) (2013) 250–261.

[33] SAE International, SAE charging configurations and ratings terminology, 2011. Available from: http://www.sae.org/smartgrid/chargingspeeds.pdf (accessed February 13, 2014).

[34] Sylvane, Space Heater Buying Guide: Power Consumption, 2013. Available from: http://www.sylvane.com/heater-buying-guide.html (accessed February 13, 2014).

[35] Silver Spring Networks, Our network smart grid infrastructure: ensuring system connectivity, 2013. Available from: http://www.silverspringnet.com/products/network-infrastructure/ (accessed February 13, 2014).

[36] Con Edison, Demand response/distribution load relief program, 2012. Available from: http://www.coned.com/energyefficiency/dist_load_relief.asp (accessed February 13, 2014).

[37] A.D. Dominguez-Garcia, C.N. Hadjicostis, Coordination and control of distributed energy resources for provision of ancillary services, in: Proceedings of 50th IEEE Conference on Decision and Control (CDC), 2011, pp. 27–32.

[38] A.D. Dominguez-Garcia, C.N. Hadjicostis, Distributed algorithms for control of demand response and distributed energy resources, in: Proceedings of 50th IEEE Conference on Decision and Control (CDC), 2011, pp. 27–32.

[39] A.D. Dominguez-Garcia, C.N. Hadjicostis, N.H. Vaidya, Resilient networked control of distributed energy resources, IEEE J. Select. Areas Commun. 30 (6) (2012) 1137–1148.

[40] R. Olfati-Saber, J.A. Fax, R.M. Murray, Consensus and cooperation in networked multi-agent systems, Proc. IEEE 95 (1) (2007) 215–233.

[41] R. Olfati-Saber, R.M. Murray, Consensus problems in networks of agents with switching topology and time-delays, IEEE Trans. Autom. Control 49 (9) (2004) 1520–1533.

[42] L. Xiao, S. Boyd, Fast linear iterations for distributed averaging, Syst. Control Lett. 53 (2004) 65–78.

[43] B. O'Hara, A. Petrick, IEEE 802.11 Handbook: A Designer's Companion, IEEE Standards Association, Piscataway, NJ, 2005.

[44] OkSolar.com, Typical power consumption, 2012. Available from: http://www.oksolar.com/technical/consumption.html (accessed February 13, 2014).

[45] R. Stamminger, Synergy potential of smart appliances: a report prepared as part of the EIE project 'Smart Domestic Appliances in Sustainable Energy Systems', 2008. Available from: http://www.smart-a.org/D2.3_Synergy_Potential_of_Smart_Appliances_4.00.pdf (accessed February 13, 2014).

[46] New York ISO, New York ISO Real Time Actual Load, Archive, 2011. Available from: http://mis.nyiso.com/public/P-58Blist.htm (accessed June 19, 2013).

Development of a residential microgrid using home energy management systems

Mehdi Ganji*, Mohammad Shahidehpour†
*Willdan Energy Solutions, Anaheim, CA, United States, †Illinois Institute of Technology, Chicago, IL, United States

1 Introduction

Electricity consumed in the residential sector accounts for slightly more than one-fifth of the total load in the United States, as shown in Fig. 1. This consumption level has increased by 10% during the last 2 years [1]. Of this total, heating, ventilation and air conditioning (HVAC) technology and air conditioner units used 27% [2]. The increased energy consumed at the residential level required the addition of generation and transmission line capacity to provide the residents with higher power at peak hours when the overall efficiency of the grid is too low. Consequently, the increase in home energy demand caused the generation and transmission line capacity cost to increase [3].

On the other hand, aggregator companies attempted to elevate their efficiency by reducing the spinning reserves of generation required for demand uncertainties, load overestimates at peak time, and errors made by intermittent renewable energy resources. This demonstrates the need for better interception of load and environmental factors such as temperature and consumer behavior. The absence of proper behind-the-meter load monitoring systems combined with the inability of residential individual optimal load scheduling that considers the customer level of comfort make aggregator companies unable to be more efficient. Using nonillustrative load monitoring adds more complication due to the presence of a huge amount of data coming to the utility server. This amount of "big data" causes the aggregator to need to invest too much on increasing server capacity in order to keep and maintain the data and run the analysis. By making submeter technology available at each building, data can be kept locally and a comprehensive load pattern that identifies the customer energy consumption behavior is sent to the aggregator, which lowers the required analysis and maintenance procedures.

Recently, a demand-side energy management system (DSEMS) has been deployed to alleviate the impacts of load uncertainties, the intermittency of renewable energy resources, and demand overestimates in the system [4]. DSEMS techniques include direct load controlling (DLC) and dynamic pricing (DP) [5,6]. As mentioned before, by using DLC, aggregator companies can centrally control the deferrable and

Application of Smart Grid Technologies. https://doi.org/10.1016/B978-0-12-803128-5.00005-2

Fig. 1 U.S. Energy Information
Administration, Annual Energy
Review 2011.

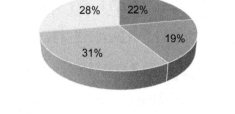

■ Residential   ■ Commercial   ■ Industrial   Transportation

curtailable appliances to maintain the balance between demand and supply at peak time. Now, different daily electricity pricing schemes are available for customers, including a fixed-rate plan (FRP), DP, and real-time pricing (RTP). Customers get charged with a constant identical fixed rate in the FRP scheme. In DP, which explicitly represents the system congestion, hourly prices are sent to customers with the hope that they will change their energy consumption accordingly. This scheme includes the day-ahead market price, which has been cleared the day before. In RTP, the energy price, which was cleared at the beginning of each hour, is sent every 5 min to the customers. All these techniques require the individual customer to manually change their behavior in response to centrally sent energy price signals, which seems unrealistic. This action needs a smart system to individually control all appliances based on price signal, appliance load profile, customer behavior, and level of comfort [7,8]. In Lee's work [9], appliances are divided into different categories based on their load elasticity, interruptibility, and deferability. Previous work conducted by Muratori [10] and Zehir [11] considered two major energy-consuming appliances and their potential for demand response (DR). Bozchalui [12] discussed DR opportunities of household appliances.

In this chapter, a fully functional home energy management system (HEMS), which helps the customer to change the aggregator's load pattern according to hourly energy price, availability of local generation, appliance profile, and customer level of comfort, is proposed and a prototypal architecture of the system is presented. The HEMS prototype built at the Galvin Center for Electricity Innovation located at the Illinois Institute of Technology is explained in detail.

2 Home energy system overview

Based on aggregated customer daily load and power generation by renewable resources, aggregators come up with a load forecast that will be used to clear the market in the day-ahead wholesale market. Forecasted day-ahead hourly loads need to be submitted by aggregators before noon. The day-ahead hourly energy price is released at 4 p.m., as shown in Fig. 2.

In the following sections, the project's technical implementation, including the specifications of components used, communication protocols, and the two main tasks of monitoring and scheduling methodologies, is presented in detail.

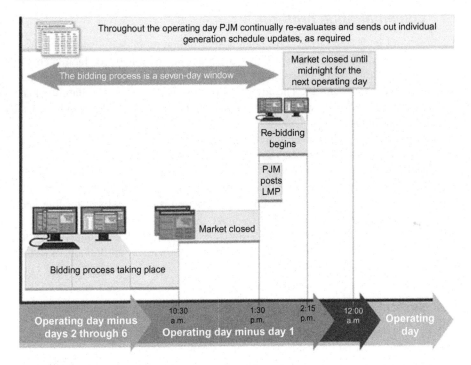

Fig. 2 PJM day-ahead market clearing price process [13].

We categorized the entire house model into the different areas with fairly similar application disciplines. Each area has its own local controller, which are all controlled by a master controller. This system has outstanding physiognomies, including hands-off interaction and a digital communication framework. Hands-off interaction frames the relation between the switches and loads in such a way that there is no conventional direct electricity feeding and, instead, it is maintained by software functions. Using a communication network in order to exchange the information in monitoring and controlling the load procedures demonstrates the efficiency of this centralized method compared to conventional hardware signaling. HEMS fortifies the end user by various operations such as instant electrical load control, environmental-sensor-based control, and 24-h control simulation. Through community control, the entire community, like a complex building, is controlled by the signal sent by the master HEMS [14]. The building manager can override all scheduling and group control by instant control capability. The emergence of more community-based renewable resources and energy storage units introduces the urgency of the community-based control. Twenty-four-hour control simulation helps the end user feel the entire 24 h of scheduling in a few seconds based on its time interval setting and gives the operator the opportunity to understand the most probable scenarios throughout the day. This capability is used mostly for training purposes. This system provides the end users with other enhanced features such as a massive messaging system for security and building announcements.

2.1 Communication protocol

As shown in Fig. 3, each local box contains controlling, monitoring, and communication hardware. All local boxes are in a mesh arrangement network facilitated by the XBee communication protocol. Each local box representing a node communicates with other nodes in this model. The advantage of a mesh network is its reliability; that is, if the communication between a pair of nodes is not working, the data can be transmitted through other nodes.

This protocol was chosen because of its short-range wireless data transmission technology, which is safe, reliable, simple, flexible, and low cost with a long battery life. The XBee module, shown in Fig. 4, is an interoperable RF module that establishes a mesh network in which all nodes need to have Zigbee as the transmitter terminal. Improved data traffic management, remote firmware updates, and self-healing and discovery for network stability capability make the Zigbee communication protocol exceptional among other protocols. In the smart community, the communication between the houses is established using the XBee communication protocol because there is clear distance between nodes.

2.2 System hardware configuration

As presented in Fig. 5, the Galvin Center Smart Home model is divided into different controllability areas facilitated by local control panels.

Each of the Smart Home area's local box, shown in Fig. 5, includes the office and garage area, the bathroom and bedroom, and the kitchen and living room. It contains numbers of current sensors and relay switches for each load and one XBee and

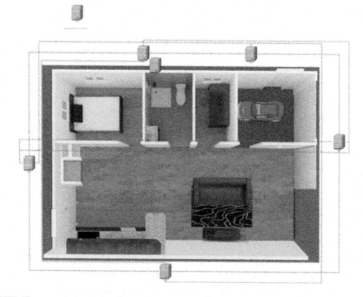

Fig. 3 HEMS communication scheme.

Fig. 4 XBee module.

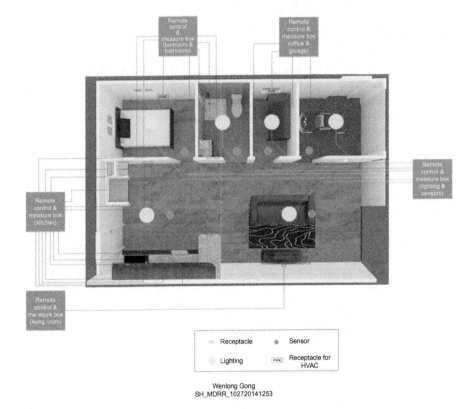

Fig. 5 Smart home layout controllability schemes.

Fig. 6 Current sensors and relay configuration.

microcontroller for each local box, as shown in Fig. 6. Lighting and HVAC systems are normally scheduled based on the optimization result and historical data. Motion and daylight sensors are used if some irregular occupancy or climate situation occurs.

This system monitors and controls all the Smart Home prototype electrical loads. In this model, all electrical loads are categorized into three different groups, including:

(1) adjustable loads such as HVAC, lighting, and the refrigerator with energy consumption levels that can be partially curtailed/increased due to the control signal.

(2) shiftable loads such as the dishwasher, washer, dryer, electrical pool pump, and plug-in hybrid vehicle (PHEV) with energy consumption levels that are considered as a load block with a constant rated power and known duty cycle. They would be on/off at the time the price of electricity is low/high and continue until its operation is completed.

(3) fixed loads such as microwaves, TVs, and computers with energy consumption levels that cannot be changed except by the tenant. Each load's real-time energy consumption can be measured dynamically every 1 s. These data are stored in the system database as historical data and are reprocessed for a self-learning operation. The database was created using the Microsoft SQL server. The controlling signal is sent to each appliance's circuit breaker initiated from the HEMS processor and can be transferred through the XBee communication protocol-based cloud within the Smart Home model.

2.3   System software configuration

Remote monitoring and scheduling of energy consumption is the main capability of this system. Smart Home tenants can monitor real-time energy consumption for specific areas and appliances and submit the schedule remotely through their portable devices. The controlling and monitoring dashboard engine, shown in Fig. 7, is a cloud-based application and enables the tenants to have access to the database and to control the electrical loads through remotely controllable circuit breakers. Each user can visualize the real-time data, including electrical load real-time energy consumption, circuit status, and scheduling.

The HEMS prototype includes two main tasks as described below:

(1) Monitoring.
(2) Scheduling.

2.3.1 Monitoring

Electrical loads including lighting, receptacles, and total electricity demand are measured and logged with resolution of every second. A data monitoring and visualization flowchart is shown in Fig. 8. In this model, all data are stored in the SQL server for research purposes, as shown in Fig. 9. More data points such as real-time current, voltage, max demand, max electricity energy consumption, and so on can be set to be downloaded according to the end user's preference. To consider the customer level of comfort, we need to log the customer's energy consumption and update the database with real-time data. This helps the scheduler to postpone/change the shiftable/adjustable loads according to historical data representing the customer's level of comfort and flexibility.

Every time the database gets updated, the system has more information about the tenant's schedules and, consequently, their preferences. Every 24 h, a new data set is added to the database and a new load pattern is released. Using this daily pattern, a probability to each appliance i at each hour t- on k-th day of the year representing the frequency of running appliance i at hour t_0 by the customer is obtained, as shown in Eq. (1) [15].

$$\pi_i(t) = \frac{\sum_{n=1}^{k} u_i(t - (24n + t_0))}{52 \times 7 \times 24} \tag{1}$$

2.3.2 Scheduling

The optimal load scheduling is implemented by considering the day-ahead market price sent by the aggregator, the customer's level of comfort using the most recent updated database and load pattern, and the load profiles of the all-electrical appliances. The objective of HEMS is to schedule the loads during the hours that the price of electricity is lower, based on the day-ahead market price, compared to the other hours with a relatively higher price representing the congestion in the system. This methodology will reduce the customer's bill as well as help the aggregators reduce the requirement of committing more fast generation units, which consequently leads to generation/procurement of power at a higher price. The drawback for this type of scheduling is the "rebound" effect, which can be eliminated through the aggregator

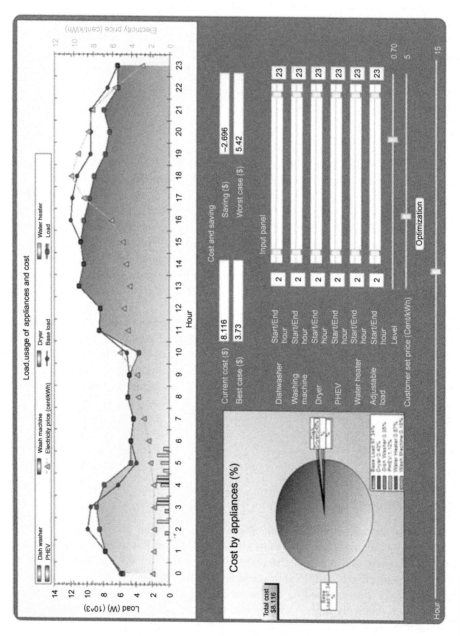

Fig. 7 HEMS dashboard.

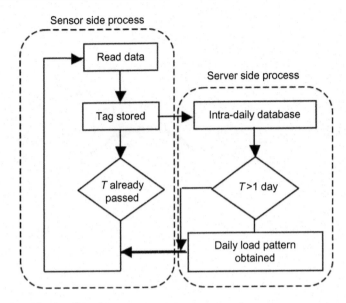

Fig. 8 Load monitoring flowchart.

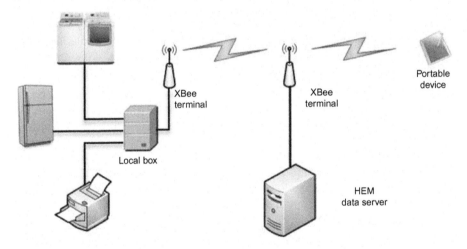

Fig. 9 HEMS monitoring map.

load limit constraint. This method forces all customers to defer their shiftable loads to the hours that the price of energy is the lowest; therefore, the new peak is introduced during those hours, as shown in Fig. 10.

As mentioned previously, one of the determinant factors in residential optimal load scheduling is the customer level of comfort, which needs to be considered seriously. In this chapter, customer comfort level is taken into consideration at two levels:

(1) Customer price setpoint (¢/kWh): customer allocates the price setpoint representing the preferred level of hourly rate to pay. During hours with an hourly price rate higher than

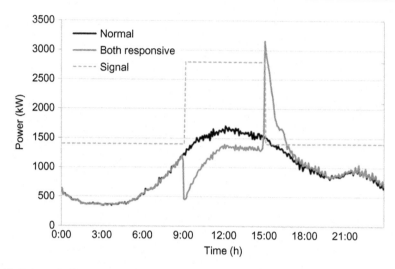

Fig. 10 Rebound effect [16].

the tenant's price setpoint, the adjustable load is set to its minimum load profile, and, during the hours with an hourly price rate lower than the tenant's setpoint, the adjustable load is set to its comfortable (base) load profile. Also, shiftable loads run during the hours with a lower energy price rate than the customer's price set.

(2) End-user predefined shiftable load operation time span: tenants can identify the time interval that they prefer to run their shiftable load such as washers, dryers, dishwashers, pool pumps, and PHEVs.

In other words, the optimal load scheduling is doable using a day-ahead price scheme (time of use) that defers shiftable loads to the hours with a relatively low electricity price while the preferred appliances' service quality (start time, finish time, operating hours) set by the customers is considered as well. Implementing both the above strategies requires that the customers keep watching the hourly electricity price and run their loads accordingly, which is unworkable. HEMS helps the tenant to set the price threshold as well as the service quality. HEMS will also implement the schedules based on the optimization result driven by the above constraints.

(3) Environmental sensors: schedules generated by area lighting and HVAC units can be overridden by occupancy and daylight sensors. Lighting is scheduled based on resident's daily behavior, but any changes in the area occupancy deactivate the already generated schedule and cause the lighting status to get updated in order to satisfy the customer's level of comfort. Daylight helps the HEMS system to keep the output lumens of the lighting at its minimum lumen level while also satisfying the resident's requirement and, consequently, helping the residents to consume the least amount of energy.

2.4 Scheduling methodology

The HEMS prototype is considered an energy hub, which is an integrated energy management system as the core of the activity with various energy entities. Thus, different energy sectors including energy resources, conversion systems, and storage should be considered in scheduling.

The optimal load scheduling is a mixed integer linear programing (MILP) model [17], which is introduced by an objective function of minimizing the net load seen by the aggregator considering different types of loads including fixed, shiftable, curtailable, and storage units that can represent positive loads at time of charging and negative loads at the time that discharging is implemented.

2.5 Case studies and results

In our system prototype evaluation, 1 day's hourly data, including fixed load, adjustable load, and shiftable load, were collected from the Smart Home prototype, as shown in Figs. 11–14. These data were considered to demonstrate the optimal load scheduling using a fully functional HEMS.

Hour	1	2	3	4	5	6	7	8
Load (kW)	2.5	2.3	1.9	1.7	1.7	1.9	2.2	1.2
Hour	9	10	11	12	13	14	15	16
Load (kW)	1.5	0.8	1.4	2.1	3.4	3.8	3.9	4.1
Hour	17	18	19	20	21	22	23	24
Load (kW)	5.6	5.8	5.3	4.4	4.1	4.3	3.4	2.8

Fig. 11 Fixed load data.

Hour	1	2	3	4	5	6	7	8
Load (kW)	3.1	2.8	2.4	2.2	2.1	2.2	2.7	3.1
Hour	9	10	11	12	13	14	15	16
Load (kW)	3.5	4	4.7	5.4	6	6.2	6.5	6.7
Hour	17	18	19	20	21	22	23	24
Load (kW)	6.9	7	6.9	6.2	5.9	5.2	4.1	3.5

Fig. 12 Adjustable load data.

Hour	1	2	3	4	5	6	7	8
Load (kW)	2.2	2.0	1.7	1.5	1.5	1.5	1.9	2.2
Hour	9	10	11	12	13	14	15	16
Load (kW)	2.5	2.8	3.3	3.8	4.2	4.3	4.6	4.7
Hour	17	18	19	20	21	22	23	24
Load (kW)	4.8	4.9	4.8	4.3	4.1	3.6	2.9	2.5

Fig. 13 Minimum required adjustable load data.

Adjustable loads		Dishwasher	Washer	Dryer	Water heater	PHEV
Rated power (kW)		0.7	0.24	1	1.4	1.5
Cycle duration (h)		2	2	2	2	4
Base case	Start	1	1	1	1	1
	End	24	24	24	24	24
User defined	Start	20	11	13	1	1
	End	24	17	19	16	8

Fig. 14 Shiftable load data.

Hour	1	2	3	4	5	6	7	8
Price (¢/kWh)	1.7	1.6	1.5	1.5	1.6	1.9	2.1	2.6
Hour	9	10	11	12	13	14	15	16
Price (¢/kWh)	3.2	3.3	5.1	4.3	4.7	4.1	4.5	4.8
Hour	17	18	19	20	21	22	23	24
Price (¢/kWh)	6	8.9	10.2	9.5	8.5	8	5.7	2.8

Fig. 15 Hourly market price.

The day-ahead market price used by HEMS to schedule the loads obtained from the local aggregator is shown in Fig. 15.

The Smart Home prototype is adjacent to a local 8 kW wind turbine and 250 kW solar units. Because typical wind and solar capacity are 1.5 and 2 kW, respectively, local generation is used partially to represent real distributed energy resources (DERs) used in residential buildings. Based on historical data obtained locally, 1.5 kW wind power generation is available in hours 1–5, 23, and 24. Solar capacity of 2 kW will generate power during hours 11–19. Below are the different load scheduling cases considering hourly price, customer level of comfort, available local generation, and storage discussed in this section:

(1) FRP price-based scheduling.
(2) Hourly electricity price based on optimal load scheduling without PHEV storage.
(3) Hourly electricity price based on optimal load scheduling with PHEV storage.
(4) Hourly electricity price based on optimal load scheduling with PHEV storage and aggregator load limit.
(5) Tenant's load reduction criteria (¢/kWh).
(6) User predefined shiftable load operation time interval.

Different cases of user-defined shiftable load time interval and adjustable reduction criteria are compared in Fig. 16.

Case 1 is the most optimal schedule. Using HEMS and considering just shiftable load and no PHEV battery brings an 18% reduction in peak load. In Case 3, a PHEV battery considered as storage in the system causes the peak load to be reduced by 10%, which was associated with charging the PHEV at hour 17. By implementing the load limit in Case 4 assigned by the aggregator, this peak load is removed and the reduction is again back to 18%. By considering the load reduction criteria in Case 5 assigned by

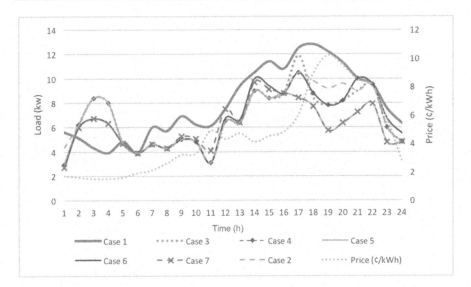

Fig. 16 Single home optimal hourly load schedule.

the tenant, the peak load reduction is increased by 30%. The tenant's level of comfort (start/end time of shiftable loads) considered in Case 6 lowers the peak load reduction to 18% compared to Case 5 and can be improved to 24% by changing the start/end time of the shiftable loads, as shown in Case 7.

As shown in Figs. 16 and 17, considering just shiftable load and no PHEV battery available (Case 2), the HEMS system brings just 11% savings in the energy bill and 18% in peak load reduction. In Case 3, the HEMS generates 14% electricity bill and 11% peak load reduction due to the extra power drawn by the PHEV at hour 17 to charge the battery. In Case 4, the aggregator assigns the load limit, which causes the PHEV not to charge its battery to maximum at hour 17; therefore, the peak load is kept lower than Case 3 and equal to Case 2 (load limit equal to Case 2 peak load).

3 Smart buildings/smart residential community

So far, using individual HEMS is not cost-efficient because of the high capital cost relative to the individual house electricity bill and the minor peak demand reduction and limited controllability on each of the houses in a community. Therefore, a community-based HEMS, CHEMS, is introduced, which includes one single master HEMS and a number of slave HEMS.

3.1 Communication protocol

As shown in Figs. 18 and 19, CHEMS mesh networks can be established in both neighborhoods and tall buildings. In both arrangements, each unit (house, apartment) is introduced to the master HEMS as a slave HEMS node. In neighborhood CHEMS,

	Case 1	Case 2	Case 3	Case 4	Case 5	Case 6	Case 7
Hour 1	5.6	4.34	2.94	2.94	2.94	2.7	2.7
Hour 2	5.1	6.24	6.24	6.24	6.24	6	6
Hour 3	4.3	8.4	8.4	8.4	8.4	6.7	6.7
Hour 4	3.9	8	8	8	8	6.3	6.3
Hour 5	4.8	4.7	4.7	4.7	4.7	4.7	4.7
Hour 6	3.9	3.9	3.9	3.9	3.9	3.9	3.9
Hour 7	6	4.6	4.6	4.6	4.6	4.6	4.6
Hour 8	5.7	4.3	4.3	4.3	4.3	4.3	4.3
Hour 9	6.9	5	5	5	5	5	5.24
Hour 10	6.2	4.8	4.8	4.8	4.8	4.8	5.04
Hour 11	6.1	3.1	3.1	3.1	3.1	3.1	4.1
Hour 12	7.5	6.5	6.5	6.5	6.5	6.74	7.5
Hour 13	9.4	6.4	6.4	6.4	6.4	6.64	6.4
Hour 14	10.5	9	9	9	9	10	9.7
Hour 15	11.4	8.4	8.4	8.4	8.4	9.4	9.1
Hour 16	10.8	8.8	8.8	8.8	8.8	8.8	8.8
Hour 17	12.5	10.5	11.9	10.5	8.43	10.5	8.43
Hour 18	12.8	9.8	8.4	8.8	7.7	8.8	7.7
Hour 19	12.2	9.2	7.8	7.8	5.73	7.8	5.73
Hour 20	11.3	9.6	8.2	8.2	6.34	8.2	6.34
Hour 21	10	9	9	10	7.23	10	7.23
Hour 22	9.5	9.5	9.5	9.5	7.94	9.5	7.94
Hour 23	7.5	6	6	6	4.77	6.7	4.77
Hour 24	6.3	4.8	4.8	4.8	4.8	5.5	4.8
Peak load	12.8	10.5	11.9	10.5	9	10.5	9.7
Peak reduction	0%	18%	7%	18%	30%	18%	24%
Daily bill ($)	10.51	8.35	8.01	8.05	7.00	8.15	7.11
Bill reduction	0	11%	14%	14%	34%	13%	33.3%

Fig. 17 Single home optimal hourly load schedule summary.

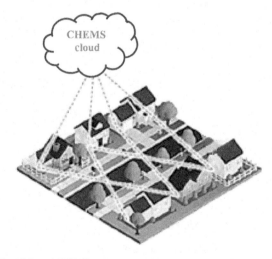

Fig. 18 Neighborhood-based HEMS.

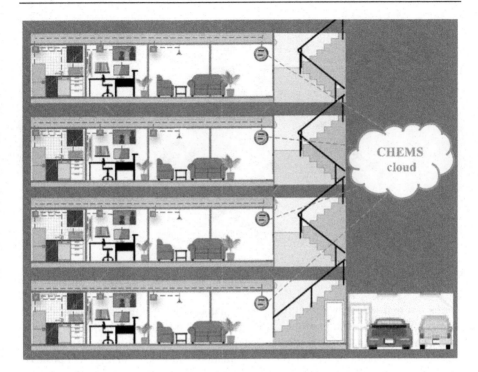

Fig. 19 Building-based HEMS.

because clear space is available, most of the nodes are connected to the network using XBee. However, in high rises, due to a lack of clear space between floors, ethernet-based communication may be used to integrate all the mesh networks deployed on each building floor.

3.2 System hardware/software control configuration

As shown in Fig. 20, CHEMS includes several slave HEMS that are numbers of apartments (houses) each containing different types of loads (residential loads) and storage units and also a community that has communal loads (lighting, pool pump) and a shared storage unit.

3.3 Scheduling methodology

Our CHEMS model is considered as an energy hub, which is a centrally integrated energy management system that serves as the core of the activity with various local energy entities. Therefore, different energy sectors including community energy resources, conversion systems, and community/mobile (PHEV) as well as community and residential loads and shared storage units are considered in this chapter.

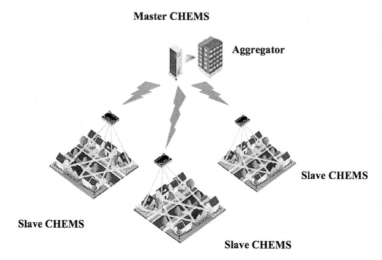

Fig. 20 CHEMS.

The optimal load scheduling problem is mathematically introduced as an MILP problem. The minimization of the net load seen by the aggregator including various types of loads such as fixed, shiftable, curtailable, and storage units (community storage unit, mobile units such as PHEV and residential units) that can represent positive loads at the time of charging and negative loads at the time discharging is implemented. Different types of loads modeled in this study are:

(1) Fixed load: those critical loads that cannot be curtailed and shifted such as TVs, computers, electrical ranges, microwaves, and emergency lighting.

(2) Shiftable load: those types of noncritical loads whose operation can be deferred to the time the price of electricity is relatively lower. Resident's level of comfort and load quality of service need to be considered. Washer and dryers, pool pumps, and dishwashers are examples of this type of load.

(3) Adjustable load: those loads that can be curtailed seeing the hourly price of electricity or an external load reduction signal. Lighting and air conditioning systems are good examples of this type of load.

(4) Storage units: this type of load can be modeled as a load when they are getting charged or as generation units when they are in discharging mode. Residential storage units and PHEVs can be considered as storage units.

(5) Community loads: include all the loads that are common between all.

3.4 Case studies and results

Two methods of DR include: (1) price-based and (2) incentive-based DR programs (limited aggregated load), and optimal load scheduling of the building in islanded mode (microgrid) are considered in this section. Also, the proposed model is tested using the real data obtained from the Smart Home system prototype located at the Illinois Institute of Technology (IIT), DERs including 200 kW solar generation, generating in hours 11–19, and 150 kW wind generation available during hours 1–5 and 23–24, and storage units are considered in this chapter. These data are extrapolated considering the real power output generated by the wind turbine and solar panels installed at IIT.

First, optimal load scheduling is implemented in the community through price-based DR (total $E(t) \leq \infty$), where $E(t)$ represents the aggregated load. Based on a tenant's daily schedules and an appliance's quality of service, a 24-h optimal schedule is generated. Then, optimal load scheduling is implemented through incentive-based load scheduling ($E(t) \leq K$). In this case, an aggregator request is delivered and, based on the amount of load curtailment submitted in the agreement, the community must curtail its load to meet that amount. Finally, the microgrid configuration of the community which is the operation of the community in islanded mode through leveraging the local generation and battery storage units to respond the building loads ($E(t)=0$) is studied. By using the data acquired from the Smart Home model, considering a community including 200 of these units, and applying our CHEMS model, the optimal load scheduling in the cases below is obtained.

(1) Case 1: community with fixed rate tariff (FRT) without PHEV and residential battery unit.
(2) Case 2: community price-based DR without shared PHEV and residential battery unit.
(3) Case 3: community price-based DR with shared PHEV and residential battery unit.
(4) Case 4: incentive-based DR in community using CHEMS.
(5) Case 5: community microgrid optimal load scheduling.

As seen in Figs. 21 and 22, Case 1 is the most nonoptimal scheduling. Because the price is fixed and load shaving and curtailing do not affect their electricity bill, there is not any encouragement for tenants to follow the price signal and change their schedule accordingly.

In Cases 2 and 3, shiftable loads are deferred to the hours with the lower hourly electricity rate (hours 1–6) and curtailable loads are adjusted to their minimum power

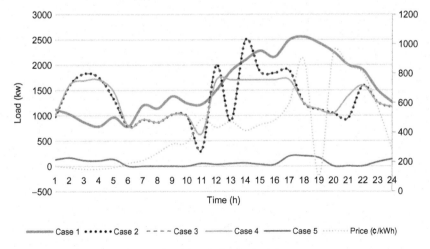

Fig. 21 Residential community optimal hourly load schedule.

Case Hour	Case 1	Case 2	Case 3	Case 4
Hour 1	1120	980.2	980.2	1008.56
Hour 2	1020	1598.4	1598.4	1604.84
Hour 3	860	1826	1826	1687.66
Hour 4	780	1746	1746	1700
Hour 5	960	1324	1324	1468.54
Hour 6	780	780	780	784
Hour 7	1200	920	920	921
Hour 8	1140	860	860	860
Hour 9	1380	1019.2	1019.2	1019.2
Hour 10	1240	979.2	979.2	979.2
Hour 11	1220	315.4	315.4	643
Hour 12	1500	1997.2	1997.2	1700
Hour 13	1880	886.4	886.4	1700
Hour 14	2100	2490.4	2490.4	1700
Hour 15	2280	1865	1865	1700
Hour 16	2160	1839.4	1839.4	1700
Hour 17	2500	1886	1886	1700
Hour 18	2560	1240	1240	1240
Hour 19	2440	1121.8	1121.8	1121.8
Hour 20	2260	1043.8	1043.8	1043.8
Hour 21	2000	946	946	1382.8
Hour 22	1900	1588	1588	1588
Hour 23	1500	1258.4	1258.4	1258.4
Hour 24	1260	1166	1166	1166
Peak load	1120	980.2	980.2	1008.56
Peak reduction	0%	3%	3%	34%
Daily bill ($)	2019.28	1404.6788	1404.6788	1425.8659

Fig. 22 Residential community optimal hourly load schedule summary.

consumption during the hours the price is higher than the tenant's setpoint. However, storage units use the available additional capacity provided by the reduced loads to charge and discharge. Consequently, although we lower the peak load by deferring the shiftable load and curtailing the adjustable load, the peak load is not decreased noticeably compared to Case 1 (about 3%). Peak load happens at hour 14 when the community storage is being charged at its maximum load to recharge the power already discharged at hour 13. So even a small oscillation in electricity price causes the storage units to charge/discharge in consecutive hours that lowers the battery lifetime as well as generates unwanted peak load. This event needs to be prevented either by considering a minimum off/on time charging/discharging consecutive cycle or assigning load limits to each community.

In Case 4, the aggregator set a load limit of 1.7 MW. This limit increases the cost by 1% compared to Cases 2 and 3. This increase happens due to the absence of demand charges imposed on residential loads. Therefore, the benefits of storage would not be noticeable compared to commercial and industrial buildings that pay the demand charge. As shown in Fig. 21, the storage unit's optimal load schedule in Case 4 was changed compared to Cases 2 and 3. Load limit schedules the storage units' charging/discharging, and consequently prevents the community from exceeding that load limit. In this case, the charging/discharging availability of different types of storage units including community, residential, and PHEV needs to be prioritized through the introduction of different \$/kW rates. Therefore, each storage unit will be charged/discharged with different rates considering the load limit set by the aggregator.

Finally, in Case 5, the model was evaluated considering the unavailability of power. As seen in Fig. 21, CHEMs will modify the loads considering the availability of DERs (residents and community) to keep the critical loads (fixed) on.

4   Conclusion

Enabling the residents and community system operator (aggregator) to monitor and control their real-time energy consumption will facilitate the deployment of the micro energy market. Through this micro energy market, residents will be enabled to sell and buy the energy from other community stakeholders and help the aggregator in predicting day-ahead data. The aggregator may play distribution system operator using distributed CHEMs.

References

[1] G. Cena, et al., Hybrid wired/wireless networks for real-time communications, IEEE Ind. Electron. Mag. 2 (1) (2008) 8–20.

[2] eia.com. Household energy consumption and expenditures by end use and energy source, selected years, 1978–2005, U.S. Department of Energy, Washington, DC, USA.

[3] P. Wattles, ERCOT demand response overview and status report, AMIT-DSWG Workshop, AMI's Next Frontier: Demand Response, 2011.

[4] M.M. Ganji, Security-Constrained Unit Commitment Reserve Determination in Joint Energy and Ancillary Services Auction, Illinois Institute of Technology, Chicago, IL, 2012.

[5] G. Strbac, Demand side management: benefits and challenges, Energy Policy 36 (12) (2008) 4419–4426.

[6] K.-H. Ng, G.B. Sheble, Direct load control—a profit-based load management using linear programming, IEEE Trans. Power Syst. 13 (2) (1998) 688–694.

[7] D.-M. Han, J.-H. Lim, Design and implementation of smart home energy management systems based on Zigbee, IEEE Trans. Consum. Electron. 56 (3) (2010) 1417–1425.

[8] D. O'Neill, et al., Residential demand response using reinforcement learning, 2010 First IEEE International Conference on Smart Grid Communications (SmartGridComm), IEEE, 2010.

[9] J.-W. Lee, D.-H. Lee, Residential electricity load scheduling for multi-class appliances with time-of-use pricing, GLOBECOM Workshops (GC Wkshps), 2011 IEEE, 2011.

[10] M. Muratori, et al., A highly resolved modeling technique to simulate residential power demand, Appl. Energy 107 (2013) 465–473.

[11] M.A. Zehir, M. Bagriyanik, Demand side management potential of refrigerators with different energy classes, Universities Power Engineering Conference (UPEC), 2012 47th International, IEEE, 2012.

[12] M.C. Bozchalu, et al., Optimal operation of residential energy hubs in smart grids, IEEE Trans. Smart Grid 3 (4) (2012) 1755–1766.

[13] *http://learn.pjm.com*. PJM's day-ahead energy market.
[14] A. Khodaei, M. Shahidehpour, in: Optimal operation of a community-based microgrid, Innovative Smart Grid Technologies Asia (ISGT), 2011 IEEE PES, IEEE, 2011.
[15] B. Stephen, et al., Self-learning load characteristic models for smart appliances, IEEE Trans. Smart Grid 5 (5) (2014) 2432–2439.
[16] J.C. Fuller, et al., Modeling of GE Appliances in GridLAB-D: Peak Demand Reduction. Pacific Northwest National Laboratory, The United States, 2012, https://doi.org/10.2172/1047418.
[17] M.M. Ganji, Optimal Load Scheduling in Commercial and Residential Microgrids, Illinois Institute of Technology, 2015.

Part Three

South America

Case studies in saving electricity in Brazil

Yuri R. Rodrigues*, Paulo F. Ribeiro†
*University of British Columbia (UBC), Vancouver, BC, Canada, †Federal University of Itajuba (UNIFEI), Itajuba, Brazil

1 Introduction—Brazilian motivation

Smart Grid (SG) deployments are motivated by the reality in which utilities are embedded along with intrinsic values such as business culture, technological and process maturity, and the current marketplace as well as the socioeconomic and environmental scenario of its concession area [1,2]. Particularly in Brazil, given the continental territorial extension, there are extremely particular socioeconomic and environmental realities in each region that require different topics to be addressed in the SG projects.

Among the motivators, saving electricity is one of the most important aspects. High levels of nontechnical and technical losses, especially in urban areas, lead to 16% of the total energy produced in Brazil not being sold, accounting for about $5 billion in annual revenue loss due to theft, billing, and metering errors [3]. This scenario drives investments in SG projects focused in loss reduction by concessionaires such as Light [4,5] and Ampla [6]. For utilities with lower losses, the investments in SG become feasible through the economic benefits provided by greater operational efficiency related to the automation of internal processes, systems management, identification of failures, and reduction of operational expenses in preventive and corrective maintenance [7,8]. Another great stimulus factor is the Brazilian potential for distributed renewable generation (DG), especially solar and wind sources, inspiring several research and development (R&D) projects [9].

Therefore, the development of SG projects in Brazil is mainly motivated by features such as an increase in operational efficiency, efficient management of assets, energy saving, monitoring, network automation, reliability, power quality, loss reduction, and DGs penetration.

These projects are financed in large part by the mandatory investments in R&D and supported by an important mechanism for the acceleration of the development of SG in Brazil: the "Inova Energia" program. This program was designed to assist in the analysis and approval of projects related to SG, integrating funding sources under the responsibility of the Brazilian Electricity Regulatory Agency (ANEEL), the Funding Authority for Studies and Projects (FINEP), and the Brazilian Development Bank (BNDES) [1].

It is estimated that more than 200 SG projects are under development, focusing on initiatives in the generation, transmission, and distribution segments. These involve

Application of Smart Grid Technologies. https://doi.org/10.1016/B978-0-12-803128-5.00006-4

Table 1 **ABRADEE investment scenarios for the implementation of SGs in Brazil (in R$ billions) [9]**

Area	Accelerated	(%)	Moderate	(%)	Conservative	(%)
Measurement	45.6	50	35.4	58	28.8	62
IT measurement	0.5	1	0.5	1	0.4	1
Telecommunication measurement	13.6	15	10.9	18	9.2	20
Automation	2.1	2	1.8	3	1.1	2
IT automation	1.5	2	1.5	2	1.4	3
Telecommunication automation	5.9	6	5.6	9	5.2	11
Distributed generation IT/electric vehicles	0.2	0	0.2	0	0.1	0
Incentives distributed generation	21.7	24	5.3	9	0	0
Total	91.1	100	61.2	100	46.2	100

approximately 450 institutions, more than 300 suppliers, 126 research centers, and 60 public service companies as well as universities, ministries, and regulatory agencies that are focusing on the deployment and implementation of aspects such as intelligent metering, distributed generation, automation, storage, telecommunication, customers services, etc. [1].

The Brazilian Association of Electric Power Distribution Companies (ABRADEE), by means of a wide-ranging study concerning distributors, institutes of science and technology (ICTs) and universities with R&D resources of ANEEL, indicates that the total investment for the implementation of SGs in Brazil may vary from R$46 billion (Reais—Brazilian currency) in a conservative scenario to R$91 billion for the accelerated one [9], as shown in Table 1.

As can be seen in the measurement area, the accelerated scenario shows a significant increase in the volume of investments, especially when compared to the conservative case, presenting an increase of R$16.8 billion. However, it should be noted that, as the scenarios evolve from conservative to moderate and accelerated, the percentage of participation tends to decline because there will be more money to apply to other areas previously without incentives due to economic restrictions.

2 Smart Grid perspective in Brazil

A comprehensive survey of the Smart Grid projects financed by the R&D Program coordinated by ANEEL, the main instrument for national research in SG, was performed by the Center for Strategic Studies and Management Science, Technology, and Innovation (CGEE) [10]. It indicates a total of 178 projects with a total investment of approximately R$411.3 million. The major concentration of resources and number of projects occur in the Southeastern and South regions of Brazil. São Paulo and Santa Catarina are the most engaged federative units in the development of Smart Grid projects, leading the number of projects. The uneven distribution of projects is driven by socioeconomic and environmental aspects, especially urbanization, industrialization, and population density indexes.

Fig. 1 illustrates the regional distribution of Smart Grid projects cataloged by ANEEL. The described numbers refer to the total projects developed in specific states while the overall costs in R$ million per region are depicted in Fig. 2.

The main areas addressed by the Smart Grid projects developed in Brazil are represented by the following topics:

- Automated metering infrastructure, including intelligent metering systems, new models, and tests of new functionalities (AMI).
- Automation of the distribution and substations, including supervision systems of the electricity distribution networks (AD&S).
- Distributed Generation, microgeneration, and microgrids (DG).
- Distributed storage and battery systems (DS).
- Plug-in Hybrid Electric Vehicles, including charging and supervision systems (PHEV).

Fig. 1 Regional distribution of SG projects cataloged by ANEEL [10].

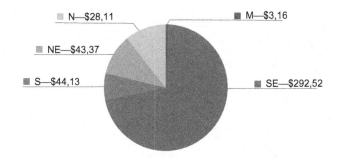

Fig. 2 Total costs of the SG projects by region, in R$ million [10].

- Telecommunications for Smart Grid systems (Telecom).
- Information Technologies for Smart Grid systems (IT).
- Intelligent buildings and residences, including consumer interaction (IB&R).
- New services for the final consumer—gas and water services measurement, security, communication services, and energy efficiency strategies (CSM);
- Others—relationship with clients, IP, cyber security, management of assets, etc. (others).

A perspective of the investment in R$ million per thematic area of the Smart Grid projects is shown in Fig. 3. The average value of the projects is R$1.89 million. The "Distributed generation" thematic is the one with the highest number of projects, the largest amount of investment, and higher than average project values; a segmentation of the investments per DG source type is featured in Fig. 4. Another thematic of

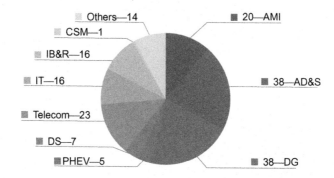

Fig. 3 Total amount of investment in R$ million per thematic area of Smart Grid [10].

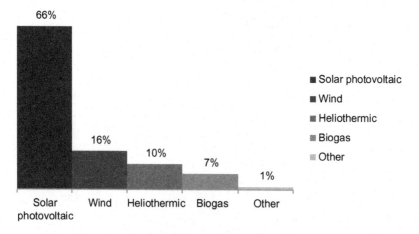

Fig. 4 Percentage of investment segmented per DG.
Based on ABDI, iAPTEL, Mapping of National Suppliers and their Products and Services for Intelligent Electrical Networks (REI): Executive Summary (Preliminary), June 2014 (in Portuguese).

great interest is the "Automation of the distribution and substations." It has the same number of projects but with about half the values invested. The themes with the largest granted values in a project are "Intelligent buildings and residences" and "Telecommunications" [10].

A summary table presenting the number of Smart Grid projects under development and the amount of investment associated per Brazilian region and thematics addressed is depicted in Table 2. One can see that the projects are focused on more immediate aspects such as smart metering and distributed generation while initiatives such as PHEVs, smart buildings, and new services are not widely covered, especially in the less urbanized middle and north regions. In addition, it is possible to note that the southeast region has the largest volume of projects, investments, and covered thematics, which is directly related to its socioeconomic reality and requirements.

3 Main Smart Grid projects in Brazil

In Brazil, Smart Grid projects are more concentrated at the distribution system level because this is the segment most affected by the changes arising from this new concept, especially by the enabling capability of new products and services for the final consumer.

In this section, the 11 main Smart Grid pilot projects in Brazil that comprise an estimated/realized investment of more than R$200 million will be discussed, evaluating the thematic developed, volume of investment, suppliers, and centers of research development and innovation (CRD&I) participation [1,11] as well as an offshore venture in the northeast region of the Archipelago of Fernando de Noronha. These projects and the responsible utilities are illustrated in Fig. 5.

3.1 Cities of the future

The pilot project Cities of the Future is an initiative of CEMIG distribution and one of the most comprehensive Brazilian programs for the implementation of Smart Grid architecture, due to the diversity of thematic and socioeconomic aspects involved. The main objective is the establishment of a functional reference model to support future large-scale deployment decisions. In this sense, technical, regulatory, financial, and customer perception aspects are equally evaluated, with emphasis on an extensive experimentation of several possible technologies. Methodologically, the project also accesses the impacts on CEMIG's business processes, the smart grid's value chain, and the training required for professionals who will deal with these new technologies.

The site selected is a great representative of the concession area of CEMIG, housing urban and rural residential, commercial, and industrial consumers, covering 8000 consumers in a region with 23,000 km of network consisting of two substations and eight feeders. The project started in 2010 with a contribution of R$25.3 million (ANEEL R&D resources). Its efforts are focused on the automation experimentation of substations and distribution networks, intelligent metering, operational communication networks, management systems, and saving electricity practices such as LED

Table 2 Summary table of the number of Smart Grid projects (NP) and invested values (R$ mi) per Brazilian region and thematic addressed [10]

Thematic	Brazilian regions										Total of projects	Total investment
	M		N		NE		S		SE			
	NP	Invest.	NP	Invest.	NP	Invest.	NP	Invest.	NP	Invest.		
AMI	1	1.145	2	4.391	5	3.854	5	2.889	7	16.763	20	29.042
AD&S			1	541	10	12.714	11	7.685	16	33.607	38	54.547
DG	1	787	1	318	6	9.631	12	13.164	18	76.708	38	100.608
DS							3	2.236	4	10.342	7	12.578
PHEV									5	9.838	5	9.838
Telecom					5	8.062	6	4.834	12	30.337	23	43.233
IT			1	1.065	3	4.322	5	2.524	7	19.903	16	27.814
IB&R					1	1.73	7	10.356	8	19.57	16	31.656
CSM									1	490	1	490
Others	1	1.232	1	21.793	3	3.057	1	441	8	74.964	14	101.487
Total	3	3.164	6	28.108	33	43.37	50	44.129	86	292.522	178	411.293

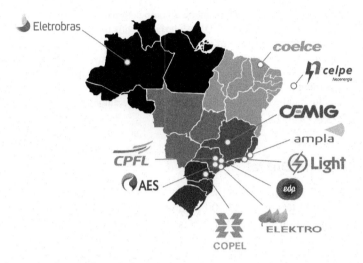

Fig. 5 Main Smart Grid pilot projects in Brazil.
Based on ABDI, iAPTEL, Mapping of National Suppliers and their Products and Services
for Intelligent Electrical Networks (REI): Executive Summary (Preliminary), June 2014
(in Portuguese).

lighting systems, management and integration of distributed generation, and the relationship with the consumer. The thematics addressed in this initiative are represented
in the following topics [7,12–14]:

- *Intelligent Measurement Systems*: There are 4200 smart meters installed in low voltage networks, including functions such as: power quality, notification of failures and instabilities,
disconnection and reconnection, energy balancing, and application of differentiated tariffs.
For last-mile communication, two strategies were adopted. The first set of smart meters
(about 3000 consumer units) was employed RF Mesh while the second group (about
1200 consumer units) was applied by PLC technology.
- *Automation of Distribution*: The project covers fault location, isolation and supply restoration (FLISR), Volt-VAr control, and temperature monitoring in distribution transformers.
The adopted FLISR solution allows remote reconfiguration of 30 points by the operator,
using the DNP3 protocol for communication with 30 automatic reclosers, five medium
voltage meters, and four fault sensors. The Volt-VAr control solution includes automation
of eight capacitor banks in the network to optimize voltage levels.
- *Distributed Generation*: The project covers two initiatives: the implementation of distributed solar microgeneration in low voltage consumer units in the city of Sete Lagoas and
the integration of the 3 MW photovoltaic solar plant.
- *Storage Systems*: Integration of an energy storage system of 100 kW with a solar photovoltaic plant of 50 kW.
- *Plug-in Hybrid Electric Vehicles*: Study of the impact of PHEVs on the electric grid by the
incorporation of another R&D project.
- *Telecommunications*: Implementation of a multiservice communications network to support
intelligent distribution and measurement automation applications. Use of a variety of technologies for the backhaul segment: radio, fiber optics, satellite, and GPRS.

- *Information Technology*: In the IT thematic a computational topology integrated with the Meter Data Management System (MDM) interfaces with the central systems to support the processes of operation and the implementation of tools for interaction with customers.
- *Others*: Public lighting—installation of LED lighting systems. Customer Relationship—development of tools for customer interaction such as websites, apps, and communication strategies according to the customer profile based on marketing research. Other initiatives include feasibility studies, process improvements, house automation, cyber security, data privacy, and SG planning.

3.2   Eletropaulo Digital

The Eletropaulo Digital project is the largest SG project in Brazil, including 60,000 customers translated to about 250,000 inhabitants impacted. With an investment of more than R$70 million, it intends to transform Barueri in the first municipality in the metropolitan region of Brazil, giving it an intelligent energy distribution network. Eletropaulo Digital is structured in a living lab concept of demonstration and deployment. The main objective is to implement infrastructures, applications, and SG functionalities that will lead to energy saving and customer satisfaction while surpassing market and strategic operational challenges.

The Barueri region is in frank expansion, possessing a diversified customer profile (residences, commerce, and industries) that provides a consistent sample of the AES Eletropaulo concession area. The site has high load density and urban complexity characteristic of a large metropolis, requiring several innovative technological solutions in measurement, automation, and telecommunication thematics, and specific metrics of evaluation to guide the concept expansion to the rest of the concession area's technological and strategic road map.

Associated with the Eletropaulo Digital SG initiative, an integrated project aiming to develop the energy metering and balance equipment to be installed in the Eletropaulo Digital project was also deployed by AES Eletropaulo. The project has an investment of about R$44 million, being based on previous R&D developments in the innovation chain and the pioneer lot of energy metering and balance systems. Another company, SG initiative, is held in the intelligent metering thematic in the Ipiranga region of São Paulo. The project proposes to employ remote measurement integrated with automation solutions for 2000 clients as well as energy balance technologies.

The Eletropaulo Digital project intends to deploy and develop the following SG thematics [15]:

- *Intelligent Measurement Systems*: Implantation of intelligent measurement solutions for 60,000 customers in the city of Barueri. These intelligent meters will be developed in association with selected manufacturers, including energy balancing capacity and last mile communication by PLC and radio-frequency technologies. This module aims to enable communication, alternatively or cooperatively, in a mesh network topology.
- *Distribution Automation*: The demonstration project foresees the installation of local fault detection solution in aerial distribution networks, able to send information about the event

to the operation center and other systems of interest. Self-healing solutions are also planned to be implemented to locate and isolate faults as well as the reestablishment of service through network reconfiguration. Other initiatives include Volt-VAr control in substations and generation units, and the development of a website and mobile app linked to the SCADA system for real-time operational information.

- *Distributed Generation*: A theoretical and comprehensive study of the impact of distributed generation on the utility distribution network is featured.
- *Plug-in Hybrid Electric Vehicles*: Preparations are taken to accommodate PHEV insertion. An associated project using PHEVs as taxi units is also presented.
- *Telecommunications*: The project uses a WiMAX network to perform the communication of distribution automation equipment and meter's last-mile integration.
- *Information Technology*: The utility is deploying a distribution management system (DMS), an outage management system (OMS), and mobile workforce management (MWM) software. Their combination allows the identification of an event's location by the analysis of the information received through the call center, optimizing the mobile workforce displacement and monitoring the service until its conclusion. The project also foresees the penetration of smart metering, executing preparations in data collection and measurement systems (MDC and MDM), billing, and hardware and software solutions for big data analytical applications.
- *Buildings and Intelligent Residences*: The project contemplates the creation of a Smart Home considering the concept of energy management and advanced measurement infrastructure (AMI), which allows bidirectional energy and information flows. It seeks a greater relationship and interaction between company and consumer as well as consumer engagement and dissemination of SG functionalities as demand response, residential microgeneration, and PHEV units' integration to intelligent devices.
- *New Services*: The Project foresees the study of new services and business models associated with SG functionalities to the end customer.
- *Others*: Customer interaction—development of tools for interaction with customers such as websites and mobile apps, based on the perception assessed through surveys. Multiutility—SG integration with other public services (water, gas, telecommunication utilities). Workforce training—development of training initiatives for technicians on SG concepts and technologies. KPIs—development of metrics for SG benefit evaluation. Road map—development of a road map for the expansion of the SG initiative through the utility concession area.

3.3 Smart Grid Light

The Smart Grid Light Program is a set of SG research and development (R&D) projects with new automation and measurement technologies applied from distribution networks to customers' homes. The initiative includes the implementation of intelligent meters with load-shedding and switch functions in approximately 400,000 consumer units, the automation of 1700 underground chambers, 1200 reclosing switches, and the implementation of a communication network and computational systems to support the applications [4].

It is composed of five related projects addressing the topics of intelligent energy metering, digital certification, new tariffs, products and services (including prepayment), distributed generation insertion, island mode operation, distribution network automation, energy saving policies such as residential automation (i.e., intelligent outlets), demand side management, channels for customer interaction, and a recharging systems for PHEVs. These products are under development to extend the concept of SG into the company's concession region. The Smart Grid Light Program initiative includes the following SG themes [4,5]:

- *Intelligent Measurement Systems*: Two types of measuring equipment were developed: a meter with a built-in reader for installation in residences and a remotely readable meter for installation of equipment on poles. The equipment perform real-time consumption monitoring, provides suggestions for energy saving, and schedules events in the electricity grid. The communication interface adopted is the ZigbeePRO open standard, with the expectation that about 2000 m will be installed by 2023.
- *Automation of Distribution*: The project aimed to develop real-time network diagnoses with autoreconfiguration functions using proprietary protocols with the implementation of 1200 automatic reclosers and 1700 underground transformer chambers.
- *Distributed Generation*: Installation of photovoltaic panels and energy storage systems in 10 consumer units.
- *Storage Systems*: Integration of an energy storage system with photovoltaic panels.
- *Plug-in Hybrid Electric Vehicles*: Development of intelligent recharging stations for PHEVs.
- *Telecommunications*: Standards with redundancy in concentrators, protocols of transport based on TCP/IP, and traditional communication technologies that must be applied according to the specificities of each region were adopted. Among the options considered are Wi-Fi Mesh, GSM/GPRS, and Satellite.
- *Information Technology*: Provide synergy through the interoperability between the standards, allowing the various products present in the project to interact efficiently and at reduced cost.
- *Buildings and Smart Homes*: Development of smart outlets that can be remotely controlled by customer interaction tools.
- *Others*: Innovative channels of interaction—in addition to the remote displays of smart metering, some consumers will be able to monitor consumption through other communication media such as cell phones, TVs, and apps providing friendly and dynamic interfaces through analytical charts and advanced reports.

3.4 Parintins Project

The Parintins Project is a Smart Grid initiative developed in the interior of the state of Amazonas by Eletrobras. Similar to the Cities of the Future project, it seeks the establishment of a functional Smart Grid project to work as a reference model for energy distribution companies.

The initiative planned to perform demand peak shaving by stimulating consumption outside peak demand periods through the application of differentiated rates during the day. For this sake, all electric power meters of Group B (residential and commercial consumers) were replaced; Group A (large customers connected at high voltage) were not altered. The project also aimed at the evaluation of distribution automation applications along with the measurement and monitoring of transformers.

The following smart grid functionalities were addressed [16]:

- *Intelligent Measurement Systems*: The project provided the deployment of 14,500 smart meter units with the measurement of electrical variables, mass memory, remote connection/disconnection and communication. Because the project seeks the establishment of an SG reference model for the Eletrobras group of concessionaires, different communication technologies were tested (Point-Multipoint 450 MHz, Zigbee 2.4 GHz, MESH 900 MHz, and PLC) with the city divided into four large areas. The project discontinued the first two technologies because of economic and technical aspects. Another topic addressed is the installation of metering systems in distribution transformers for energy balance, fault identification, and determination of quality indicators. These devices are expected to be installed in about 300 transformers with DigMesh, Zigbee, and Satellite communication methods.
- *Automation of Distribution*: The project accounts for the installation of 16 automatic reclosers.
- *Distributed Generation*: The project provides for the installation of a 120 kWp photovoltaic system, equivalent to 40 consumers, with remote connection and disconnection control to monitor the impacts of distributed generation on the grid.
- *Telecommunications*: The last-mile data are transported by the Internet, with its expected migration to the Backhaul radio network. To transport data to the measurement center, a satellite link is employed.
- *Information Technology*: Development of a Measurement Center with the implementation of the Measurement Data Collection System (MDC) and MDM.
- *Buildings and Smart Homes*: It was planned to install a consumption management system, particularly air conditioning control. This initiative was abandoned due to the incompatibility of the equipment installed and clients' technologies.
- *Others*: Public Lighting—installation of LED technology to evaluate its feasibility in the replacement of conventional lighting. Customer Relationship—development of a website and mobile application for energy bill monitoring.

3.5 Búzios Intelligent City

The Ampla utility demonstration project Búzios Intelligent City is held in a coastal tourist region of Brazil. It contemplates a substation with four feeders supplying 10,000 consumer units. The project aimed at learning about the operation, infrastructure, costs, socioenvironmental, and service quality impacts as well as distribution automation, renewable generation, electric mobility, public lighting, energy storage, smart buildings, and citizen awareness.

The initiative of the Búzios Intelligent City project contemplates the following Smart Grid thematic [6]:

- *Intelligent Measurement Systems*: Installation of 10,000 smart meters, with last-mile communication by GPRS, PLC, and RF MESH.
- *Automation of Distribution*: Use of 17 motorized switches distributed in the four feeders.
- *Distributed Generation*: Installation of nine solar panels of 5 kWp, four wind turbines with vertical axis of 2 kW, and one with horizontal axis of 1 kW.
- *Storage Systems*: The installation of a 200 kW storage system for two-stage testing is proposed: an internally controlled network and the Ampla network.

- *Plug-in Hybrid Electric Vehicles*: This considers the participation of 4 PHEVs, 40 electric bicycles, and a small electric boat.
- *Telecommunications*: To transmit the data from last mile-networks, a optical fiber link is used. The automation data is transmitted by point-multipoint radio communication up to the substation level and then by a optical fiber link to the Operation Center.
- *Information Technology*: Implementation of automation system and medium voltage network telecontrol.
- *Buildings and Intelligent Residences*: Intelligent building pilot with the provision of power and energy control platform for 20 consumers.
- *Others*: Public Lighting—installation of 130 LED luminaries, of which 40 have intensity control and an experimental hourly rate for low voltage.

3.6 Fernando de Noronha Archipelago Smart Grid project

The Archipelago Fernando de Noronha Smart Grid project developed by CELPE is held in the Island of Fernando de Noronha, an ecological reserve with several environmental restrictions. The objective of the project is to develop and implement an SG concept including technological resources for network automation, telecommunication, measurement, distributed microgeneration, energy quality, recharge of PHEVs, and differentiated tariffs. A particularity that must be highlighted of this SG initiative is the evaluation of the thematic viability under the sustainability aspect.

The SG thematic developed at Fernando de Noronha Archipelago initiative are discussed as follows [17,18]:

- *Intelligent Measurement Systems*: Installation of intelligent metering in 831 consumer units of Group B (residential and commercial consumers) featuring measurement in four quadrants, remote disconnection and reconnection, energy balance, load profile acquisition, and identification of failures. The adopted solution uses an open communication protocol, multivendor interoperability. and PLC for last-mile communication.
- *Distribution Automation*: Deployment of decentralized automation solution in three feeders through the installation of reclosers and fault locators.
- *Distributed Generation*: Installation of solar and wind distributed generation.
- *Storage Systems*: Association of storage system to a microgeneration.
- *Plug-in Hybrid Electric Vehicles*: Implantation of PHEV recharging service by renewable generation disconnected from the grid.
- *Telecommunications*: A hybrid network using optical fiber and radio communications for automation and intelligent metering, and WiMAX and RF MESH technologies for unlicensed frequencies.
- *Others*: Public Lighting—control and remote management system for LED and induction lamps. Submeasurement—assessment of submeasurement technology and identification of the active consumer practices.

3.7 InovCity project

The InovCity project is a partnership between EDP Bandeirante and the Portuguese intelligent city of Évora, an associated of the EDP Group. The project is located at the city of Aparecida do Norte—SP aiming to analyze the feasibility of a set of energy-saving technologies that will provide greater efficiency and quality to customer services. It works as a living lab for the testing of intelligent measurement, the vehicle recharging process, microgeneration, energy efficiency, network automation, telecommunications, customers' interaction, and regulation measures such as dynamic tariffs, prepayment, micro and mini tariffs, and residential automation practices. A second phase, the InovCity II project, takes place at the federative unit of Espirito Santo being held by ESCELSA, a utility of the EDP Group. It seeks an improvement in customer services such as smart metering, efficient public lighting, microgeneration with renewable energy sources, electric transportation, and energy efficiency actions. The following Smart Grid themes are covered [19,20]:

- *Intelligent Measurement Systems*: EDP Bandeirante in partnership with Ecil is committed to the development of smart metering equipment scheduled to be deployed in 5 years and three phases. Phase 1, initial deployments. Phase 2, expansion of the initially deployed and integration of the web portal and new tariff options. Phase 3 will address actions related to energy storage and energy saving. It is projected that the project will include the installation of 15,300 intelligent electronic meters at low voltage, (2000 single-phase, 12,000 biphase, and 1300 three-phase). Approximately 120 concentrators equipped with ZigBee technology should centralize customer measurements.
- *Distribution Automation*: Network automation with remotely controlled reclosers.
- *Distributed Generation*: Development of solutions that allow customers with microgeneration to expand their benefits.
- *Plug-in Hybrid Electric Vehicles*: Promotions and incentives for the use of PHEVs associated with the installation of recharging stations.
- *Telecommunications*: The topology implemented employs a combination of GPRS and the WiMAX network.
- *Information Technology*: A management system responsible for telemetry collection, solution management, and presenting a component's status of operation is featured.
- *Others*: Public Lighting—use of LEDs and energy management systems. Consumer Interaction—interface with other R&D projects aimed at evaluating consumer reactions to new technologies such as energy management systems. Differential rates—test and research with consumers on different tariffs including prepayment, dynamic tariffs, and tariff flags.

3.8 CPFL Smart Grid

The CPFL Group is focused in the development of automation projects in its feeders and substations in order to reduce the displacement of teams and the promotion of faster supply restoration. Also of interest is smart measurement for remote data collection of the energy consumed by distribution customers of Group A (large customers connected at high voltage) and Group B (residential and commercial consumers), along with distributed generation, PHEVs and workforce management. The main objective of this project is to enable remote operation by the company's distribution systems, such as automatic reconfigurations by the displacement of telecontrolled switching devices.

The initiative addresses the development of the following Smart Grid thematic [8]:

- *Intelligent Measurement Systems*: A proof-of-concept phase proposes the installation of 25,000 intelligent meters of Group A and Group B meters connected by an RF Mesh communication network.
- *Distribution Automation*: It seeks the standardization of the feeder and substation automated systems. These systems will operate in synchronization with mobility teams to increase the efficiency of occurrence attendance by the company as well as allowing faster and more efficient maneuvers to restore supply. This includes the installation of 5000 telecontrolled switches enabling the network to be divided into 250,000 microgrid schemes with automatic reconfiguration.
- *Distributed Generation*: Implementation of a solar power plant of 1.6 GWh/year with investment provided by ANEEL R&D.
- *Plug-in Hybrid Electric Vehicles*: Several PHEV initiatives through R&D projects are featured.
- *Telecommunications*: A hybrid communication network with proprietary RF MESH and public GPRS network is adopted.
- *Information Technology*: Implementation of MDC, MDM, and workforce management systems.
- *Others*: Workforce management—allows service orders to be issued electronically in order to dispatch the best positioned team to perform network services.

3.9 Aquiraz Smart City

The COELCE demonstration project Aquiraz Smart City seeks the automation of the local electric system. The site selection was influenced by the existence of an automated substation, the network structure, and load displacement. In the context of smart metering, a pilot project was deployed in Fortaleza with the installation of 100 intelligent meters and data concentrators at transformers.

The following Smart Grid themes are addressed [21]:

- *Distribution Automation*: All reclosers installed in the region are automated.
- *Telecommunications*: Radio communication network deployed and tested.
- *Information Technology*: Implantation of SCADA system associated with an automatic recomposition system and automatic protection settings.

3.10 Paraná Smart Grid pilot

The particularity of this project developed by COPEL is the link to a state program. The Smart Energy Paraná program was defined by the State Decree n. 8842/2013 [22], including the participation of a governmental entity along with the water and gas companies Compagas and Sanepar for an integrated measurement cooperation. The project was establishment in an area of high density of load and visibility, with the following Smart Grid functionalities under development [23]:

- *Intelligent Measurement Systems*: The project foresees the integrated measurement of water and gas in two implementation phases. The first phase has approximately 2000 m including 487 gas and 64 water meters. This is followed by the completion of installation of approximately 10,000 m.

- *Distribution Automation*: The pilot includes remote-controlled switches for automatic reconfiguration and a self-healing strategy with four switches and two reclosers installed. Other subthemes include fault sensing, DNP3 application, Volt-VAr control, and network reconfiguration.
- *Distributed Generation*: Installation of solar panels with a capacity of 1.4 kW.
- *Storage Systems*: Testing of a storage system in association with a solar photovoltaic generation.
- *Plug-in Hybrid Electric Vehicles*: On an experimental basis, a PHEV taxi was developed in an R&D project involving the utilities COPEL and Itaipu Binacional, Institute of Technology for Development, Fiat, and other private companies. This project employs the recharging station located at Curitiba International Airport to perform studies of the impact of PHEVs and battery use on the electric power system.
- *Telecommunications*: Optical fibers were used for communication between the company's headquarters, field switches, and metering concentrators. Also, GPRS technology for Group A consumers and multipoint RF technologies and MESH for Group B consumers reading was employed.
- *Information Technology*: Implementation of a network recomposition system and cybersecurity studies were considered.
- *Buildings and Intelligent Residences*: Test of the house of the future including automation, microgeneration, energy storage, and electric mobility.

3.11 Elektro Smart Grid project

The objective of the pilot is to deploy and test SG technologies in the touristic city of São Luiz do Paraitinga in order to evaluate the main impacts on the technical-operational processes and changes in the patterns of consumption implied by the new services and products enabled. At the end, it is expected that this will achieve better energy savings by the rational use of energy and energy efficiency policies [24].

In addition to this Smart Grid demonstration project, Elektro has several R&D projects related to SG in the areas of distribution automation, distributed generation, PHEVs, and customer interaction.

3.12 Summary of the 11 Smart Grid projects

A summary of the thematics covered in the main Smart Grid projects developed in Brazil and discussed previously is shown in Table 3. Further, a comprehensive perspective of these projects is illustrated in Table 4.

The presented table indicates for each project the responsible utility, the main thematic covered, the total investment, the involved centers for research development and innovation (CRD&I), and the suppliers as well as partnerships and related projects. In association with this wide-ranging perspective of Smart Grid deployment in Brazil, another important aspect is related to the level of maturity of these utilities. Hereafter, a classification into four groups of utilities based in terms of SG operations is presented in Fig. 6. Where:

- Beginner—utilities starting SG activities working up to three thematics.
- Researcher—concessionaires working in more than three SG thematics but that do not have demonstration projects.

Table 3 Summary table of the thematic covered in each project

Thematic	Cities of the future	Paraná Smart Grid	CPFL Smart Grid	Eletropaulo Digital	Búzios Intelligent City	Smart Grid Light	Parintins	Fernando de Noronha SG Project	Inovcity	Aquiraz Smart City	Elektro Smart Grid project
Intelligent Measurement Systems (AMI)	x	x	x	x	x	x	x	x	x		
Automation of Distribution (AD&S)	x	x	x	x	x	x	x	x	x	x	x
Distributed Generation (DG)	x	x	x	x	x	x	x	x	x		x
Storage Systems (DS)	x	x	x		x	x		x			
Plug-in Hybrid Electric Vehicles (PHEV)	x	x	x	x	x	x		x	x		x
Telecommunications (Telecom)	x	x	x	x	x	x	x	x	x	x	x
Information Technology (IT)	x	x	x	x	x	x	x	x	x	x	
Buildings and Intelligent Residences (IB&R)		x			x	x	x				
New Services (CSM)		x		x					x		
Others	x	x	x	x	x	x	x	x	x		x

Table 4 Comprehensive perspective of the main Smart Grid projects developed in Brazil [25]

Project name	Utility	Thematic	Investment (R$ mi)	CRD&I	Technology solutions suppliers	International partners	Related international projects
Cities of the Future	CEMIG	AMI, AD&S, DG, DS, PHEV, Telecom, IT, Others	215	CPqD, FITec, FAPEMIG, UNIFEI	Landis+Gyr, Cemig Telecom, Ativas Data Center, Axxiom, Concert, Solaria (Spain)	N/A	
Paraná Smart Grid	COPEL	AMI, AD&S, DG, DS, PHEV, Telecom, IT, IB&R, CSM, Others	350	UTFPR, UFPR, PUC-PR	Arteche, Cooper, ABB, Siemens, Exalt, Ecyl, Lupa, FIAT, GE, Itron, Landis+Gyr. Chantex, CP Eletrônica, Emera, Smartgreen	N/A	
CPFL Smart Grid	CPFL	AMI, AD&S, DG, DS, PHEV, Telecom, IT, Others	215	(Note: without CPD&I's)	IBM, Silver Spring		
Eletropaulo Digital	AES Eletropaulo	AMI, AD&S, DG, PHEV, Telecom, IT, IB&R, CSM, Others	89 (18 R&D Eletropaulo+72 ANEEL)	FITec, USP, (ENERQ), CPqD	Synapsis Brasil Ltda.	AES Grupo AES Brasil Corporation -	

Continued

Table 4 Continued

Project name	Utility	Thematic	Investment (R\$ mi)	CRD&I	Technology solutions suppliers	International partners	Related international projects
Búzios Intelligent City	AMPLA	AMI, AD&S, DG, DS, PHEV, Telecom, IT, IB&R, Others	41 (18 R&D ANEEL+22 AMPLA)	COPPETEC, UFF, UFRJ, LACTEC, UERJ	Landis+Gyr	ENDESA (Spain), ENEL (Italy)	Smart City Málaga, Smart City Barcelona, Smart City Santiago
Smart Grid Light	LIGHT, CEMIG	AMI, AD&S, DG, DS, PHEV, Telecom, IT, IB&R, Others	35	CPqD, Lactec, Inmetro and Universities	Axxiom Soluções Tecnológicas, CAS Tecnologia S.A.		
Parintins	ELETROBRÁS	AMI, AD&S, DG, Telecom, IT, IB&R, Others	21	CPqD, UFF, UFMA, CETEL (UFRN)	Elo	N/A	
Fernando de Noronha Archipelago Smart Grid project	CELPE	AMI, AD&S, DG, DS, PHEV, Telecom, IT, Others	18	CPqD, UFPE, UPE (POLI)	RAD Data Communications (international supplier)	IBERDROLA	Projects STAR and BIDELEK (Cities of Castellón and region of Bilbao, Portugalete e Lea-Artibai)

				USP, FUSP	ECIL Energia	EDP	Évora InovCity
Inovcity	EDP BANDEIRANTE	AMI, AD&S, DG, PHEV, Telecom, IT, CSM, Others	10			ENDESA (Spain), ENEL (Italy) IBERDROLA (Spain)	Only DA parcel (self-healing) Projects STAR and BIDELEK (Cities of Castellón and region of Bilbao, Portugalete e Lea-Artibai)
Aquiraz Smart City	COELCE	AD&S, Telecom, IT	1.66	UFC, IFCE, UNIFOR,	Synapsis Brasil Ltda.		
Elektro Smart Grid project	ELEKTRO	AD&S, DG, PHEV, Telecom, CSM	18	UNESP, USP São Carlos, PUC Rio, FITec	Siemens, Beckhoff, Elipse, Advantech		

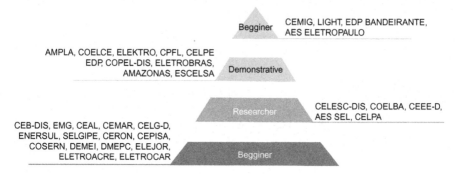

Fig. 6 Smart Grid maturity level of distribution utilities operating in Brazil [1].

- Demonstration—concessionaires that have demonstration projects in progress integrating most of the SG areas of operation.
- Pioneer—utilities with demonstration projects completed or in progress and that are planning to continue SG initiatives, including road maps and new areas for deployment and/ or R&D projects.

The segmented classification emulates a natural flow where beginners start to research, researchers implement demonstration projects to validate their propositions, and, if motivation is found, a smart grid becomes a natural business strategy that is continued by pioneers [1]. In this sense, acceleration mechanisms such as the "Inova Energia" program are essential for a faster transition of utilities between these segments.

4 Centers for research development and innovation (CRD&I)

The main developments of Smart Grid initiatives in Brazil have been performed by R&D projects financed by the ANEEL R&D program resources. According to the Brazilian Industrial Development Agency (ABDI), these projects involve 450 institutions in the execution of activities ranging from basic research aimed at demonstration projects and equipment development in various stages of the innovation chain as pioneer and market insertion [26]. Among these institutions, there are 126 entities that can be characterized as centers of research, development and innovation (CRD&I). These entities are nonprofit organizations comprised of institutes, foundations, associations, and private and public academic entities of the federal, state, and municipal level [1] that work autonomously and in partnerships with specialized consulting institutions in the development of R&D projects.

Their participation has been determinant in the evolution of the application of the Smart Grid concept in Brazil because, among the 11 main SG demonstrative projects in the country, 10 have the participation of at least one national CRD&I [26]. Among the Brazilian research institutions, at least 80 are involved in R&D activities in SG, accounting for about 64% of the country's CRD&Is.

Their fundraising portfolio contemplates government resources at the federal, state, and municipal level, that is, the Financier of Studies and Projects (Finep) ANEEL R&D program and the National Council for Scientific and Technological Development (CNPq) as well as indirect financing through foundations, investment banks, tax benefits, and investment from other nations [25,26].

CRD&I actuation covers all research areas of SG, with an emphasis on Photovoltaic Distributed Generation (DG) and Distribution Automation, which have approximately 56% and 48% of the country's entities acting in these research lines. Instead, important research lines such as intelligent measurement, intelligent buildings, consumer services, and storage systems have incipient deployment with indexes below 25% [26,27].

In Fig. 7, a summary of the CRD&I participation index in Smart Grid research lines is presented. A picture of the number of CRD&Is involved in the main SG thematic developed in Brazilian projects is shown in Fig. 8.

In order to assess CRD&I significance in the development of the SG concept in Brazil, an investment-raising classification based on their annual revenues is presented in Table 5. It is possible to observe that there are a reasonable number of large and medium-sized research centers with a significant amount of investment raised to the development and deployment of SG thematics.

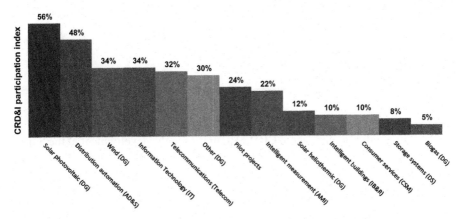

Fig. 7 Percentage of CRD&Is acting on the lines of research in SG.
Based on ABDI, iAPTEL, Mapping of National Suppliers and their Products and Services for Intelligent Electrical Networks (REI): Executive Summary (Preliminary), June 2014 (in Portuguese).

Fig. 8 Number of CRD&Is working on
SG projects.
Based on ABDI, iAPTEL, Mapping of
National Suppliers and their Products and
Services for Intelligent Electrical
Networks (REI): Executive Summary
(Preliminary), June 2014 (in Portuguese).

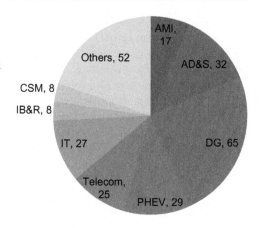

Table 5 CRD&I classification by size and annual revenues [26]

Specialized research centers in SG (CRD&I)	Annual revenues (in R$ mi)	Size (BNDES)
Center for Research and Development in Telecommunications (CPqD)	300	Large
Electrical Energy Research Center (CEPEL)	210	Large
Institute Eldorado	103	Large
Institute of Technology of Paraná (TECPAR)	100	Large
Institute of Technology for Development (LACTEC)	71	Average
Foundation Centers of Reference in Innovative Technologies (CERTI)	60	Average
Cesar Institute	60	Average
Foundation for Technological Innovations (FITec)	50	Average
Atlantic Institute	25	Average
Renato Archer Information Technology Center (CTI)	12.5	Small

To contextualize the impact of CRD&I activities in the development of the Smart
Grid concept in Brazil, a summary of the nationalization index of some of the main
Smart Grid thematics and their related products is presented in Table 6. One may
note that AMI, AD&S MSD, and Residential & Commercial management solutions
are significantly developed by national suppliers and CRD&I institutions. For generation and transmission automation (AG&T) and prosumer access network (PAN)
thematics, a medium rate of nationalization is featured, with an expected accelerated
increase in development while DG initiatives are in the beginning of the nationalization process.

Table 6 **Smart Grid products nationalization indexes.**

Supply chain	Products	Nationalization index (Brazil)
Advanced Measurement Infrastructure (AMI)	Group A Smart Meter Group B Intelligent Meter Intelligent Border Meter and Free Customers Meter Communication Device Router/Data Hub (network) Meter Communication Gateway (MDC) Communication Equipment (GPRS, Mesh, PLC, Satellite, Radio, Fiber, WiMax, etc.)	High
Automation of Distribution and Substation (AD&S)	Field Recloser/Feeder Circuit Breaker Capacitor Bank Voltage Regulator Voltage-Current Sensors Field Transformer Power Transformer (substation) Volt/VAr Control Disconnect Switch Automatic Switch PMU Device (Synchrophasor) Field Communication Equipment (GPRS, Mesh, PLC, Satellite, Radio, Fiber, WiMax, 3G, LTE, etc.) Substation Communication Equipment (routers, switches, media converters, etc.) Devices/Substation Elements (IED, sensors, IR cameras, etc.)	High
Business/ Commercial Management (software)	Power Management System (PMS) SCADA Distribution Management System (DMS) Energy Interruption Management System (EIM) Geographic Information System (GIS) Mobile Service Dispatch System (MSD) Network Asset Management System (NAM) Distributed Energy Resources Management System (DERM) Wide Area Monitoring and Control (WAMC)/Synchrophasor Management System Meter Data Management System (MDM) Applications (commercial losses, technical losses, etc.)	High

Continued

Table 6 **Continued**

Supply chain	Products	Nationalization index (Brazil)
	Analytical Data (mining, processing, display/dashboard)	
	Information Management System of the Subscriber (IMS)	
	Billing	
Mobile Service Dispatch (MSD)	Mobile Communication Device	High
	Tablet/Notebook	
	MSD Management Software installed on the Mobile Device	
Prosumer Access Network (PAN)	Residential Energy Management (HEMS)	Medium
	Commercial/Building Energy Management (CEMS)	
	Industrial Energy Management (IEMS)	
	Smart Gateway/Gateway	
	Smart Display	
	Smart Thermostat	
	Energy Services Device (ESI) (interface, router, gateway)	
	Power Management Applications by Smartphone, Tablets, etc.	
	Submeter	
	Smart Charges	
	Home Appliances	
	Demand Response Control Devices (interface, communication, etc.)	
	Connected Business Devices (elevators, motors, etc.)	
	Industrial Displays (Engines, etc.)	
	Prossumidor Communication Equipment (Mesh, PLC, WiFi, etc.)	
Generation and Transmission Automation (AG&T)	PMU (device)/PDC (collector)/ Synchrophasor	Medium
	Sensors	
	High Voltage Direct Current Systems (HVDC)	
	Flexible AC Transmission Systems (FACTS)	
Distributed Generation (DG)	Residential Grid-Tie Inverter (DG)—250 W–10 kW	Low
	Commercial Grid-Tie Inverter (DG)—up to 10 kW (up to 1 MW)	
	Industrial Grid-Tie Inverter (DG)—up to 50–100 kW (up to 1 MW)	
	Intelligent Inverter for Electric Vehicle (PHEV)	
	Smart Battery	
	DC Combination Box (DG)	

Table 6 **Continued**

Supply chain	Products	Nationalization index (Brazil)
	Smart Meter Net Metering Intelligent Load Controller (DG, DS, IPHEV) Communication Devices (DG, DS, IPHEV) Network Controller (DG, DS, IPHEV) Communication Devices (DG, DS, IPHEV) Intelligent Load Control Devices/Demand Response (DR)	

Based on ABDI, iAPTEL, Mapping of National Suppliers and their Products and Services for Intelligent Electrical Networks (REI): Executive Summary (Preliminary), June 2014 (in Portuguese).

5 Smart Grid roadmap—Brazilian case

In recent years, the SG concept has been implemented in several scales and initiatives in Brazil, with much of the hardware already tested and approved, albeit in isolation. In this context, the adoption of a framework or unifying architecture of interfaces and protocols based on norms and standards is essential. Thus, a reference architecture for the exchange of information between devices and electrical systems must be defined to allow all existing products, services, protocols, and interfaces to relate with one another. This framework, commonly defined as a road map, is of great interest to the Brazilian institutions and government [1,28].

In this perspective, the Brazilian Government by the figure of the Ministry of Mines and Energy (MME) instituted a working group to analyze and identify necessary actions to subsidize the establishment of public policies for the implementation of the "Brazilian Program of Intelligent Electrical Network" [28]. The program mainly addressed a proposal for the adequacy of regulations and general rules for public electricity distribution services, the identification of sources of funds for financing, and incentives for the deployment of national equipment and new possible prosumers-consumers that are able to generate their own consumption and sell their surplus.

Associated with this initiative, the Brazilian Industrial Development Agency (ABDI) articulated the creation of a government working group to structure the base for the strategic development of the SG supply industry in Brazil. It seeks to evaluate the sector as well as discuss and promote strategic actions presenting a proposal for the creation of the "Brazilian Program for the Development of the Intelligent Electrical Network Supply Industry." The actions proposed in this program for the development of the SG and industry are described by the strategic areas and themes described in Table 7.

At the end of 2013, the first proposal of the Referential Architecture of Intelligent Electrical Networks of Brazil (REI-BR-2030) was created by AGX Energia in partnership with ABDI. This is the first work to define the "Brazilian Road Map"

Table 7 Strategic areas of the Brazilian road map for the development of the SG and supply industry [28]

Strategic area	Themes
Management	Mapping/Articulation of the productive chain.
	Development of an observatory.
	Evaluation of new business models.
Financing and Taxation	Creation of financing line (Inova Energia).
	Special Taxation Regime.
	Expansion of the national capital offer for Smart Grid/City.
New Opportunities	Promotion of startups.
	Incentives for design projects.
Incentives for National Production and Technologies	Development of national software integrators.
	Definition of PPB for smart meters.
	Strengthen existing companies and solutions.
Legislation and Regulation	Adequacy of legislation and regulation.
	Definition of the Brazilian model for standards.
	Interoperability standards and certification.
	Criteria for cyber security.
Research, Development, and Innovation	Develop and strengthening of CRD&Is.
	Encouraging the deployment of ICT products.
	Encouraging the creation of demonstration cities with innovative technologies.
Internationalization	Accomplishment of international insertion and training missions.
	Creation of certification for Brazilian products.
	International collaboration.
Promotion and training	Development of technical-vocational courses.
	Development of postgraduate courses.
	Training of labor abroad.
	Teacher training.
	Program for the concept dissemination.
Infrastructure	Definition of criteria for sharing of the electric network and the telecommunication network.
	Integration of actions with PNBL and Digital Cities.

containing all the main concepts and elements of Smart Grids customized for the Brazilian case, in a complete inclusive, and agnostic way to technological implementations [28].

The REI-BR-2030 road map has been validated in interactions with SG industry specialists, the academy, utilities, and government entities. It addresses in a logical and simplified way the basic block diagrams constituting the Brazilian Smart Grid. The Brazilian Smart Grid Reference Architecture that presents a schematic summary of the proposed arrangement, the relationships between the agents, and the main thematics and technologies to be implemented is shown in Fig. 9.

Brazilian Smart Grid: Reference Architecture

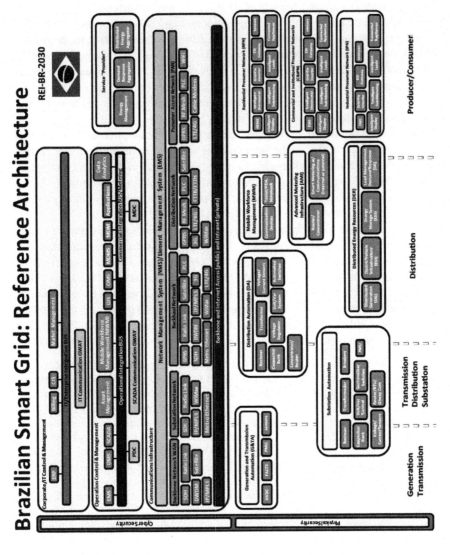

Fig. 9 Brazilian Smart Grid Reference Architecture—REI-BR-2030 [28].

6 Lessons learned, diagnostics, and barriers

Smart Grid implementation imposes a series of significant changes to the model of the electric sector. The current model based on great investments in large infrastructure has been limited in the face of the new challenges, and the boost of technological evolution, digitalization, and connectivity of all the elements that surround life [1]. One of the most emblematic features of the changes promoted by SGs is the mass insertion of distributed generation, where the sector moves from a captive market model with a single supplier to a market with multiple sources of energy and a customer's actuation as prosumers. Given these characteristics, it is clear the need for changes in the sector way of making business, regulation, politics, and industry because the current model does not seem to accommodate the new characteristics presented by this new technological model of the electric sector and society [29].

This scenario is not Brazil's privilege. The barriers to the adoption and evolution of SG have also been identified in the international scenario, for example in the United Kingdom, where technical and regulatory conditions did not encourage its adoption. The United States is still seeking regulatory improvements to accommodate issues such as market restructuring, renewable insertion, net metering policies, tariff decoupling, and the provision of new services [1].

In Brazil, the Smart Grid domestic market is still incipient when compared to its potential. Among the 11 main SG projects, only one presents the concessionaire's own investments as the main financing source. All others have, as a common characteristic, most financing resources through the R&D Program regulated by ANEEL [26]. The incipience of this market is justified by the existence of barriers to the adoption of these technologies, which also impacts the evolution of the SG industry in the country.

In these terms, a diagnosis was made on the current barriers to the adoption of SG in Brazil in the context of industry, regulation, technology, and science [1–9,29–32].

- *R&D resource limitation*: Given the resource constraints scenario in Brazil's economy, it is important that the new initiatives are articulated to maximize the available instruments, including innovations that have significant limitations of funding for R&D activities.
- *Deployment of the regulatory model*: The evolution of SG adoption is directly related to the regulatory improvements toward a new model for the sector. The main problems include the uncertainty about the recognition of investments in telecommunication networks and IT, which in SGs is inherently and account for 21%–36% of SG project expenditures. Another aspect is the regulation on "Other Revenues" of the concessionaire, limiting the services and compelling the benefits to be shared with the society through tariff modality.
- *Low market maturity*: Most of the resources spent with SG come from research support instruments. The industry presents difficulties in positioning itself in relation to investments in SG, avoiding the realization of relevant investments and the insertion of new products and technologies.
- *Industrial policies*: There is a lack of clear signals about the creation of an SG market and the difficulty of competition between national companies and large multinational players that already have solutions with great maturity.
- *Difficulty of synergy with the telecommunications sector*: The distributors have opted in their private projects to invest in their own communication networks. However, these companies

Table 8 **Strengths and weaknesses of Smart Grids in Brazil [9]**

Strengths	Weaknesses
• Pilots in progress, some with the provision of national technology. • Existing innovation policies and funding sources (Aneel R&D, BNDES, Finep) and institutional support. • Existence of companies with national technology and CRD&Is with technical qualification for surpassing SG implementation challenges. • Demand to create specific solutions due to the particularities of the Brazilian market.	• Company's size with national technology inferior to their world peers. • Absence of integrated solution providers. • Low articulation in R&D and introduction of innovations in the market. • Regulatory mismatch on the cost of the smart grid. • High cost of production and local R&D. • Lack of national agenda: late entering into the SG market.

do not have the right or authorization to operate communication services. In this way, the implanted infrastructure is underutilized, either from the point of view of installation of antennas or the use of frequencies.

• *Lack of synergy with the other utilities and public services*: The underutilization of telecommunication networks dedicated to SG can be enlivened with partnerships with other utilities (water and gas) interested in performing remote measurements. These companies have difficulty implementing these systems because their infrastructure is less than that of energy companies.

• *Qualified labor force*: The creation of more complex solutions entails the need for higher qualified labor. Electricians may have to deal with more complex knowledge, such as the configuration of a firmware, because they will be required to install digital equipment.

• *Certification and standardization process*: The approval process for equipment poses some important difficulties for the adoption of SG and for the development of the industry. Three obstacles have been identified in this respect: difficulty in following the creation of ordinances in relation to the regulation of the electricity sector, long periods for homologation of meters, and radio frequency equipment and standardization.

• *Taxes, financing, and incentives*: The last barrier identified refers to taxes, financing, and incentives for the development of technologies in the country. In the pioneer batch phase, the production cost is very high, mainly in terms of the lack of scale gain. This favors production in other countries, notably China. Possible discounts at different stages of the production chain could help the development of national solutions.

At last, a summarized diagnosis of the Brazilian strengths and weakness in the development of Smart Grids is presented in Table 8. Also, in order to contextualize the opportunities and threats that this concept faces on its implementation, a comparative analysis of possible opportunities and barriers based on the lessons learned is shown in Table 9.

Table 9 **Opportunities and threats of Smart Grids in Brazil [9]**

Opportunities	Barriers
• Development and local production of integrated circuits, reduction of the impact on the CRD&Is trade balance and training. • Strengthening of local players. • Export of typical solutions to emerging markets (high losses, low consumption, etc.). • Potential use of the state's purchasing power (regulation of Law 12.349/10) and regulatory instruments. • Specific tax regime with emphasis on local technological content. • Smart metering: potential integration with other utilities (telecommunication, gas, water).	• Attention to the main suppliers worldwide, who are buying companies with national technology (risks of denationalization), caused by the size of the Brazilian market for SG. • Implantation of SG with low/very low penetration of technology developed in the country and without technology trade. • Possibility of increasing the technological and business gap in Brazil due to the priority treatment of developed countries.

7   Conclusions

Brazil's incipient framework for the implementation of SGs is expected by, among other things, the continental physical dimensions, institutional policies, and complexities inferred into the management of the electricity grid. However, paraphrasing CS Lewis, "If things can improve, this means that there must be some absolute standard of good above and outside the process to which that process can approximate. There is no sense in talking of 'becoming better' or ´smarter´ if better or smarter means simply 'what we are becoming'—it is like congratulating yourself on reaching your destination and defining destination as 'the place you have reached.'" This perspective defines a Smart Grid's framework as a coordinated effort not in assuming a better or smarter electrical system, but in the nature of the development, which must be performed by a continuous and integrated effort of government, industry, utilities, and research/academic institutions in establishing dynamic standards of good and acceptable enactments within all dimensions.

It is also necessary to recognize that in designing and establishing a framework for complex technological systems such as a smart grid, the phases of "analysis" and "creative design" are not successive steps but strongly interwoven stages [29]. The unfolding of Smart Grids must happen within the dynamics presented in Fig. 10, in which the development is reached by a continuous improved development.

Therefore, neither personal prejudices nor vested interests can permanently keep in favor a model of the grid that cannot be technically and economically justifiable. The grid of the future, call it: Smart, Smarter, Intelligent, Flexible, Modern, Integrated, Virtual, etc., will triumph over the traditional one. If it does, it will be not because the current model is a failure, but only if it proves to be a better one. As well as in the future, when the Smart Grid model gets discarded, the same principle will be in operation. The post-Smart Grid model will depend on the technical developments along with human will and preferences.

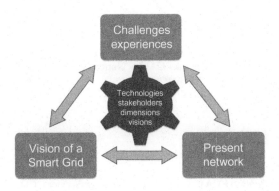

Fig. 10 Continuous learning dynamics applied to Smart Grids.

Some aspects still need more debate and improvements such as stimulating mass migration to SGs, addressing regulatory and technological issues (standardization, interoperability, etc.), and consumer responsiveness as well as new market models to foster investments. Although Brazil has already executed some initial steps in the process of Smart Grid implementation and energy saving goals, comprehensive demonstration projects were/are under development for different scenarios (metropolitan regions, environmental restriction, difficult access, high load density, diversity of customers, etc.). Alternatively, initiatives for the national deployment of SG innovative technologies and product nationalization through CRD&Is and industry partnerships must continue.

An overall picture of Smart Grid deployment in Brazil presenting the main thematic areas of development with the number of utilities, CRD&Is, and industry suppliers involved in its development and improvement are illustrated in Fig. 11.

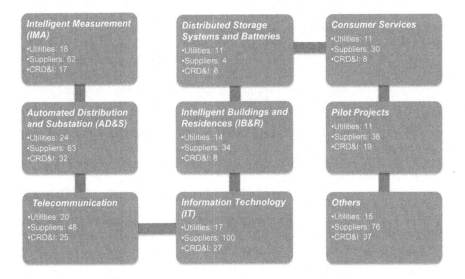

Fig. 11 Overall picture of the Smart Grid deployment in Brazil [26].

Acknowledgments

The authors thank to CAPES, FAPEMIG, CNPq, and INERGE for partially supporting this work. The first author, Yuri R. Rodrigues, especially thanks CAPES Notice No. 18/2016 of the Full Doctoral Program Abroad/ Process n° 88881.128399/2016-01.

References

[1] J.R.C. MCTI, Intelligent Electric Networks Brazil-European Union Sector Dialogue, November 2014 (in Portuguese).

[2] Japan International Cooperation Agency (JICA), Tokyo Electric Power Services CO. (TEPSCO), Study on Collection of Basic Information Concerning Smart Grid/Smart Community Introduction in the Federative Republic of Brazil, 2014.

[3] Bloomberg, Energy Smart Technologies—Digital Energy—Research Note, New Energy Finance, 2012.

[4] R.J. Riella, Experiences in Distributed Generation and Smart Grids, LACTEC Institute, 2013 (in Portuguese).

[5] D. Ribera, Revenue Recovery Project, Smart Grid MME, January 2014 (in Portuguese).

[6] W. Correia, Advances of the Búzios Intelligent City project, 6 Latin America Smart Grid Forum, São Paulo, November, 2013 (in Portuguese).

[7] A. Diniz, Cities of the Future Project Cemig Sete Lagoas, IWSGCom, Juiz de Fora, MG, 2013.

[8] P. Bombassaro, Smart Grids and the interfaces between energy and telecommunications, LETS—Infrastructure Week, São Paulo, May, 2014 (in Portuguese).

[9] R. Rivera, A.S. Esposito, I. Teixeira, Smart Grids: opportunity for local productive and technological clustering, BNDES J. 40 2013 43–84 (in Portuguese).

[10] CGEE, Intelligent Power Grids: National Context, Center for Strategic Studies and Management Science, Technology and Innovation, 2012 (in Portuguese).

[11] C.Q. Pica, D. Vieira and G. Dettogni, An overview of Smart Grids in brazil opportunities, needs and pilot initiatives, In: International Conference on Smart Grids, Green Communications and IT Energy-Aware Technologies, Venice, May 2011.

[12] D.S. Guimarães, in: Intelligent energy networks cemig distribution achievements 2011–2012, IV Smart Grid International Seminar, Campinas, May 2012 (in Portuguese).

[13] P.L. Cavalcante, J.F. Franco, M.J. Rider, A.V. Garcia, L.L. Martins, M.R.R. Malveira, R. J. Franco, P.F.S. Carvalho, D.S. Guimarães, L.J. Hernandes Junior, Advanced network reconfiguration system applied to CEMIG-D system, IEEE PES Conference on Innovative Smart Grid Technologies Latin America, São Paulo, April 2013.

[14] T. Batista, Project Cities of the Future Convention D423—Cemig D, Minas Gerais Renewable Energy Forum, Belo Horizonte, June 2014 (in Portuguese).

[15] P. Pimentel, in: AES Eletropaulo Smart Grid program, V International Smart Grid Seminar, São Paulo, May 2013 (in Portuguese).

[16] J.C. Medeiros, Eletrobras Amazonas Energia Parintins Project, Smart Grid MME, January 2014 (in Portuguese).

[17] N. Mincov, L. Zanatta, J.A. Lopes, M. Lôbo, in: Communication technologies for differentiated Smart Grid environments, IEEE PES Conference on Innovative Smart Grid Technologies Latin America, São Paulo, April 2013.

[18] J.A. Lopes, in: Development and implementations of intelligent network concept tests in a pilot location with high environmental restrictions—case Island of Fernando de Noronha Paraná Smart Energy, International Conference on Intelligent Energy, Curitiba, May 2014 (in Portuguese).

[19] J.B. Martins, in: InovCity project, 6° Latin America Smart Grid Forum, São Paulo, November 2013 (in Portuguese).

[20] J. Marcondes, Projetos InovCity EDP, Maio, Rio de Janeiro, 2014 (in Portuguese).

[21] COELCE, Implementation of a Smart Grid Pilot for Electric Systems Automation, Smart Grid MME, January 2014 (in Portuguese).

[22] BRAZIL, State Decree n° 8.842, from September 24th of 2013. On the Creation of the Smart Energy Project Paraná and Other Measures, Official Gazette Curitiba, PR, 26th September 2013 (in Portuguese).

[23] J.S. Omori, Projetos Pilotos de Redes Inteligentes, Smart Grid MME, January 2014 (in Portuguese).

[24] Elektro, PD—0385-0062/2013—SMART CITY—Reference Model for the Implementation of Intelligent Electrical Networks (Smart Grid), http://www2.elektro.com.br/html_pd_13/projetos/0385-0062_2013.html. Accessed April 2017 (in Portuguese).

[25] ABDI, iAPTEL, International Mapping of the Intelligent Electrical Network Product and Service Chain: Executive Summary (Preliminary), ABDI, December 2014 (in Portuguese).

[26] ABDI, iAPTEL, Mapping of National Suppliers and their Products and Services for Intelligent Electrical Networks (REI): Executive Summary (Preliminary), June 2014 (in Portuguese).

[27] A.L. Ferreira Filho, J.J. Coura, P.F. Correia, in: The development of Smart Grid in Brazil and the role of universities, IEEE 13th Brazilian Power Electronics Conference and 1st Southern Power Electronics Conference (COBEP/SPEC), Brazil, 2015.

[28] ABDI, iAPTEL, Technical Standards, Standards and Regulations Applied to the Intelligent Electrical Network (REI) Product and Service Chain: Executive Summary (Preliminary), ABDI, October 2014 (in Portuguese).

[29] S. Suryanarayanan, P.F. Ribeiro, M.G. Simões, Grid modernization efforts in the USA and Brazil—some common lessons based on the Smart Grid initiative, IEEE PES General Meeting, 2015.

[30] F.C. Maia, Intelligent Electric Networks in Brazil: Subsidies for a Deployment Plan, Synergia, Rio de Janeiro, 2013 (in Portuguese).

[31] N. Kagan, Intelligent Electric Networks in Brazil: Cost and Benefit Analysis of a National Implementation Plan, Synergia, Rio de Janeiro, 2013 (in Portuguese).

[32] ANEEL, Tariff Regulation Procedures. Module 2—Periodic Tariff Review of Electricity Distribution Concessionaires, Submodule 2.7—Other Revenues, ANEEL, 2016 (in Portuguese).

Further reading

[33] P.F. Ribeiro, H. Polinder, M.J. Verkerk, Planning and designing Smart Grids: philosophical considerations, IEEE Technol. Soc. Mag. 31 (3) 2012 34–43.

Part Four

Europe

Automation for smart grids in Europe

7

Davide Della Giustina*, Ferdinanda Ponci†, Sami Repo‡
*Unareti SpA, Brescia, Italy, †RWTH Aachen University, Aachen, Germany
‡Tampere University of Technology, Tampere, Finland

1 Introduction

1.1 Distribution system operators' challenges and needs in the EU

Distribution system operators (DSOs) operate in a scenario defined by a national and international regulatory framework that, on one hand, vigorously promotes and rewards customers who install distributed generation (DG) and, on the other hand, pressures the DSO's operating margin. In most European countries these operating margins are contingent upon regulated prices (a price cap), which tend to decrease over time. They are also subjected to remuneration or penalization mechanisms related to the number and the length of outages for both medium voltage (MV) and low voltage (LV) customers. This growing complexity, brought about by DG, makes it increasingly difficult to meet those quality of service standards. Therefore the only sustainable way for DSOs to cope with this trend and keep their economic model in balance is to enhance the technological level of the grid and the efficiency of their operations, moving toward the smart grid [1].

1.1.1 Regulations about service continuity

The power distribution grid is a dependable asset of the modern society. For this reason the DSO—the body in charge of its management—is constantly assessed to determine the quality of the service delivered to its customers. In the scope of the quality of service domain, one of the most relevant criteria is the continuity of service, which relates to voltage interruption. A voltage interruption is defined as the decrease in the voltage supply level to $<10\%$ of nominal for up to 1 min duration [2]. Many national authorities set output-based incentive schemes aimed at improving the continuity of service, based on the measurement of key performance indicators (KPI) related to voltage interruptions. Among them, two of the most commonly used are:

- the System Average Interruption Frequency Index (SAIFI), the average number of interruptions that a customer would experience.
- the System Average Interruption Duration Index (SAIDI), the average interruption duration for each customer served, defined according to the following formulas:

Application of Smart Grid Technologies. https://doi.org/10.1016/B978-0-12-803128-5.00007-6

$$\text{SAIFI} = \frac{\sum_{i=1}^{n} U_i}{U_{tot}}, \quad \text{SAIDI} = \frac{\sum_{i=1}^{n} \sum_{j=1}^{m} U_{i,j} t_{i,j}}{U_{tot}},$$

where n is the number of interruptions in a year; m is the number of customer clusters affected by the same duration of interruption; U_i is the number of customers involved in the ith interruption; $U_{i,j}$ is the number of customers involved in the ith interruption, belonging to the jth cluster; $t_{i,j}$ is the duration of the interruption for the cluster $U_{i,j}$; and U_{tot} is the number of customers at the end of the year.

Voltage interruptions are classified according to their duration t in: transient interruption, when the event lasts <1 s ($t \leq 1$ s); short interruption, when the event duration is between 1 s and 3 min (1 s $< t \leq 180$ s); and long interruption, when the event last more than 3 min ($t > 180$ s) [2].

As an example, in Italy [3] an output-based incentive scheme aimed at improving the continuity of service has been set based on those indicators: each year, a DSO receives a bonus or pays a penalty if the SAIFI—when evaluated for short and long interruptions only—and the SAIDI—when evaluated for long interruptions only—exceed a specific threshold value, varying with the territorial density of customers. At present, the regulatory framework does not foresee any incentive/penalty mechanism for transient interruptions.

1.1.2 Regulations about voltage quality

Besides the continuity of service, the quality of service is characterized by the level of power quality (PQ). The growing presence of distributed energy resources (DERs) is one of the most impactful causes affecting the quality of the energy provided by DSOs. In Europe, this topic is monitored by the Council of European Energy Regulators (CEER). In [4], the working group on the quality of electricity supply of the CEER provides a comparison among quality levels, standards, and regulation strategies for electricity supply in European countries. The benchmark about the quality of the electricity supply in 2011 [5] reports that in several EU countries (among them the Czech Republic, Hungary, Norway, Slovenia, Italy, Austria, Lithuania, The Netherlands, Portugal, etc.), the DSO is required to perform PQ measurements. Different phenomena are monitored in different countries; however, the EN 50160 [2] is used as a reference in most cases. Among the many PQ events, the short-duration RMS variations—also called voltage dips—are the most important.

As an example, in Italy the national authority mandates that DSOs monitor the PQ on MV bus bars in PS-using devices compliant with IEC 61000-4-30 [6], and periodically prepare reports about the number and the origin of voltage dips measured on the MV network.

1.2 Smart grid automation demos in Europe

The European Commission plays a major role toward the realization of the smart grid in Europe. The current progress toward the Energy Union is laid out in the 2nd Report on the State of the Energy Union 1.2.2017 [7] and it indicates positive trends in the direction of achieving the 2030 energy and environmental goals.

The research, development, and demonstration program to accelerate the evolution of the energy sector is laid out in the European Electricity Grid Initiative (EEGI) [8] road map, the Strategic Energy Technologies Plan (SET-Plan), and, eventually, assessed in Transforming the European Energy System Through INNOVATION—Integrated SET Plan—Progress in 2016 [9].

The latest legislative proposals to the implementation of the Energy Union have been released in the "Clean Energy For All Europeans," known as the Winter Package 2016 [10], pointing to a system where the automation of the electricity system in the direction of the smart grid is clearly unavoidable. The research, development, and demonstration projects funded by the EU Commission point in this direction. Noticeably, the role of demonstrations is evident in the relevant portion of the H2020 funding program: Demonstrations involve smart grids, storage, and system integration technologies for distribution systems with increasing share of renewables.

An overview of the status of European smart grid projects, from 2004 to 2014 from all funding sources, has been compiled by the Joint Research Center of the EU Commission in [11], for projects up to 2011 and in [12] or a comprehensive overview up to 2014. The new outlook report is expected to be published in Summer 2017.

Data show an increasing investment, with 221 ongoing projects totaling 2 billion € (with an average of 9 million € per project) and 238 completed projects totaling 1.15 billion € (with an average of 5 million € per project).

Furthermore, research and development projects show a total budget of about €830 million whereas demonstration and deployment projects total about 2320 million €.

In the context of EU Commission funding, the reference broad topic for smart grids is the H2020-EU.3.3.4—A single, smart European electricity grid. In this context, the most relevant program is the societal challenge of competitive low-carbon energy, in particular the research initiative "LCE-02-2016—Demonstration of smart grid, storage and system integration technologies with increasing share of renewables: distribution system." Here the targeted technology readiness levels typically range between 5 and 8. It is required that the proposed solutions are demonstrated in large-scale pilots and validated in real-life conditions.

In the FP7 program, the antecedent of H2020 and whose projects ended in 2016, the project Grid4EU Large-Scale Demonstration of Advanced Smart Grid Solutions with wide Replication and Scalability Potential for EUROPE [13] stands out for its 2011–16 €54 million cost, six DSO demonstrators, and 51 month total testing period in six European countries. Most of the demonstrations focused or were based on automation of the MV and LV distribution system.

A success story in terms of development and demonstration of smart grid automation is the IDE4L Ideal Grid for All project, rewarded with the excellence stamp of the EU Commission. Some results of the project are included in the current chapter. Development and demonstrations of the smart grid can be found also among the IDE4l cooperation projects: INCREASE, which focuses mainly on increasing the penetration of renewable energy sources into the grid; EvolvDSO, which has defined new roles of the DSOs and also encompasses planning, operational scheduling, and maintenance; DREAM, which builds and demonstrates an industry-quality reference solution for DER aggregation level control and coordination based on available ICT components.

The new research program H2020 is bringing a new, demonstration-oriented thrust in the smart grid domain, in particular through the low-carbon energy calls LCE-07 ("Innovation Initiatives") and LCE-02 ("Demonstration of smart grid, storage and system integration technologies with increasing share of renewables: distribution system.").

In project INTERFLEX, five DSOs and a total of 20 industrial partners, including utilities, manufacturers, and research centers, are involved in the project, which has a budget of €23 million. Partners will run six demonstrations for 12–24 months with the goal of validating the role of DSOs in calling for flexibility. Demonstrations run numerous use cases of energy storage technologies (electricity, heat, cold), demand response schemes with two network couplings (electricity and gas, electricity and heat/cold), the integration of electric vehicle owners, and advanced automation and microgrids. The use cases address the enhancement of the distribution network flexibility itself, demonstrate the role of IT solutions to speed up automation of distribution for local single or aggregated flexibilities, and combine automation and an increased level of aggregation to validate the plausibility of local flexibility markets. INTERFLEX [14] runs from 2017 to 2019, on a €22 million budget.

Project FLEXICIENCY [15] targets the accessibility of metering data, close to real time, made available by DSOs in a standardized and nondiscriminatory way to all the players in electricity retail markets (e.g., electricity retailers, aggregators, ESCOs, and end consumers). The aim is to facilitate the emergence of new markets for energy services, enhancing competitiveness and encouraging the entry of new players, which benefits the customers. Economic models of these new services will be proposed and assessed. Based on the five demonstrations, while connecting with parallel projects funded at EU or national levels, a total of about 16.000 customers involved. The project runs from 2015 to 2019 on a €19 million budget.

The project UPGRID [16] develops and demonstrates new solutions for advanced operation and exploitation of LV/MV networks. The context is the smart grid environment with increased capacity for hosting DERs, involving the customers and creating new business opportunities. The demonstration is primarily carried out in four field pilots in Portugal, Poland, Spain, and Sweden. They cover monitoring and operation, advanced grid maintenance, awareness of the consumers, and their cost-effective participation. The project runs from 2015 to 2017 on a budget of €15 million.

Project GOFLEX [17] aims to enable the flexibility of the automatic trading for the capabilities of the virtual power plant and virtual storage. Data services for local estimation and prediction support data-driven decision making. GoFlex will start with three European demonstration sites involving >400 prosumers from industry, buildings, and transport—in Cyprus, Germany and Switzerland 2016–19 on a budget of €11 million.

For an overview on regulation, refer to CEER Status Review on European Regulatory Approaches Enabling Smart Grids Solutions ("Smart Regulation," C13-EQS-57-04 February 18, 2014). This status review represents a cross-section of smart grid demonstration projects in Europe. It also contains a summary of the status (and links to detailed reports) of the smart grid demonstration projects, including funding and assessment.

Projects that are closer to commercial implementation are funded in the program FTIPilot-01-2016 [18]—Fast Track to Innovation Pilot. Among these, ADMS (Active Distribution Management Systems) develops and demonstrates on the premises of DSOs new low-cost solutions for monitoring and automation based on innovative sensing technologies, data-driven monitoring functions, and cloud implementations.

Finally, the large, typically cross-border Projects of Common Interest (PCI) [19] support the field implementation of new technologies. In particular, currently active PCI in the smart grid area are the following:

The North Atlantic Green Zone Project (Ireland, United Kingdom/Northern Ireland) aims at lowering wind curtailment by implementing communication infrastructure, enhanced grid control, and interconnection and establishing (cross-border) protocols for demand-side management. The commissioning is expected in 2018. The underlying technology is that of intelligent distribution networks, overlaid with high speed communications.

Green-Me (France, Italy) aims at enhancing RES integration by implementing automation, control, and monitoring systems in HV and HV/MV substations, including communication with the renewable generators and storage in primary substations as well as new data exchanges for a better cross-border interconnection management. The enabling technologies are load and generation forecasting and improved communication between TSO (Transmission System Operator) and DSO (Distribution System Operator) automation systems. As a result, the hosting capacity of further RES will grow while quality and system reliability will be guaranteed. The realization of this connection is under consideration for 2025.

SINCRO.GRID (Slovenia/Croatia) aims at solving network voltage, frequency control, and congestion issues via a virtual cross-border control center involving two neighboring TSOs and the two neighboring DSOs. The dedicated IT infrastructure and software will support algorithms for Volt-Var Control (VVC) optimization, secondary reserve, managing battery storage, advanced real-time operation of the grid with advanced forecasting tools, and using dynamic thermal rating. The feasibility stage is expected to be completed in 2021.

1.3   IDE4L at a glance

The IDE4L project was focused on the investigation and development of new solutions and architectures to plan, manage, and control future distribution networks. In detail, the project covered the integration of DERs (e.g., DG, electric vehicles) in distribution networks in order to reduce emissions, save energy, reduce network losses, improve network monitoring and controllability, utilize existing networks more efficiently, and improve visibility of DERs to transmission system operators and aggregators. The project also proposed innovative smart solutions to design and operate distribution networks in order to guarantee the continuity and quality of electricity supply.

The conceptual development of the active distribution network and automation architecture as well as the network management have been the basis of the whole project. The main goal was to create a comprehensive concept and map for DSOs to start deploying and developing the active distribution network, based on existing technologies and solutions, to meet future requirements of distribution grids. The project has addressed some major challenges of future electricity distribution networks, which will be much more complex than today's passive networks. The hosting capacity of passive networks can be increased by huge investments, but by applying the IDE4L approach and concept, the active network management will dramatically reduce these investments.

Another aspect taken into consideration in the IDE4L project has been the coordination of network and energy market operations together with customers, which requires extended analysis of electricity systems at all levels. Possible conflict of DERs participating in the electricity market and distribution network technical constraints can be solved using algorithms and functionalities developed within IDE4L through the smart active network management.

A smart automation infrastructure has been designed and realized in order to monitor LV and MV grids, to mitigate the fault effects in the distribution network based on innovative fault location, isolation, and supply restoration (FLISR) logics. State estimation and load-production forecasting solutions have been implemented as well in order to provide correct inputs to a power control system that is able to regulate the voltage in the network feeders by controlling several resources, such as controllable photovoltaic (PV) inverters or transformer online tap-changers.

2 Architecture

2.1 DSO control hierarchy

Fig. 1 illustrates the hierarchy of the controllers and their interactions at the conceptual level. The control system has three hierarchical levels (primary, secondary, and tertiary controllers) that interact with each other by providing monitoring and estimation data to an upward direction and by sending optimized reference values to downward direction. Note that that figure does not include the monitoring part of the complete automation system. The primary controller operates continuously while the secondary controller may update reference values periodically (5 min in the demonstration), and the tertiary controller operates primarily in day-ahead basis, but may participate in real-time operation in case of emergency (after a fault and required supply restoration or while a secondary controller requests help). Secondary and tertiary controllers are responsible for controlling voltage within control areas (background color codes of squares in the figure). A typical selection of control areas would be according to voltage levels, that is, one secondary controller for each LV grid where control is needed or enabled; one secondary controller for the MV grid below a primary substation; and a tertiary controller for the whole MV grid including multiple primary substations. The color codes of lines indicate which part of the complete system has been utilized

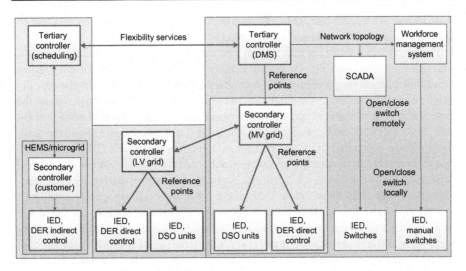

Fig. 1 Hierarchy of controllers (shaded areas indicate control areas, portions involved in the demonstration listed in the text).

in the demonstrations: Demonstration of Secondary Controller (LV grid), IED DER direct control and IED DSO Units in Unareti; Demonstration of Tertiary Controler (scheduling), Tertiary controller (DMS), Secondary Controller (MV grid), IED DSO Units, IED DER direct control in integration laboratory extension for the field Demonstration.

Primary controllers (intelligent electronic devices, IEDs) operate autonomously with fast response based on local measurements only, and a secondary controller dedicated to that control area may adjust the set points remotely. The information that control area primary controllers belong to is maintained in the grid model needed for the computation of state estimation and the optimal power flow of secondary and tertiary controllers. Status changes of automated circuit breakers and switches is immediately known at the secondary controller level as well as at the tertiary control level, but the operation of manually controlled switches must be updated from distribution management system (DMS). DMS interacts with the workforce management system to send and receive work orders and to update the status information of switches to/from field crews.

Secondary controller coordinates (optimizes) the operation of primary controllers (IEDs) within a control area based on state estimation results of the control area. Secondary controllers are located at primary or secondary substation automation units (SAUs), depending on which network, MV, or LV grid they are managing. A tertiary controller (DMS) manages the whole distribution grid. It is located in DMS and it communicates with commercial aggregators via a market operator to validate and to request flexibility services (scheduled reprofilling (SRP) and conditional reprofiling (CRP) products to indirectly control DERs via microgrids or home energy management systems (HEMS) [20]), and it adjusts network topology remotely via

SCADA and workforce management systems. It may also advise secondary controllers from system and day-ahead viewpoints.

The secondary control for distribution grid congestion management optimizes the settings of primary controllers to enhance the hosting capacity for DERs and to solve occasional congestion problems. The secondary controller most benefits the cases in which congestion occurs occasionally, network reinforcement costs are high, and the control area includes several primary controllers such as the automatic voltage regulator of the DG unit and/or production curtailment (DER direct control) and the automatic voltage controller (AVC) of the on-load tap changer (OLTC) (DSO unit). DSO should naturally have a control contract or obligation based on grid code to control DER units. In a weak distribution grid, the hosting capacity provided by the secondary controller is typically three to four times higher than the hosting capacity of the passive network, and two to three times higher than the hosting capacity of the active network utilizing only primary controllers [21–23]. A secondary controller is running in real time and it is based on optimal power flow to define set points for primary controllers to minimize operational costs such as grid losses within the control area, production curtailment, demand response, tap change action, and voltage deviations. As an input, it needs the state estimation results of the control area. Details and thorough explanations of the secondary and tertiary controllers and the operation of all of their functionalities are given in [24].

SAU collects real-time monitoring data from smart meters (SMs) and IEDs as described earlier. In principle, two clients are used for that purpose: IEC 61850 MMS and DLMS clients. Because client poll data from a single meter or IED at time, the process of going through all units takes some time and secondly the latest measurements received from units are not from the same time instant. These practical issues have impacts on the performance of state estimation and voltage control. These issues may be partly removed by enhancing the communication applied, but the choice of communication is always a compromise between technical performance and overall costs. After receiving data, clients will write it into a local database of SAU. State estimation and secondary voltage control are run periodically, utilizing available data on that time in the database. After finalizing the computation, algorithms write results to the local database to be used by other algorithms as inputs and clients as changes of the primary controller reference points.

Both the state estimation and the secondary voltage control (optimal power flow) require an exact distribution grid model. SAU maintains information about grid topology and grid parameters in the database. The data model of grid presentation is based on the common information model (CIM) [25]. The CIM data model is utilized to present grid topology by status of switches, grid parameters such as impedances, component types and control characteristics, and customer/prosumer data like location, contracted capacity, annual energy, consumption/production load profiles, etc. Real-time measurements, historian values, and calculation results of state estimation, forecasting, and secondary voltage control are modeled utilizing the data model of IEC 61850. Therefore, a mapping between these two standard-based data models has been created [26,27].

The update of the distribution grid model in SAU is realized in a few steps. The most frequent updates are received from IEDs and SMs that belong to the control area of SAU. These data packages include, in addition to real-time measurements of electrical quantities, the status of a device or customer (grid connected or not) and parameters of primary controllers (setting of reference value, control mode, delays, etc.). Therefore, if a switch or distribution grid has an IED read by SAU, the topology of the grid model in SAU may be updated immediately after opening or closing a switch. If a supply restoration is utilized in the grid operation, then the grid model in the SAU should also include grid elements of possible backup connections. SCADA, DMS, and the workforce management system in DSO's control center are managing the status of all switches. The update of these switches could be realized by CIM difference model exchange from DMS to SAU. The CIM difference model should be exchanged immediately after a topology change in the control area of SAU has been updated in DMS. In a similar way, grid and customer parameters require updating from time to time due to grid extension. These changes are less frequent and are always well planned, and therefore a batch update of the grid model in SAU is enough. Scalability of such a solution requires, however, full automation of the information exchange from DMS to SAU. These two updates of difference models have not been implemented in the demonstration because DMS with necessary CIM interface was not available.

2.2 Commercial aggregator control hierarchy and interaction with DSO

Fig. 2 does not completely represent the IT and automation systems of market operator and commercial aggregators, which were not in the focus of the IDE4L project although information exchange between these stakeholders was considered. Fig. 2 represents the control hierarchy of the commercial aggregator, which has a similar three-level structure as DSO had in his IT and automation system. The tertiary controller (scheduling) optimizes the utilization of aggregated flexibility resources in different energy and flexibility markets such as day-ahead, intraday, intrahour, TSO's control and reserve markets, and DSO's local flexibility market. Typically, scheduling is realized in an energy management system (EMS) of the commercial aggregator control center.

The secondary controller (customer) optimizes and manages the local flexibility resources of the customer, based on price or volume signals coming from the commercial aggregator and the real-time monitoring data of the flexibility resources. The customer might be an energy community (virtual microgrid), microgrid (local physical microgrid), building/commercial service/industrial facility, or a single end-customer such as a household. The secondary controller in these cases might be, for example, a microgrid central controller, a building/factory/home EMS located at a customer site, or a cloud service depending on what kind of functional and nonfunctional requirements the IT and automation systems should perform.

The IED (DER indirect control) is the primary controller of the flexibility resource. The control is indirect from DSO's perspective because the IT and automation system of the commercial aggregator is making the control decision instead of DSO although DSO requests a control action via a flexibility service marketplace. The IED is, for

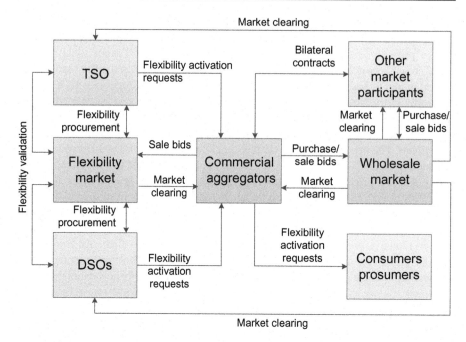

Fig. 2 Interactions of electricity and flexibility market stakeholders.

example, a thermostat of the heating/cooling device, which has advanced functionalities such as the capability to remotely adjust the reference point or other parameters of the primary controller, or has advanced functionalities such as the frequency or voltage-dependent control characteristics [28,29]. Many grid codes already mandate frequency and voltage-dependent control for DG units. The draft harmonized grid code of ENTSO-E would extend that kind of requirements down to an 800 W unit size [30].

The control of the customer-owned DERs may also be realized by direct control (e.g., grid code or bilateral contract), but the flexibility of DERs may be utilized only by a DSO. Direct control is deterministic for a DSO and rather straightforward to realize compared to indirect control of the local flexibility market. Nonfirm connection contract between a customer and DSO is an example of bilateral contract, where DSO has the chance to curtail production or shift controllable loads to off-peak periods. Grid code might be used to mandate control and communication capabilities for DERs (e.g., reactive power control capability of a DG unit, which is available for DSO).

A commercial aggregator, in principle, may operate as a third party, but that would create a serious conflict of interest between electricity market stakeholders such as retailers, balance responsible parties and TSO/DSO without proper co-ordination, open information exchange, and clarification of responsibilities of each party. In order to simplify this process, the IDE4L project considered only a case where the balance responsible party or retailer acts as the commercial aggregator, making the chain of balance responsibility remain intact and delivering simple arrangements such as one main contact point for the customer [31]. In a similar way to ensure safe, secure, and cost-efficient distribution, transmission network operation, and development, commercial aggregators will have to coordinate with DSOs and TSOs that must also coordinate with each other and have access to flexibility services and all technical relevant data needed to perform their activities, both at prequalification and operation stages.

The IDE4L project adopted and further developed the concept of flexibility products called SRP and CRP, originally developed in the ADDRESS project [29]. SRP is an obligation for a specified demand modification at a given time. CRP is the capacity for a specified demand modification at a given time with an activation coming from a buyer as a control signal. The SRPs are formed by means of price incentives, triggering the load reprofiling of the prosumers. In response to the signals, the consumers change their consumption level at specific time intervals. Therefore, the aggregator sells a deviation from the forecasted level of demand and not a specific level of demand. The CRPs demand a more immediate and direct control, so their activation is performed by means of direct control signals (power setpoints).

Fig. 2 represents the interaction of the commercial aggregator and other market stakeholders of electricity (wholesale) and flexibility markets. Flexibility markets include balancing and ancillary service markets organized by TSOs today and local flexibility markets aimed at congestion management and voltage control, which are more localized issues compared to system balancing and frequency control. Both TSO and DSO have access to a common flexibility market to avoid fragmentation of markets and ensuring the effectiveness of the system operation [30]. The bidding for the flexibility market is realized on a day-ahead and intraday basis, as in the wholesale market. The DSO/TSO sends to market flexibility requests that are required to alleviate congestion and other system needs. At the same time, the commercial aggregator sends its bids to different markets. When the gate closes, the market clearing process starts to match the flexibility requests received from the DSO, TSO, or any other interested entities with the available bids. After market clearing process, the DSO offline validation will check if distribution grid violations occur due to SRPs. Validation is realized with a sequential load-flow of 15-min time slots based on market clearing and forecasted data of distribution grid configuration and operating conditions. If a distribution grid violation is detected, then an optimal power flow (OPF) is performed to find a minimum set of SRP changes, production curtailment, etc. Once the offline validation is done, DSO sends one of three signals to the flexibility marketplace regarding each bid (approved, rejected, or need for curtailment). The assumption here is that the DSO has the right to reject or demand curtailment of the bids that will cause a violation of network constrains with no compensation for the commercial aggregator. When the offline validation process has finished, the marketplace sends contracts to the relevant flexibility buyers and sellers.

DSO uses real-time validation before giving its consent to the request of a flexibility product by the TSO or other market stakeholder, which is activated close to real time and therefore refers to the actual configuration and operating conditions of the distribution system. Real-time validation checks distribution grid technical limitation violations in a similar way to offline validation, but this function runs automatically and periodically, every 15 min. It will focus on validation of CRP products going to be activated during the next period.

The business role of DSO calculates and determines how much flexibility is required to alleviate congestion in the DSO grid or for network optimization as well as what kind of flexibility is required (SRP or CRP). It communicates this to the market place as a request for flexibility. The second role is the validation role. In this role, the DSO performs the offline validation and real-time validation. From the DSO perspective, flexibility should be integrated as part of the tertiary controller (DMS) where new functionalities should be integrated, aiming to realize the new roles of DSO: flexibility procurer and validation of distribution network-located flexibility products.

3   IDE4L demo

3.1   Unareti field demonstrator

The demonstration phase of the IDE4L project revolved around three field demonstrators and four laboratory demonstrators. Field demonstrators were located in Italy, Spain, and Denmark (Fig. 3): different countries and different network conditions to prove the scalability and replicability of the proposed solutions at the European level via specific KPI. As a whole, the demonstrators included large- and small-scale PV panels, wind turbines, heat pumps, and EVs located in urban and rural areas. In the following, the description will be focused on the Italian demonstrator. Among the laboratory demonstrators, the ones of Tampere University and Aachen University modeled this demonstrator to extend the set of testing conditions and to compare simulations versus real field results.

The Italian field demonstrator is located in the city of Brescia (north of Italy) in a district called Il Violino. This district was recently designed to promote an ecocompatible lifestyle: it is characterized by a high percentage of customers

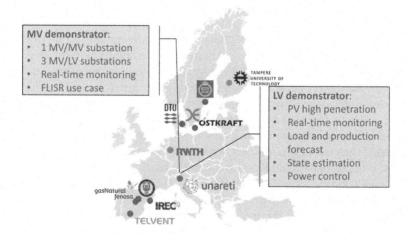

Fig. 3 IDE4L lab and filed demonstrators, with a focus on the features tested in the Italian demo.

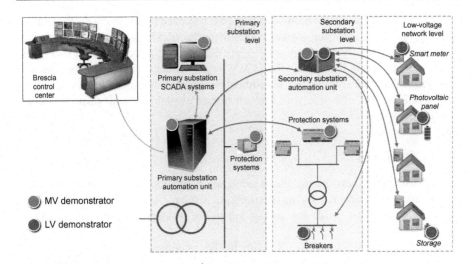

Fig. 4 Conceptual scheme of the MV and LV demonstrators.

equipped with high PV penetration, which is about 40% of the peak demand panels and uses a district heating system. The demonstrator covers the MV and the LV networks up to the final customer connection points, and is general enough to be a concrete test bed for experimenting with the new concepts proposed by IDE4L, in reference to both the MV and LV networks (Fig. 4).

3.1.1 MV demo

The MV network demonstrator consists of 1 MV/MV substation, 3 MV lines, 40 MV/ LV substations, and 9 MV customers. Of the three MV lines, two will be fully automated with monitoring, control, protection, and simulation systems while the third one will be mainly involved in simulations and for the LV field trial. To enable the MV automation services, a proper communication network will be implemented by using a mix of technologies, such as fiber optics, broadband power lines (BPL) over MV cables, and Wi-Fi. This communication network is an extension of the one implemented for the FP7 European Project INTEGRIS.

The use of protection functions on the MV network is one of the most relevant application to improve the management of the distribution network. Despite of that, this application is still not applied on a large scale. Across Europe, there are some pilots in which the coordination of IEDs in secondary substations and IEDs in primary substations has already been tested. In some cases, the 61,850 standard has been used to implement a peer-to-peer coordination mechanism. The main limitation of this application, commonly known as logic selectivity, is that the protection settings and logics are based on the standard grid configuration. As a consequence, every time the grid changes, the protection settings and logics are no longer valid. To address this issue, IDE4L proposes an adaptive FLISR solution—based on the logic selectivity—in which the protection system is always configured in the proper way, independently from the power grid configuration. This feature is achieved by modeling the protection settings and logics by using logical nodes from the 61,850 standard and updating values in real time. Those logical nodes were defined by the standard, but they have never been applied in real cases. UNR's MV demo site is the first operating environment where this concept is tested: investments in smarter solutions for the FLISR are made available and effective with a much higher availability with respect to non-adaptive approaches.

3.1.2 LV demo

The LV distribution network includes a secondary substation with 10 feeders, 299 customers, and 118 residential PV installations As is typical in Italy (Fig. 5), LV backbones are three-phase while the connection to the residential customers is single-phase.

Each of the 10 LV lines of the substation has been equipped with new breakers with remote monitoring and control features. Those breakers have the Modbus RTU interface so Modbus RTU-to-61,850 gateways were used. The first gateway (Schneider G3200 A) interfaces the breaker of four LV lines and—furthermore—the temperature and radiation sensors. The second gateway (Schneider G3200 B) interfaces all other LV lines (Fig. 6).

Two extra breakers (without advanced features) were installed to allow a manual bypass of the DVR installed on the LV line 7. This is a further protection besides the automatic DVR bypass already available in series to line 7. Those protections were needed, especially because the DVR is a prototype developed by a university during the Smart Domo Grid project.

Each smart breaker provides the following measurements with a frequency of 10 s: real-time RMS values of phase voltage, current, active power, reactive power, power factor, frequency, voltage imbalance, current imbalance, and voltage total harmonic distortion.

All the monitoring, control and protection devices installed inside the secondary substation were connected to a local switch/router. Among them, there are also:

- The SSAU PC, where interfaces, the database, and algorithm were installed,
- The power supply box, which has an ethernet interface allowing for a remote supervision of batteries and the power electronic.

In this area, 60 SMs were installed to monitor in real time the power exchange between the customer, the network and the power produced by PV panels. The installed SMs use the BPL to communicate with the secondary substation where the measurements are collected. Among the real-time RMS values are: phase voltage, current, active power, reactive power, power factor, with a frequency of 1 min. Those SMs use the DLMS/COSEM protocol.

Data coming both from SMs and smart breakers are stored in a database to perform cross analysis, such as those reported later on.

LV functionalities: despite the fact that in some EU countries electronic meters have been already deployed, the LV network cannot be monitored or controlled by

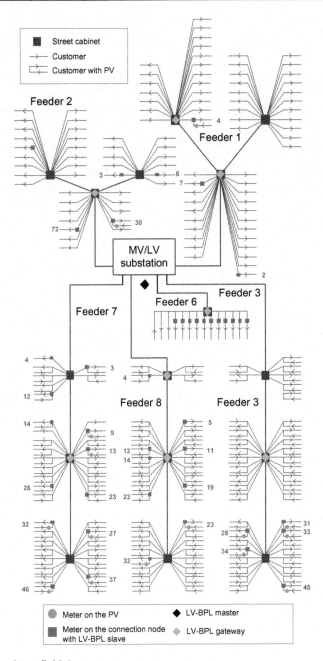

Fig. 5 Low voltage field demonstrator.

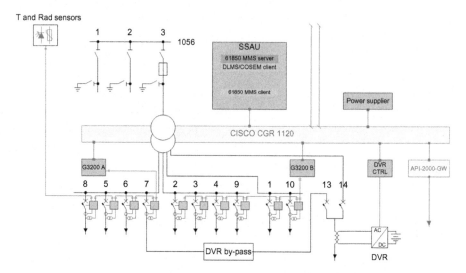

Fig. 6 Secondary substation schematic representation.

the DSO. This limitation is becoming more and more critical because of the increasing presence of DERs. The use of SMs for the technical management of the LV grid has been proposed as a proof of concept in some EU projects (e.g., INTEGRIS), but these projects have mainly focused on monitoring services. In the UNR's demonstrator, in addition to these functionalities, one uses monitored and estimated data to feed brand new algorithms to condition the LV status, such as load and generation forecast, state estimation, and power control algorithms. Moreover, one scales up the infrastructure put in place by UNR during the FP7 European Project INTEGRIS and the Italian Smart Domo Grid project, in order to cover the whole LV grid of a secondary substation. Among the benefits derived from these new LV functionalities are:

- electronic meters are also used for value-added services related to LV network management.
- the ease of exploitation in the EU market due to the usage of DLMS/COSEM specifications for meter communication.
- highlight the future role of DSOs as local dispatchers.

3.2 TUT laboratory demonstrator

Laboratory demonstrations had multiple aims within the IDE4L project, but the ultimate aim was to shorten the initialization duration of field demonstrations. This aim was achieved by fixing software and configuration bugs at as early in the demonstration as possible and by testing the performance of the system to understand if the developed implementations meet specified technical requirements. Also, the searching and solving of bugs from complex automation systems in field conditions is not an easy task, because the real reason might be hidden and the field conditions may produce additional complexity for the performance analysis of the automation system. Therefore, as many hardware and software components and combinations of those were tested in the laboratory environment prior to field demonstrations.

After finalizing the testing of each individual component (e.g., interface, hardware, software component, etc.), it was time to integrate them together and test the interoperability of those components. The test environment of the complete automation system consists of hardware devices, software components, and interfaces between them as described in Fig. 7. The testing of individual components and interfaces separately does not quarantee the performance of the complete automation system and does not provide enough understanding about the complex interactions within the automation system and between different use cases. Interactions are naturally dependent on the implementation of the automation system and therefore the testing of the complete automation system requires integrated hardware and software in the loop simulations.

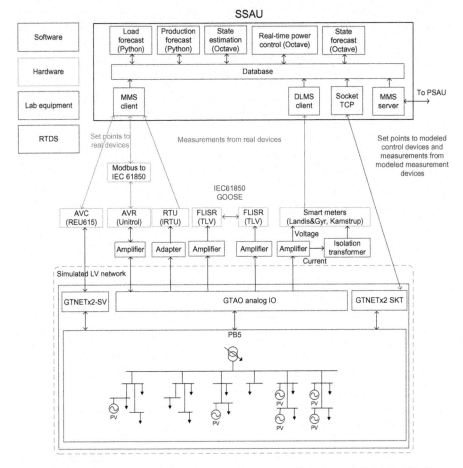

Fig. 7 TUT's hardware and software in the loop test setup for a decentralized distribution automation system.

The advantage of a real-time testing environment is the ability to do testing as close to field conditions as possible. Also the testing can be realized in controlled environments where exactly the same test may be easily repeated or multiple different types of tests may be run in sequence to understand the performance of the system in normal conditions but also in emergency and very rare conditions, which might not be able to be realized in a real distribution grid due to safety or other reasons. Real-time simulation also enables the analysis of such grid conditions easily, which would require expensive investments for the field demonstration. For example, the analysis of the OLTC in the secondary transformer for congestion management functionality is one example that was simulated instead of realizing real investments in the demonstration site. A second example is the increment of DG within the demonstration grid to create challenges for grid management, which existing grids typically cannot have due to the service responsibility of DSOs.

The integation testing is also the testing phase when the delays in information exchange or processing and other unwanted characteristics of the automation system become visible for the performance of the complete system. The aim of automation system testing is therefore searching limits of acceptable behavior of the automation system in the space of possible operation conditions and events. Realization of such testing requires combining a priori knowledge about automation systems, ICT, and electrical grid management. Therefore, the design of interoperability testing is not a trivial task and requires good imagination to find the problematic "corners" of the tested system. Application of the black-box testing approach might be too time consuming because the system is running in real time and some functionalities might require relatively long simulation periods (from minutes to days) to find out the functional and nonfunctional performance of them. Therefore, the gray-box testing approach is recommended, which may utilize the internal structure of the automation system and combine it with ideas of black-box testing while searching for weaknesses in the system without the need to know everything exactly, as in white-box testing. Commercial components such as IEDs are like black-boxes anyway for the complete automation system although there is some knowledge about them via technical manuals and contacts to vendors.

The laboratory demonstration test setup of a decentralized distribution automation system and functionalities will be discussed next. Two test setups and their roles in the demonstration project will also be described. The vital parts of TUT's test setup (Fig. 7) are the physical and commercial IEDs for network monitoring and control. This provides credibility for the test results, the opportunity to test the interoperability of the complete automation system, and the realistic capability and performance of IEDs. IEDs are interfaced with the real-time simulator system as hardware in the loop test devices utilizing analog and digital signals. IEC 61850 sampled values are used when possible; otherwise analog signals and signal amplifier are used. SAU collecting measurements from IEDs as well as storing data and processing data for reporting and control purposes is interfaced with the test setup as software in the loop test systems. Only standard-based information exchange interfaces such as the IEC 61850 MMS for IED reading and writing and the DLMS/COSEM for SM reading were utilized. Additional measurement and control signals between the SAU and real-time digital simulator (RTDS) simulation model are exchanged with RTDS socket interface, which is fast enough for the purpose of secondary control. In this case, the SAU, interoperability of interfaces, and automation functionalities were especially the target of the testing. Detailed information about the test setup may be found [31].

Fig. 7 describes the integration testing environment of the TUT utilized for functional and nonfunctional testing of the IDE4L automation system. The color code of the figure indicates different parts of the system. The purple color (RTDS) represents the distribution grid, DERs, and other elements existing in physical reality, which are real-time simulated in this case. Simulation may also include control systems that are outside the management of DSO, such as HEMS managed by a prosumer or aggregator. A digital real-time simulator is a dedicated simulator for power system simulation combined with the possibility to integrate hardware and software components as part of the simulation. The blue color (lab equipment) is needed to physically interface simulators to IEDs. The orange color (hardware) includes the hardware devices used in the implementation of the automation system. Typically those are IEDs, SMs, and protocol converters, as indicated in the figure. Configuration and performance of these devices is an important part of the complete system and therefore an essential part of integration testing. The red color (software) includes, in this case, all software components needed in the SAU or multiple SAUs. The SAU itself is running on a Linux computer. The testing of SAU includes testing of each individual interface, the performance of the database, functional testing of algorithms, and nonfunctional testing of the complete SAU (Fig. 8).

3.3 RWTH laboratory demonstrator

The RWTH test setup (Fig. 9) resembled the field demonstration in all its components. The UNARETI MV grid has been modeled, node by node, in an RTDS environment. A portion of the LV grid has also been accurately modeled whereas the other parts are modeled as a concentrated load. The RTDS computes a real-time simulation of several grid scenarios, including ones with high renewable generation and high loading conditions. The numerical outputs are converted into digital samples, provided in the DNP3 protocol via the GTNET card, and into analog samples by the GTAO card. Some commercial measurement devices, such as phasor measurement units and SM can be fed with these analog signals thus emulating field conditions for the related measurements. Such measurement devices provide the SAUs with real-time measurements in, respectively, the IEEE c37.118 and DLMS/COSEM protocols. RWTH has also developed some virtual intelligent electronic devices (VIEDs) that provide measurements and accept control setpoints. In this way, it is possible to simulate several classes of IEDs in the grid with few hardware and software expenses. Similarly to the commercial devices, they communicate regularly with the SAUs. The VIEDs exploit the IEC 61850 MMS protocol. The SAUs, developed in the version to be installed in the secondary and primary substations, work as in the field demonstration with full coverage of automation functionalities. In fact, the SAU run the common version of the IDE4L database, the IEC 61850, DLMS/COSEM, IEEE C37.118 software interfaces and state estimation, power control, and load forecast algorithms.

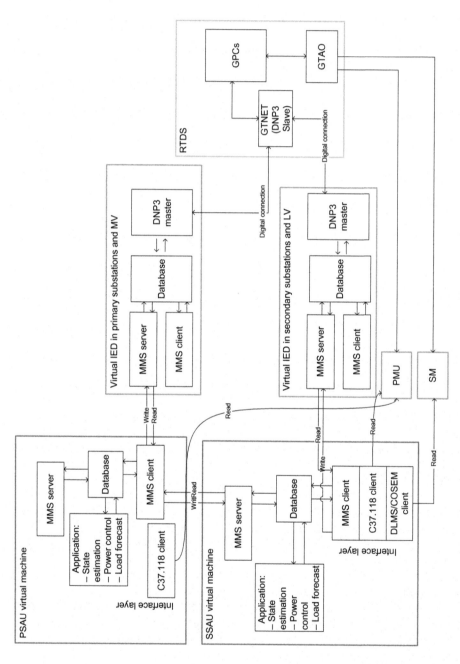

Fig. 8 Details about software components used in the TUT laboratory demonstrator.

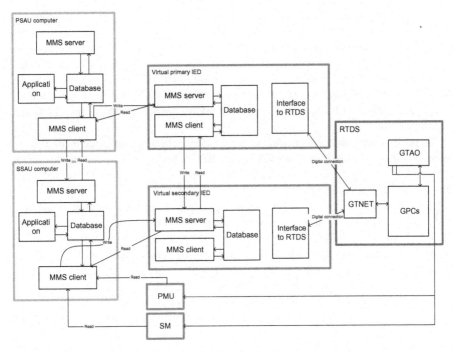

Fig. 9 Details about software components used in the RWTH laboratory demonstrator.

One functionality that was tested only on the RWTH premises, was the communication among SAUs. Such communication exchange permitted the exchange of state estimation results at the connection point between MV and LV. In this way, both the estimation results can be further optimized. Fig. 9 RWTH shows the RWTH laboratory test set up for the IDE4L project.

4   Monitoring and forecast

4.1   Performance of the communication network for the LV monitoring

As explained in the above sections, the LV field demonstrator is based on the use of the BPL technology to collect data from SMs. The present section reports some indicators about the performance of this network based on the data collected through an intensive experimental characterization.

It's worth recalling that traditionally, smart metering applications haven't been mandated for a performing communication infrastructure because they have been mainly focused on the billing and customer relationship management (e.g., remotely activate/deactivate a contract, change contractual power, etc.). Requirements for such nontechnical services were easily fulfilled by many technologies. One of the most used was power line communication (PLC) such as the IEC 61334-5-1 [32], offering a limited bandwidth (approximately a few kilobits per second). Still remaining in the PLC domain, more advanced technologies were then proposed to overcome some of the limitations of the IEC 61334-5-1. Among them, the PRIME and G3-PLC [33] were the most relevant ones. As an example, PRIME has been tested in Spain, France, and Switzerland. Technical features of PRIME allow the reaching of a theoretical performance of 128 kbit/s, which is a good improvement with respect to previous standards.

To further boost the performance of the data collection, a leap in the technology is needed. For this reason it was decided to use a BPL solution, which—taking advantage of a wider bandwidth for the communication—provides a theoretical performance higher than 100 Mbit/s [34].

Fig. 1 describes how this BPL network was deployed in a real field demonstrator. SMs are connected to a BPL modem (API-2000-GW), working at 2–12 MHz. Repeaters were installed in some central nodes of LV feeders (e.g., LV street cabinets) to group together the slave modem connecting the meters. The master node was located in the secondary substation in order to collect the data (depicted in Fig. 6).

This network was experimentally characterized in order to measure two important parameters affecting the performance of a distributed measurement system: the round trip time—which is related to the latency of the system—and the time offset of the synchronization system. Details about the measurement procedure can be found in [35]. Table 1 reports those two quantities for some of the nodes of the BPL network.

4.2   Analysis of data from the LV monitoring system

In the next part of the section, the analysis will revolve around data collected by SMs.

As a sample of the data collected by the SMs installed in the demo area, Fig. 10 reports active power profiles from two LV customers with a contractual power of 4.5 and 3 kW, respectively. For the same customers, the bottom part of the picture shows the production of their PV panels (5.6 and 1.29 kW).

SM data were stored in a database for 1 year and then further processed to derive average load and production profiles. The result of the aggregation is reported in Fig. 11 where a comparison between offline average profiles and real measurements is shown for the sake of comparison. It's worth pointing out that the average profile—despite the fact that their accuracy is not perfect—is extremely important because it can be used on behalf of measurements when those are missing to calculate the overall state of the network through a state estimation algorithm, like the one described later.

An analysis of the data reported by smart breakers installed in the substation involved in the demo is reported—as an example—in Figs. 12 and 13, showing respectively the current/current imbalance[1] and the voltage/voltage imbalance for one LV feeder (FD 1) in the period between October 10–16, 2016.

---

[1] The current imbalance index is, for each phase A, B, and C, the difference between the phase current magnitude and the average current magnitude $(I_A + I_B + I_C)/3$.

Table 1 **Summary of the network infrastructure, round trip time (RTT) (ms) 95-percentile (64 byte) and the time offset statistics evaluated using NTP configurations 1 (observation interval: 2 weeks) [35]**

	RTT (ms)	Time synchronization		
Node	95-perc. (ms)	Mean (ms)	Std (ms)	Jitter (ms)
N_1	2337	9.1	109	884
N_2	2085	20.7	105	933
N_3	–	9.2	104	880
N_4	2090	28.8	95	1049
N_5	2323	14.9	97	1098
N_6	1849	−4.6	89	1010
N_7	916	−38.7	59	939
N_8	1725	15.4	87	943
N_9	2372	−26.4	77	884
N_{10}	1923	35.7	80	1055
N_{11}	2013	1.2	82	647
N_{12}	–	7.5	92	943
N_{13}	2356	15.9	100	1132
N_{14}	1956	22.4	87	819
N_{15}	2062	15.4	72	599
N_{16}	2383	12.3	98	1036
N_{17}	2094	12.7	95	864
N_{18}	1790	−0.1	84	825
N_{19}	2313	5.9	98	925
N_{20}	2160	2.0	94	878
N_{21}	913	−31.6	56	830
N_{22}	2419	123.7	114	1639

In Fig. 12, a periodic behavior over the day with peaks after sunset due to a lack of production from the PVs and the increase of consumption during dinner, and dips over the night, can be appreciated. Furthermore, the current appears to be very unbalanced with a constant value of neutral current over 10 A, and the C phase is normally more loaded than the others.

Observing Fig. 13, two general considerations can be made. First, the voltage at the substation level has a variation of about 10 V during the day. This may be due to voltage variations from the MV. Second, the voltage imbalances are, at the substation level, considerably small. This can also be inferred from Fig. 13 where the voltage imbalances, equal to the difference between the phase voltages and the average voltage $(V_A + V_B + V_C)/3$, are shown. Also, it may be meaningful to mention that the measurement chain values used by IEDs are fed through sensors with an accuracy class (0.2), therefore an error of approximately 0.4 V may be present. However, it can be observed that the voltage imbalance is, in general, constant; whereas if the sensors introduce the error, it is expected to be statistically distributed over positive and negative values.

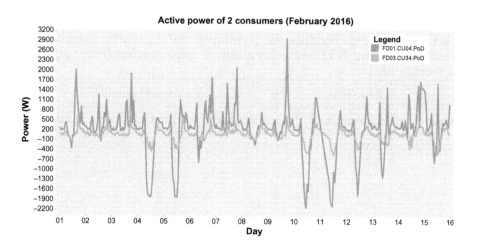

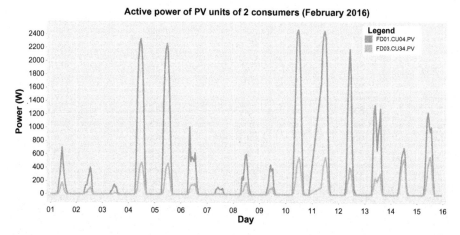

Fig. 10 Active power profiles monitored by the smart meters in the first half of February 2016: (up) two customers with 4.5 kW (FD01.CU04·PoD) and 3 kW (FD03.CU34·PoD) of nominal peak power; (bottom) domestic PV plants of two customers with, respectively, 5.6 (FD01. CU04·PoD) and 1.29 (FD03.CU34·PoD) kilowatt-peak PV units [36].

The RMS voltage magnitude of feeder 1—phase B is reported in Fig. 14, where it is compared with the values measured by two SMs (namely CU04 and CU07) located on the same phase in two distinct nodes on the LV network. As can be appreciated, in a typical midseason sunny day, the SM CU04 reports a significant voltage rise during the afternoon because of the presence of a PV installation. At the same time, the voltage increase of a customer without a PV is smaller (SM CU07). From this observation it can be derived that a significant part of voltage variation takes place on the short connection between the cabinet and the customer's premises whereas the voltage in the main feeder is probably going to have smaller variations (in the range of a

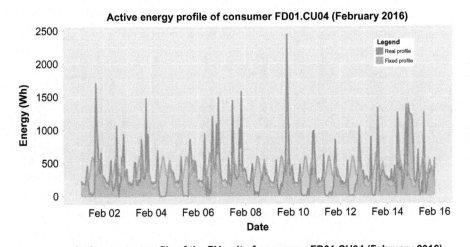

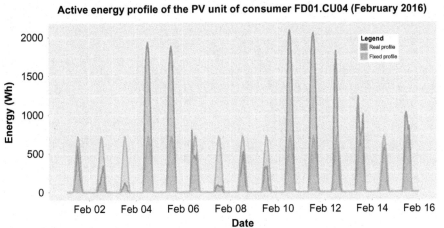

Fig. 11 Real active energy profiles and fixed active energy profiles of customer FD01.CU04 (up) and the PV unit of the same customer (down) during the first half of February 2016 [36].

few volts). A second effect that can be observe is the LV value at a customer's premises during the night and the evening, due to the zero power production of PVs and the higher energy demand of customers.

5   State estimation and voltage control

5.1   State estimation results

State estimation is a key enabler for an active network control (e.g., power control algorithm in IDE4L project) because the state of the network must be known in order to efficiently control a network with distributed resources. The IDE4L LV network

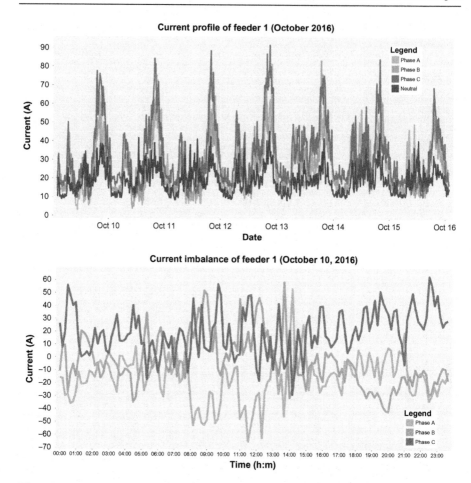

Fig. 12 Current of feeder 1: phase A, B, and C and neutral (up) and current imbalance index of feeder 1: phase A, B, and C (bottom). October 10, 2016 [36].

state estimator uses network data, real-time measurements, load and production forecasts, and fixed load and production profiles as inputs and, based on this information, estimates what is the most likely state of the network at a given moment. The estimated quantities are node voltage magnitudes and node power injections (i.e., load and production) and current flows in all network nodes and lines. The state estimation algorithm was demonstrated in two laboratories (TUT and RWTH) and in three electric utilities (UFD, OST, and UNR). The following results are from Unareti's field demonstration.

The demonstration of the LV grid state estimation proved to be more difficult than expected, but an appropriate solution was finally found. Problems in monitoring were frequent and, because the state estimator relies heavily on input measurements, the state estimator performance was often affected by issues in the monitoring systems.

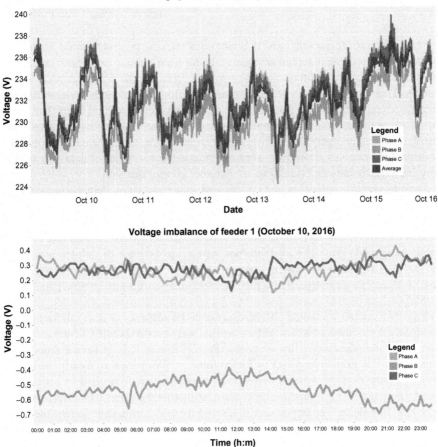

Fig. 13 Voltage of feeder 1: phase A, B, and C with the average of the three phases (up) and voltage imbalance index of feeder 1: phase A, B, and C (bottom). October 10, 2016 [36].

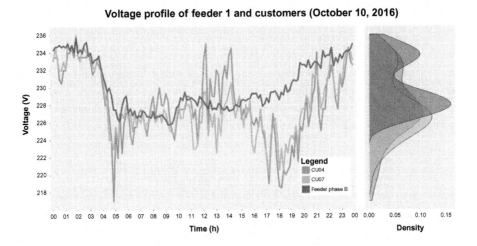

Fig. 14 Voltage of feeder 1—phase B compared with the voltage measured on customer 4 and 7 connected on the same phase. October 16, 2016 [36].

As learned from demonstrations, a bombproof backup solution should always be available to overcome potential problems coming from missing or incorrect measurements. The robustness of the state estimator is critical for the online secondary controllers. Therefore, the focus of development should be given to the availability and accuracy of monitoring data instead of the state estimation algorithm itself. The handing of bad or missing data is extremely important because of the radial structure of the distribution grid and very few real-time measurement points compared to the number of estimated variables. Therefore, even one missing or bad piece of data may lead to very incorrect results and finally to an incorrect control decision.

Despite several challenges, the LV network state estimation demonstrations were successfully finalized. It was proven that the developed state estimation algorithm using the proposed automation system works quite properly and provides an improved view of the state of the network without a need to extend real-time monitoring to every corner of the LV grid. The average root means square error of voltage was 2.4 V (about 1% of nominal voltage), which is very low compared to the resolution of SMs (1 V) and the coverage of real-time measurements on customer connection points (<20%).

Fig. 15 gives more detailed information about the accuracy of state estimation during different moments of a day. Measurements were realized during September 2016 when solar radiation was still strong enough in Northern Italy to have a remarkable impact for LV grid voltages. The uncertainty of PV output is clearly visible in Fig. 15 where daytime error is higher than average error. It is good to realize that the location of SMs is not optimized for state estimation, but other factors have instead been affecting those. The importance of a secondary substation (SS) voltage measurement for the accuracy of state estimation is visualized in the figure too. It is the most

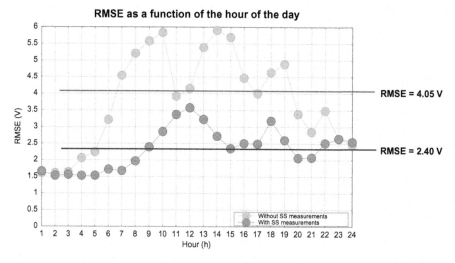

Fig. 15 State estimation accuracy during different moments of a day for 1 month (September 2016).

important single measurement for the state estimation, which must be available. Therefore, if voltage measurement is not available at a secondary substation, the nearest SM measurement might be used instead, but the addition of a more accurate PQ meter for the secondary substation is preferred. The trade-off between the benefits of more accurate state estimation and the required investments on more accurate and a higher number of measurements and more frequent reading of meters needs to be done for each DSO/investment case separately. Our demonstration in three different DSOs showed that state estimation of a European-style LV distribution grid is possible to realize today with existing algorithms, automation, and communication technology, and with reasonable investment and operational cost.

Fig. 16 provides similar results to Fig. 15 with an additional view by linking the node location (impedance between the node and secondary substation) and the accuracy of state estimation. It represents the case where secondary substation measurements (voltage and feeder currents) are available for the state estimation. The figure clearly indicates how the uncertainty of estimation results increases toward the tail parts of the radial feeders. This kind of analysis is needed to decide the location of additional measurement points if the accuracy is not acceptable. In this specific case, an additional voltage measurement from a PV unit located in the tail part of the feeder(s) would benefit the voltage estimation accuracy.

Fig. 17 gives an example of state estimation results compared to real-time measurements as a time series. Estimated voltage follows very well with the real-time measurement, although the estimation errors are also clearly visible.

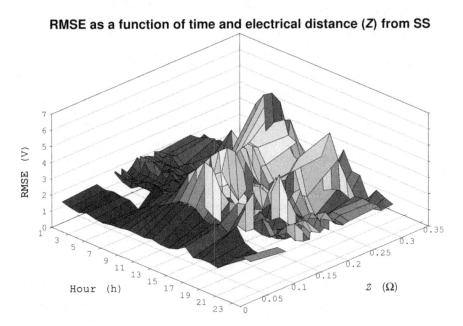

Fig. 16 State estimation accuracy during different moments of a day and in locations.

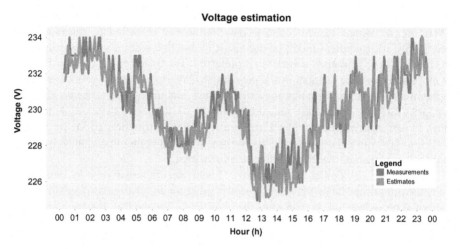

Fig. 17 Example of time domain behavior of state estimation results compared to measurements not utilized in the estimation.

5.2 Secondary voltage control results

The demonstration results confirm that the proposed and implemented secondary controller for congestion management operates in all the demonstration cases as expected, and that no adverse time domain operation occurs. The same controller is able to operate for MV and LV grids. Also, the results show that the algorithm is able to improve the defined KPIs (network losses, curtailed production, and over-voltage volume) with an increased number of control actions in all the demo cases. In general terms, it can be said that the weaker the network and the larger the amount of generation connected to it, the larger the benefits of the secondary controller. The annual benefits of the secondary controller have been evaluated in [22], which shows that the secondary controller is able to both prevent voltage and current congestions and to decrease the annual network losses. The integration of an MV secondary controller and a LV secondary controller was also effective and resulted in correct operation interactions of MV and LV secondary controllers. Graded time delays of the AVC relays seem to be an adequate measure to prevent back-and-forth operations of control in most cases [37].

In the Unareti case, an extract of the secondary voltage control operation during the 15-day test-period is presented here. Fig. 18 presents network voltages on a given test day. It can be noticed, on this day as well in the remaining testing campaign of the secondary voltage control, that no voltage violations outside the acceptable voltage range ($\pm5\%$) have been observed due to the current penetration of the PV generation in the network. As a consequence, during the demonstration of the available controllable resources in the network, six PV inverters have been utilized with the goal of reducing reactive power flows and, thus, network losses. Fig. 19 reports an example of the reactive power profiles of controlled resources during the testing campaign. Utilization of a small scale of reactive power has a minor effect on network voltages due

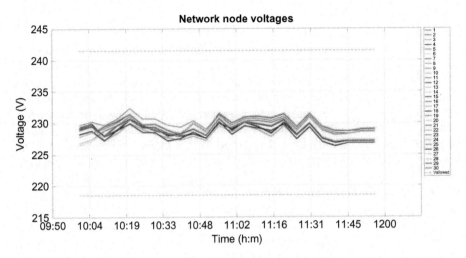

Fig. 18 RMS voltage for 30 nodes on September 10, 2016 in UNR demo site.

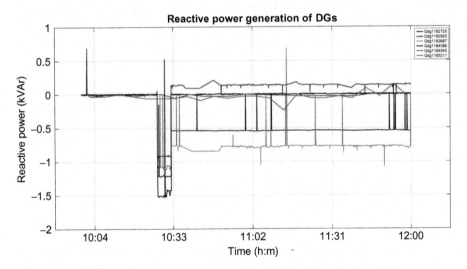

Fig. 19 Reactive power output of PV units during 15-day demonstration period.

to the reactance/resistance ratio of lines in LV networks. In the Unareti case, the cost function was set to minimize network losses where the effect of reactive power flows can be seen. The secondary voltage control realized totally 43 setpoint changes for six PV inverters during the whole 15-day test period. This demonstration confirms that the secondary voltage control has been successfully installed, and setpoints of available controllable resources have been successfully altered in real LV distribution network operations.

Several hardware and software in the loop tests were realized in the integration laboratories of RWTH and TUT. These laboratories were utilized to test the functionality of algorithms prior to field demonstration, to realize nonfunctional tests for a complete automation system, and to extend the testing of developed control methods beyond the capabilities of a field demonstration site. The secondary voltage control algorithm, which is based on OPF, implemented in both MV and LV grids is based on [23]. A weighted least squares-based state estimation [38] is used to provide necessary input, which includes phase voltages of each network node, branch currents, and real and reactive powers of each load and production node. Full network topology information is available in the SAU database. The secondary voltage control can utilize multiple types of controllable resources that have a primary controller (IED) and an interface to the SAU database implemented. Real and reactive power setpoints of DG units, reactive power setpoints of reactive power compensation devices, real power setpoints of controllable loads, and voltage reference setpoints to AVC relays of OLTCs are the controllable resources that can be currently utilized. The algorithm aims to minimize the objective function in Eq. (1), where C_{losses} is the cost for network losses P_{losses}, C_{cur} is the cost for generation curtailment $\sum P_{cur}$, C_{tap} is the cost for tap changer operations n_{tap}, C_{DR} is the cost for load control $\sum P_{DR}$, and C_{vdiff} is the cost for voltage deviation from nominal on all nodes, $\sum (V_{i,r} - V_i)^2$. Network constraints and load flow equations are taken into account in the optimization. The problem is a mixed-integer nonlinear programming problem, which is solved by sequential quadratic programming in Octave due to demonstration constraints.

$$f = C_{losses}P_{losses} + C_{cur}\sum P_{cur} + C_{tap}n_{tap} + C_{DR}\sum P_{DR} + C_{vdiff}\sum (V_{i,r} - V_i)^2 \qquad (1)$$

In real-time simulations, the algorithms are executed once a minute. Real-time monitoring of a distribution network provides necessary measurement data from selected SMs and IEDs. [31] State estimation is run first, followed by a voltage control algorithm. In order to avoid oscillation between the simultaneous real-time control of MV and LV networks, some sort of coordination is required among the secondary controllers. This can be either implemented using blocking signals or using graded time delays of AVC relay control loops of OLTCs. When using blocking signals, the reason for voltage variations needs to be originated first. For instance, with a blocking signal from an upper-level SAU to a lower-level SAU, the operation of OLTC and the voltage-control algorithm can be delayed until the upper-level algorithm has finished its control actions; thus unnecessary control actions can be avoided. In these studies, graded time delays alone proved to avoid unnecessary OLTC actions but clashing voltage reference setpoints were found to be set simultaneously on the MV and LV levels, which needed to be reverted later on. To prevent this from happening, the execution of algorithms is staggered within the 1-min timeframe.

The simulation network model represents a real MV and LV distribution network of Unareti's demonstration site. The model consists of one primary substation, two MV feeders, a secondary substation, and six LV feeders. Controllable resources in the network include HV/MV OLTC at the primary substation and MV/LV OLTC at the secondary substation, each controlled by an AVC relay, six controllable PV units in the LV network, and one DG in the MV network. Static constant power loads represent customers in all LV network nodes and in some MV network nodes.

The automation system consists of two SAUs, both running on the same Linux computer. With RTDS, all calculations are performed in real time, which enables the connection of commercial monitoring and control IEDs to the simulation environment. The AVC relay at the secondary substation, and two SMs measuring two critical load nodes in the LV network, are connected to the simulator. A dedicated Windows computer runs an instance of MATLAB to provide two-way communication between the databases and algorithms regarding measurements and setpoints of the modeled AVC relay and PV units within the RTDS model.

As an example, two simulation sequences are presented. The first simulation sequence represents varying loading of the feeding network and thus voltage variations that take place downstream in the network. In the simulation, the primary substation feeding network voltage was varied with the following sequence (time instant and voltage): 0:00 (23.0 kV), 1:45 (23.5 kV), 3:45 (23.0 kV), and 5:15 (23.3 kV). The combined operation of the MV and LV secondary voltage control with incorporated graded OLTC delays were simulated.

In Fig. 20, the MV and LV simulation sequence graphs are reported. The changes made by the sequence to the network are highlighted with numbers in the figures

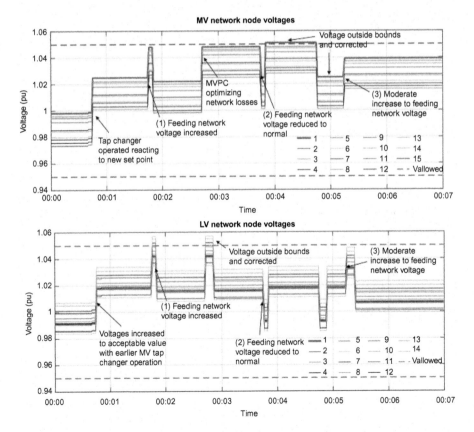

Fig. 20 MV and LV simulation sequence graphs in TUT.

Continued

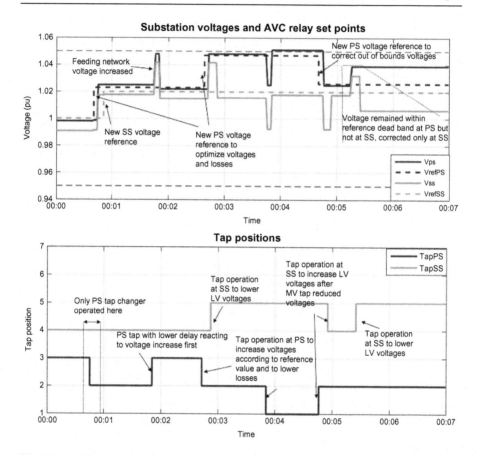

Fig. 20, cont'd

explaining the sequence. Other visible voltage variations are due to control actions in the network. When the voltage remained outside the reference values at the substation, the optimization was first blocked in order to allow tap changer operations to take effect. Tap changer delays at the primary substation were set to a half value of the secondary substation's tap changer delay. The algorithms were executed sequentially every minute, LV algorithms first at the beginning of every minute, followed by MV algorithms 30 s later. The control actions made by both secondary voltage controllers were therefore visible for both the OPF algorithms via real-time measurements collected before the run of the OPF.

In the second simulation scenario, the output of all PV units is varied according to the real data of the PV output throughout a 24-h scenario, depicted in Fig. 21. The nominal powers of the real network's PV units have been multiplied by a factor of 3 in order to create congestions in the network.

From Fig. 22, the conclusion can be drawn that the voltage variations are handled at the LV network portion where the effect of PV output in voltages is mainly seen.

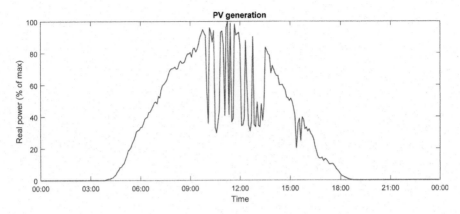

Fig. 21 Scenario 2—Typical daily PV generation curve. Measured in Tampere, Finland.

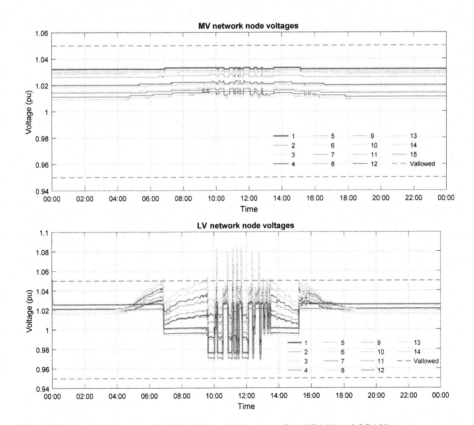

Fig. 22 Impact on large-scale PV in LV grid for cascading PSAU and SSAU.

Continued

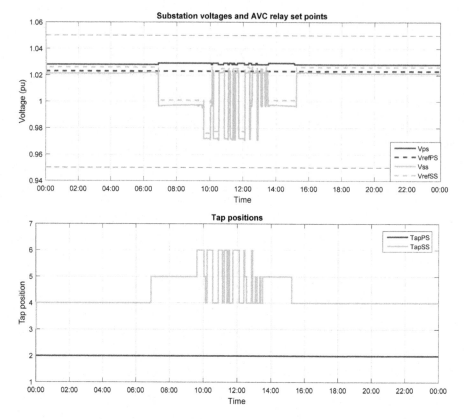

Fig. 22, cont'd

With the higher, artificial penetration of the PV varying between 30% and 100% of output, many changes to the secondary substation AVC relay voltage reference setpoint can be observed during mid-day. The reactive power of PV units is not controlled in order to see better how OLTCs perform. Therefore, tap changer actions according to the fourth graph in Fig. 22 have also been made.

The algorithm is by default tuned to have a cost for voltage difference from nominal set to zero, and thus to optimize voltages closer to the upper voltage limit. Therefore, with 30% PV output during the middle of the day, and following a substantial increase to the 100% output, over-voltages are observed. These over-voltages, however, are always corrected after the next algorithm execution by setting a new, lower voltage reference value. This increases the number of tap changer operations and the need for maintenance. By changing the algorithm setting parameters to aim for voltages closer to nominal, usually at the cost of network losses, the over-voltages in the scenarios could be avoided.

6 The role of the aggregator in the IDE4L automation architecture

The possible roles of the aggregator in support of the network operation, beyond the business models and features of aggregators currently operating in some European energy markets, are implied in [10]. Various European projects have investigated and developed schemes of operation and business models for the aggregator. From the technical point of view, a detailed development was produced by the project IDE4L and demonstrated in the lab.

An example of operation of the aggregator is presented in [39,40].

The aggregator manages the individual consumers with a two-layer communication-based control architecture. The lower layer is located on the level of the EMS of the customer. Upon request, such EMS sends the baseline and flexibility availability to the aggregator, timely enough to fairly allocate the flexibility burden, in the form of power setpoints, depending on individual flexibility. The upper layer instead only checks the point of common coupling of the customer. This is where the aggregator's central manager corrects the aggregated power consumption by reallocating individual target flexibilities from consumers incapable of delivering to those who are. The simple flexibility activation scheme is pro-rata, and curtailment is allocated proportionally to customers' volume availabilities.

The main challenge from the operative point of view is to reduce the gap between the requested flexibility and the actual delivered one. In our implementation we considered the flexibility successfully delivered if it is included in a 20% tolerance band. Right now the flexibility is divided evenly among the customers, only based on the declared flexibility available. However, it could be optimized, for example, assigning the flexibility based on the customer's "thrust score," meaning how much the customer followed the requested setpoints in the past (this would require an evaluation of the customer's past performances), thus incentivizing virtuous customers. Another way to optimize the flexibility activation among the customers would be assigning the biggest amount of flexibility to the "cheaper" customers first. This would require the definition of a cost function that could be based on the actual cost of (maximizing the aggregator revenue) or other kinds of costs, such as a performance-related cost or, for example, trying to give an advantage to customers who have smaller delays in actuating the required setpoints.

Because the performances of the customers, like delays in actuating the setpoints or how much they move away from the requested setpoint, have an effect on the final result (the one that concerns the DSO), this strategy could improve the overall performance of the aggregator in reaching flexibility delivery and thus improve the benefits for the DSO.

From the implementation point of view, the communication between customers and aggregators has a major role. Because the whole strategy is based on a continuous exchange of data between the aggregator and its customers, it is crucial to have a reliable communication infrastructure. On the other hand, interactions between the aggregator and DSO are less intensive. However, in this case the timing and intervals of such interaction have an impact on the overall performances of the flexibility delivery. The aggregator strategy includes a certain time window, defined as the flexibility delivery check window, which is the elapsed time between two energy measurement requests sent to the customers. These two measurements are used to calculate the actual energy consumed by the customer during this time window and compare it with declared base energy consumption, expected if no flexibility is activated.

In our simulations we analyzed how different configurations of the flexibility allocation strategy timing, obtained by changing the width of the flexibility delivery check window, affect the performances. Furthermore, we repeated the simulations, adding delays in the response of customers in actuating the required setpoint, thus evaluating the effect of such delays in the performances both at the customer and aggregator level.

The test cases were the following:

Case	Window (min)	Delay
1.a	3	No delay
1.b	3	With delay
2.a	4	No delay
2.b	4	With delay
3.a	5	No delay
3.b	5	With delay
4.a	6	No delay
4.b	6	With delay

As shown in Figs. 23 and 24, the different delivery check window configurations affect both the profile of the active power consumption and of the activated flexibility. Moreover, the overall performances of the aggregator decrease with the wider windows, with the 3 min window giving the best performances and with the 2 min window being still able to deliver the required flexibility within the 20% tolerance band. Finally, the percentage of performing customers decreases with wider windows.

In the presence of delays, the performances decrease for each flexibility check window configuration and only in case 1.b are the performances still good enough to consider the flexibility successfully delivered, as reported in Fig. 25.

Another implementation challenge is the modeling of the behavior of the customers. For our implementation we created, power profiles for each customer based on real data coming from a DSO, and eventually added some delays in the acceptance and execution of the setpoints sent by the aggregator. However, the modeling of the customers could be improved by adding some variability factors on such profiles and on the acceptance of the flexibility by the customers. By doing so and running a series of simulations, thus analyzing a wider set of possible scenarios, it could be possible to better validate the effects of different timing strategies on the flexibility activation and execution as well as the effects of the delays.

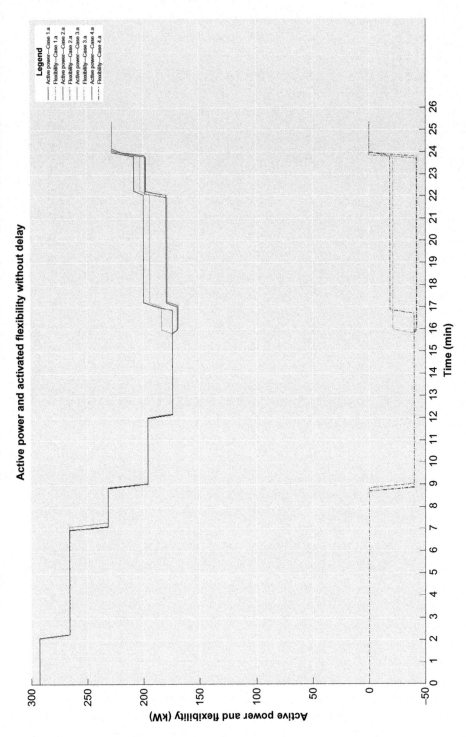

Fig. 23 Active power and activated flexibility without delays.

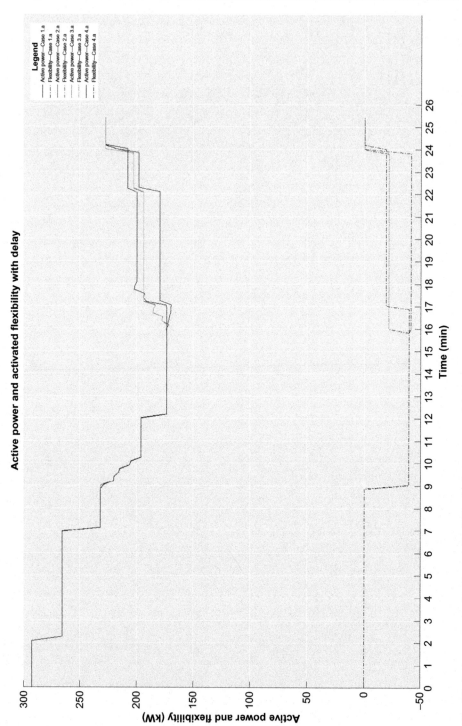

Fig. 24 Active power and activated flexibility in the presence of delays.

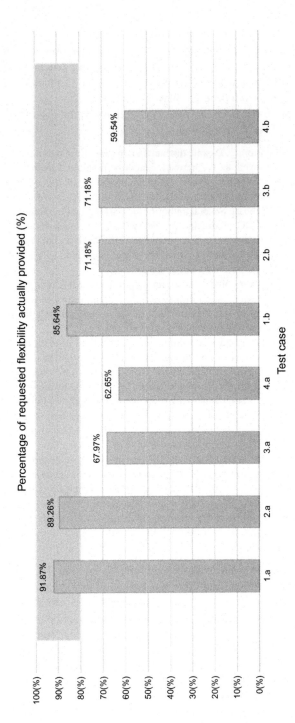

Fig. 25 Overall performances of aggregator for different test cases.

7 Conclusions

Based on three successful field and six laboratory demonstrations, we may conclude that the proposed concepts, automation architecture, and functionalities work as expected. The following paragraphs highlight the general conclusions of the IDE4L project demonstrations, from different players perspective.

In reference to what has been explained in Section 1.1, the most relevant exploitable results for a DSO are the following.

- Network real-time monitoring. Despite the fact that, in some EU Countries electronic meters have been already deployed, an LV network cannot be monitored by the DSO. Within the IDE4L project, SMs were utilized as sensors to monitor and control the LV network by the DSO to increase the hosting capacity of the LV network for RES/DER.
- Load and production forecast. Several authorities are asking DSO to forecast—on the short and medium term—the DG production on the MV and LV levels to understand the real impact of renewable generation the distribution network. The algorithm tested in the project has been proven to have a balanced trade-off between complexity of use and accuracy and can be used for such a purpose on a larger scale.
- State estimation and power control. Those are two important bricks in the supervision on today's distribution network. Based on the real-time monitoring and the forecast of load and productions, they allow the determination of the overall network status—identifying congestions—and mitigate them through control actions.
- FLISR and MV monitoring. FLISR solutions—because they have an immediate impact on quality of service indicators and though incomes for the DSOs—are in the stage of moving from a pilot project to deployment projects. Tests like those done during the IDE4L project make the business case more solid. The feature of remote reconfiguration of the protection device when the network topology changes allows for a maximization of the investment because the protection system can be used not only in the standard configuration, but even after a fault. We expect to buy IEDs like those tested during the project.
- Standards. Standards like the 61,850 and the CIM are becoming more and more a core part of our business. We are currently redesigning the data model for our asset management system and the new one will be based on the CIM standard (to be completed by December 2016). We recently released (May 2016) the technical specification for the primary substation automation systems and those specifications mandate full compliance with the 61,850. The use of the DLMS/COSEM makes the application of the "smart meters as a sensor" concept easily exploitable.

It is worth pointing out the replicability of the proposed solution. The three DSOs partners of IDE4L represent urban and rural networks in different parts and countries of Europe. Demonstration grids also include relatively high penetration of RESs connected to a LV grid. The same IDE4L automation system was implemented in all field demonstration sites. Results prove it effective in the different configurations (functionalities chosen by the DSO) and hardware implementations that were tested. Because the IDE4L solutions work in this variety of conditions, they are expected to be suitable for most of the EU.

References

[1] S. Repo, D. Della Giustina, G. Ravera, L. Cremaschini, S. Zanini, J.M. Selga, P. Jarventausta, in: Use case analysis of real-time low voltage network management, 2011 2nd IEEE PES International Conference and Exhibition on Innovative Smart Grid Technologies, Manchester, 2011, pp. 1–8.

[2] EN 50160: 2011-05, Voltage characteristics of electricity supplied by public distribution networks.

[3] AEEGSI, 646/2015/R/eel—Testo integrato della regolazione output-based dei servizi di distribuzione e misura dell'energia elettrica, per il periodo di regolazione 2016–2023 [Online]. Available from: http://www.autorita.energia.it/it/docs/15/646-15.htm.

[4] Council of European Energy Regulators, Quality of electricity supply: initial benchmarking on actual levels, standards and regulatory strategies, April [Online]. 2001. Available from: www.autorita.energia.it/pubblicazioni/volume_ceer.pdf.

[5] Council of European Energy Regulators, 5th CEER benchmarking report on the quality of electricity supply 2011, April [Online]. 2012. Available from: www.energy-community. org/pls/portal/docs/1522177.pdf.

[6] EN 61000-4-30:2008, Electromagnetic compatibility (EMC)—Part 4-30: testing and measurement techniques—power quality measurement methods.

[7] https://ec.europa.eu/commission/publications/2nd-report-state-energy-union_en.

[8] http://www.gridplus.eu/eegi.

[9] https://ec.europa.eu/energy/sites/ener/files/documents/set-plan_progress_2016.pdf.

[10] http://ec.europa.eu/energy/en/news/commission-proposes-new-rules-consumer-centred-clean-energy-transition.

[11] https://ses.jrc.ec.europa.eu/sites/ses/files/documents/smart_grid_projects_in_europe.pdf.

[12] http://ses.jrc.ec.europa.eu/sites/ses.jrc.ec.europa.eu/files/u24/2014/report/ld-na-26609-en-n_smart_grid_projects_outlook_2014_-_online.pdf.

[13] http://www.grid4eu.eu/overview.aspx.

[14] https://ec.europa.eu/inea/en/horizon-2020/projects/H2020-Energy/InterFlex.

[15] http://www.flexiciency-h2020.eu/.

[16] http://upgrid.eu/.

[17] http://cordis.europa.eu/project/rcn/206239_en.html.

[18] http://smartgrid-adms.com/.

[19] http://ec.europa.eu/energy/infrastructure/transparency_platform/map-viewer/main.html.

[20] Deliverable 7.2, Overall final demonstration report, IDE4L, 2016.

[21] Deliverable 5.2/3, Congestion management in distribution networks, IDE4L, 2015.

[22] A. Kulmala, S. Repo, P. Järventausta, Coordinated voltage control in distribution networks including several distributed energy resources, IEEE Trans. Smart Grid 5 (4) (2014) 2010–2020.

[23] A. Kulmala, M. Alonso, S. Repo, H. Amaris, A. Moreno, J. Mehmedalic, Z. Al-Jassim, Hierarchical and distributed control concept for distribution network congestion management, IET Gener. Transm. Distrib. (2017) 665–675.

[24] S. Lu, S. Repo, D. Della Giustina, in: Standard-based secondary substation automation unit – the ICT perspective, IEEE PES ISGT Europe 2014 Innovative Smart Grid Technologies, 13–15 October, 2014, Istanbul, Turkey, 2014.

[25] IDE4L—IDEAL GRID FOR ALL, Deliverable 3.2: Architecture design and implementation [online], September 2015. http://ide4l.eu/results/.

[26] A. Rautiainen, S. Repo, P. Järventausta, in: Using frequency dependent electric space heating loads to manage frequency disturbances in power systems, IEEE Bucharest PowerTech Conference Proceedings, Bucharest, Romania, June 28–July 2, 2009.

[27] A. Rautiainen, S. Repo, P. Järventausta, Using frequency dependent charging of plug-in vehicles to enhance power system's frequency stability, European Conference Smart Grids and Mobility, June 16th/17th, Würzburg, Germany, 2009.

[28] http://www-gridplus.eu/eegi.

[29] ADDRESS project, www.addressfp7.org.

[30] ENTSO-E, Towards smarter grids: developing TSO and DSO roles and interactions for the benefit of consumers, Tech. Rep., March 2015.

[31] V. Tuominen, H. Reponen, A. Kulmala, S. Lu, S. Repo, Real-time hardware- and software-in-the-loop simulation of decentralized distribution network control architecture, IET Gener. Transm. Distrib. (2017). 18 pp. (accepted).

[32] Distribution automation using distribution line carrier systems—Part 5-1: lower layer profiles—the spread frequency shift keying (S-FSK) profile, IEC 61334-5-1 ed2.0, IEC, 2001.

[33] M. Hoch, in: Comparison of PLC G3 and PRIME, Proceedings of IEEE ISPLC, 3–6 April, 2011, p. 165. 169.

[34] D. Della Giustina, P. Ferrari, A. Flammini, S. Rinaldi, E. Sisinni, Automation of distribution grids with IEC 61850: a first approach using broadband power line communication, IEEE Trans. Instrum. Meas. 62 (9) (2013) 2372–2383.

[35] S. Rinaldi, D. Della Giustina, P. Ferrari, A. Flammini, E. Sisinni, Time synchronization over heterogeneous network for smart grid application: design and characterization of a real case, Ad Hoc Netw. 50 (1) (2016) 41–57.

[36] A. Barbato, A. Dedè, D. Della Giustina, G. Massa, A. Angioni, G. Lipari, F. Ponci, S. Repo, Lessons learnt from real-time monitoring of the low voltage distribution network, Sustainable Energy Grids Networks (2017).

[37] H. Reponen, A. Kulmala, V. Tuominen, S. Repo, in: Distributed automation solution and voltage control in MV and LV distribution networks, IEEE PES ISGT Europe 2017 Innovative Smart Grid Technologies, 26–29 September, Torino, Italy, 2017. (accepted).

[38] A. Mutanen, S. Repo, A. Löf, in: Testing low voltage network state estimation in RTDS environment, Proceedings 4th IEEE PES ISGT Europe Copenhagen, 2013.

[39] G. Lipari, G. Del Rosario, C. Corchero, F. Ponci, A. Monti, A real-time commercial aggregator for distributed energy resources flexibility management, Sustainable Energy Grids Networks (2017). (under review).

[40] IDE4L Project deliverable D6.3, http://ide4l.eu/results/.

Smart distribution networks, demand side response, and community energy systems: Field trial experiences and smart grid modeling advances in the United Kingdom

8

Eduardo A. Martínez Ceseña, Pierluigi Mancarella
The University of Manchester, Manchester, England

Acronyms

ADDRESS	Active Distribution networks with full integration of Demand and distributed energy RE SourceS
APS	Autonomic Power System
BAU	Business As Usual
C₂C	Capacity to Customers
CBA	cost-benefit analysis
CHP	combined heat and power
CI	customer interruptions
CML	customer minutes lost
CVR	conservative voltage reduction
DER	distributed energy resource
DIMMER	District Information Modeling and Management for Energy Reduction
DNO	distribution network operators
DSR	demand-side response
EHP	electric heat pumps
ENWL	Electricity Northwest
EV	electric vehicle
NOP	normally open point
NPC	net present cost
Ofgem	Office of gas and electricity markets
PV	Photovoltaic
RIIO	Revenue = Incentives + Innovation + Outputs
TSO	transmission system operator

Application of Smart Grid Technologies. https://doi.org/10.1016/B978-0-12-803128-5.00008-8

Nomenclature

Transactive   Economic and control techniques to manage energy transactions between end users and the grid

1 The UK electricity context

As with many other countries, the UK electricity sector is facing significant changes to integrate smart grid solutions that are meant to facilitate the production and consumption of energy in an economic, reliable, sustainable, and socially acceptable manner [1]. However, the particular regulatory framework and energy source requirements of the country as well as the interest of various stakeholders and policy makers on introducing innovation and a sustainable future make the UK energy context particularly appealing for smart grid applications.

This chapter provides a brief overview of smart grid research, particularly from projects that involve field trials in the United Kingdom. Specific focus is placed on smart grid applications aimed at: (i) enhancing distribution networks to facilitate the integration of distributed energy resources (DERs) that are emerging at the community level; and (ii) enabling the use of this community-based energy system flexibility to provide economic and environmental benefits to end customers and, potentially, different energy, capacity, reserve, and other services throughout the value chain.

Two UK projects were selected to showcase smart grid research on distribution networks, namely the "Capacity to Customers" (C_2C) [2,3] and "Smart Street" [4] projects. These projects address smart grid solutions based on innovative operation and nonasset-based network upgrades, especially considering demand-side response (DSR), which was recently put through trials in real UK networks. Three British or European projects were chosen to highlight novel smart grid applications of energy systems for a community, namely the "District Information Modeling and Management for Energy Reduction" (DIMMER) [5,6], the "Autonomic Power System" (APS) [7], and the "Active Distribution networks with full integration of Demand and distributed energy RESourceS" (ADDRESS) [8] projects. The projects demonstrate smart grid applications that enable community energy flexibility and allow the community to offer different services based on DSR, including distribution network support.

The rest of this section provides an overview of the specific characteristics of the UK energy system, taken into consideration by the above-mentioned projects for the development and testing of smart grid applications. This includes discussion on the current and expected future energy mix in the United Kingdom as well as existing energy markets and mechanisms, current distribution network practices, and characteristics of the consumption side. Afterward, in Section 2, the smart grid features developed in the above-mentioned projects for applications in community energy systems and distribution networks are discussed in detail. Relevant to smart grid research developments, particularly on the development of mathematical tools for the modeling and planning of smart systems are presented in Section 3. Section 4 provides several examples of smart grid trials and studies in real UK networks and communities. Finally, a summary of the chapter is presented in Section 5.

1.1 Overview and future scenarios

Broadly speaking, the UK power sector can be divided into generation, transmission, distribution, and consumption levels (see Fig. 1). Traditionally, most electricity is produced at the generation level using large-scale fossil fuel-based technologies located far from the consumption sites (e.g., offshore). The transmission is carried out by using high voltage level energy transport (e.g., 275–400 kV) from the generation stations to locations throughout the country. The energy is then carried by the distribution network (at lower voltages ranging from 6.6 to 132 kV) from the transmission substations to the locations of industrial, commercial, and domestic users.

The UK's energy mix has significantly changed during the last few years due to increased social awareness of environmental issues, new policies, and the emergence of more affordable low-carbon technologies, particularly DERs, which can be distributed at the community level. At the beginning of this decade, roughly 90% of UK electricity was generated through nuclear power and large-scale coal- and gas-fired generation while renewable sources only provided 3% (mostly coming from hydroelectricity). Nowadays, nuclear and large-scale coal- and gas-fired generation still provide about two-thirds of the electricity supply while 24% of the demand is now met with renewable energies. Among renewable technologies, there has been a dramatic integration of wind and photovoltaic (PV) power, as onshore wind capacity has increased from 0.2 GW in 2000 to 8.2 GW in 2015 (offshore wind capacity provided 4.9 GW of additional wind power in 2015) while installed PV capacity increased from 0.02 to 9.6 GW during the same period [9]. This represents a significant change in the structure of the power system, as all the PV capacity is installed at the distribution level (e.g., on rooftops of buildings). In fact, 23% of the generation capacity in the United Kingdom is now connected at the distribution level, specifically to the low voltage distribution network or to domestic, commercial, and industrial buildings.

The integration of distributed renewables, particularly PV, is creating new challenges to balance electricity generation and demand. This can be attributed to the intermittency of renewables as well as the seasonal variation of demand and the renewable resource. For example, PV penetration can cause severe issues to the UK

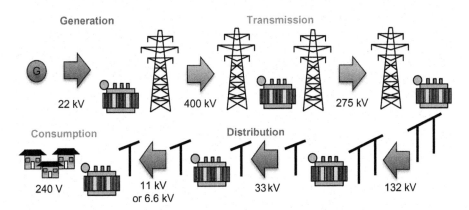

Fig. 1 UK electricity generation, transmission, and distribution levels.

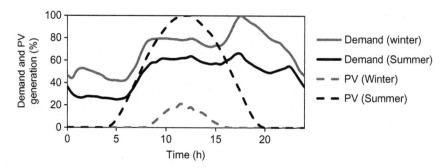

Fig. 2 Representative daily UK demand and PV profiles for different seasons.

electricity sector when considering the demand and solar profiles. As shown in Fig. 2, demand peaks in winter during evenings and is relatively low during summer at noon, which is not a good match with the relevant solar radiation profiles. As a result, without smart technologies to balance PV generation and demand mismatches, PV integration will introduce large volumes of power during periods when it is not required. This significant surplus of PV generation does not contribute to reducing peak demand while it can lead to significant balancing and network challenges at the system and local levels, respectively.

The future electricity generation mix is uncertain as it depends on energy policy, the economy, environmental concerns, and other factors (e.g., social acceptance of different technologies and security concerns). Nevertheless, it is possible to formulate future energy mix scenarios based on current conditions and practices and future targets. In this respect, National Grid, the transmission system operator (TSO), produces an annual report on future UK energy scenarios. The latest scenarios produced are based on assumptions ranging from (A), maintaining Business As Usual (BAU) practices, or (B) enabling smart community energy systems and providing strong governmental support to meet all environmental targets (Gone green scenario). An overview of these scenarios is presented in Fig. 3 [9].

These scenarios highlight the importance of the active involvement of smart community level systems on the management of the energy sector. However, significant changes to the existing energy markets and networks (particularly the distribution network) are required to cater to this emerging flexibility consumption side.

1.2 Energy markets and key actors

Liberalization of the electricity sector began in the United Kingdom in 1990 and, as a result, several actors have emerged to take the roles of retailers, TSOs, and others to trade different services in emerging markets and mechanisms. A brief description of some of these mechanisms and relevant actors is provided below:

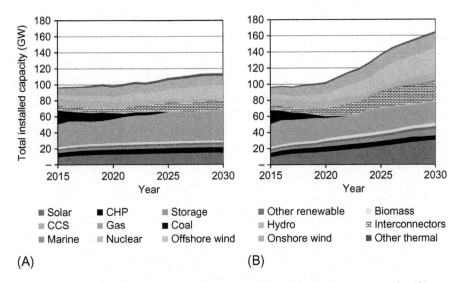

Fig. 3 UK electricity generation mix in the (A) BAU and (B) Gone green scenarios [9].

1.2.1 Electricity markets and mechanisms

- *Wholesale market*: The UK wholesale market is based on direct and unrestricted bilateral trades between different actors (e.g., generators and retailers). This mechanism allows the trade of electricity from several years in advance until 1 h before delivery (gate closure). As bilateral trading is done without regard for network constraints, National Grid, which takes the role of both system operator and transmission network operator, is responsible for assessing and approving or rejecting every transaction [10].
- *Balancing market*: As some actors may not deliver or consume the exact volume of energy agreed upon in the wholesale market, additional energy services are traded in the balancing market to match electricity supply and demand. The balancing market, which is also operated by National Grid, operates between gate closure and the time of delivery. In this market, generators offer to increase or decrease their output and retailers offer to increase or decrease the consumption of their customers if requested by the TSO.
- *Imbalance settlement*: As mentioned above, the balancing market allows the energy market to cope with actors that do not generate or consume the exact amount of power that they traded. As this results in additional costs for the system, some actors (i.e., balancing responsible actors) might be penalized for introducing imbalances to the market. Balancing penalties are calculated ex-post based on the position of the market (i.e., long or short) and the balancing responsible actors as well as the costs for balancing the market [11].
- *Ancillary service market*: As part of the grid code, some actors are obliged to provide specific ancillary services (e.g., reserve, frequency response, and so forth) [12]. Nevertheless, these (or other actors) may be flexible enough to provide services beyond their obligation, which can be traded in the ancillary service market. This market is managed by the TSO, which is the single buyer [13].

In addition to these markets, there are mechanisms in place to manage the electricity transmission and distribution businesses. The Office of gas and electricity markets

(Ofgem), which is the UK regulator, imposes price controls on the TSO as well as on the different distribution network operators (DNOs) across the country. These mechanisms are revised regularly (roughly every 5 years). The latest iteration of the price controls, namely Revenue = Incentives + Innovation + Outputs (RIIO), sets the network prices and incentives from 2015 to 2023 [14]. These prices are defined based on a wide range of mechanisms, tools, and considerations, such as uncertainty mechanisms, quality of supply incentives, broad measure of customer satisfaction, loss reporting and incentive mechanisms, an innovation stimulus package, operation and capital expenditure incentives, and workforce renewal incentives, among others [15].

1.2.2 Actors

- *TSO*: The UK TSO, namely National Grid, is a regulated entity that is responsible for the management and development of the grid transmission and for the operation of most electricity markets and mechanisms (without selling electricity to customers). National Grid provides physical access to the electricity market to different actors (e.g., generators) according to nondiscriminatory and transparent rules, and takes the role of market operator. In addition, the TSO ensures the security of supply, the safe operation and maintenance of the system, and a generation-consumption balance via services provided by different actors in accordance to the grid code or contracted in the balancing and ancillary markets [13]. In addition, balancing service and network costs are passed to the relevant actors based on specific mechanisms and system charges [16].
- *DNOs*: DNOs are regulated entities that are responsible for the transport of the electrical power on the distribution networks. DNOs provide physical access to the distribution network to customers according to nondiscriminatory and transparent rules. In addition, they are also responsible for the safe, reliable, and economic operation of the network and for investments in new infrastructure. Distribution costs are externalized to customers in the form of system and connection charges [17]. Currently, there are 14 licensed DNOs (and several independent DNOs) in the United Kingdom.
- *Retailer*: The electricity retailer acts as an intermediate agent between the wholesale and balancing energy markets and customers. The normal operation of the retailer involves buying energy from the wholesale market at a variable price and selling the energy to customers at flat prices. These retail tariffs include a profit margin that retailers charge for "protecting" customers from the variable market prices and signals. Retailers are balancing responsible actors and are penalized for introducing imbalances to the markets (i.e., from consuming a different amount of power than contracted), and are allowed to participate in the balancing market.
- *Generators*: Bulk generation comprises traditional large generators connected to the transmission system, such as nuclear and large-scale coal- and gas-fired generation. These generators can participate directly in the wholesale, balancing and ancillary markets. Due to economies of scale, this type of generator generally has an advantage in the wholesale market over DER. However, this is not necessarily the case on the balancing and ancillary markets where specific types of technologies may have the advantage regardless of economies of scale.
- *Customers*: Customers can comprise a variety of entities such as households, residential and commercial buildings, and small businesses, among others. In a traditional context, these customers passively consume electricity. However, as will be further discussed below,

customers in a smart grid context may possess DERs such as PV systems, combined heat and power (CHP) boilers, and so forth, which can enable customers to actively participate in the different energy markets and mechanisms.

1.3 Distribution networks

Under current preventive security standards such as traditional N-1 criteria or P2/6 engineering recommendations in the United Kingdom [18], the distribution networks are configured as open rings. That is, groups of two or more distribution networks at the 6.6 and 11 kV levels are configured as open rings, interconnected through normally open points (NOPs), as shown in Fig. 4A. Accordingly, if a contingency occurs, all customers connected to the affected feeder are immediately disconnected from the feeder (i.e., C1, C2, and C3 in Fig. 4B). Afterward, as shown in Fig. 4C, the protections system actuates to isolate the contingency (alongside some customers, i.e., C2) and restores supply to the customers that can be reconnected to the original feeder (i.e., C1). This automatic network reconfiguration procedure is usually done within 3 min as interruptions that last less than 3 min are not regulated in the United Kingdom and, thus, do not lead to financial penalties for DNOs.

At this stage, manual operations are required to restore supply to other affected customers (i.e., C2 and C3). These operations may include the manual operation of the NOP, which takes roughly an hour and, in the example, can be used to restore C3 (see Fig. 4D). Other alternatives required for customers directly connected to the contingency (i.e., C2) include sending a crew to the site to manually isolate the fault and restore customers by installing a mobile generator, or connecting them to the original or a neighboring feeder. The latter option is illustrated for C2 in Fig. 4D.

Based on these preventive operation and restoration practices, each distribution feeder must be significantly oversized so that it can be used to supply customers from

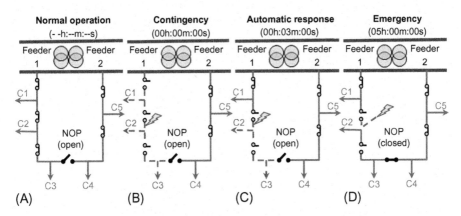

Fig. 4 Traditional distribution network operation and restoration practices – (A) normal operation conditions, (B) actuation of protection devices after a contingency occurs, (C) automatic network reconfiguration, (D) emergency conditions while the contingency is manually cleared.

neighboring feeders during emergency conditions without violating any technical constraints (dictated by voltage and thermal limits [19]). This practice can be deemed highly expensive considering that distribution networks seldom experience interruptions (e.g., roughly once every 3 years or even less frequently in urban areas [20]) while no interruptions have ever been recorded in some distribution networks). Nevertheless, regardless of how unlikely contingencies may be, without sufficient spare feeder capacity, customers could not be restored within acceptable time frames if the network were to enter emergency conditions. Accordingly, these preventive practices are reasonable in a traditional context where customers are passive and there are no smart means to actively manage voltage and thermal limits that may arise during emergency conditions.

1.4   The consumption side

Customers have traditionally been seen as passive actors that do not have the flexibility to get involved in the active management of the system. As shown in Fig. 5, the only alternative that customers have to provide active DSR to support the energy system would be to forego comfort by curtailing or shifting their energy consumption. Accordingly, these customers are generally "protected" from the variable nature of the energy system by retailers who only send them flat tariffs. DSR applications are still possible but mainly for services with low frequency and high benefits, as identified in the ADDRESS project [8].

This vision of passive energy customers made sense in a traditional context where demand was highly predictable and supplied mainly by flexible generation where customers had little or no distributed generation or other forms of energy infrastructure. However, this fundamentally changes in a smart grid context where customers may possess different energy devices (e.g., PV panels and storage devices), information and communication technologies to interact with these devices, and smart software to customize their operation [12,21,22]. These smart technologies provide significant consumption side flexibility, particularly at the community levels where diversity in energy needs and available infrastructure can be exploited through multienergy exchanges between buildings [5].

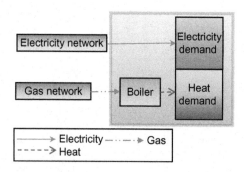

Fig. 5 Traditionally inflexible customers.

2 Smart grid features

In a smart grid context, significant flexibility is expected to shift from the generation to the consumption side. As a result, the manifold energy services that are currently provided by large-scale, flexible, and mostly fossil fuel-based generation units will now be provided by a large number of smaller, sometimes intermittent and uncertain DERs. This transition of flexibility from the generation to the consumption side changes the way the energy system must be planned and operated. In particular, the distribution networks, which have been historically designed to transport unidirectional flows from substations to the customers, must now cater to the integration of DERs and allow constant bidirectional power flows associated with the trade of energy services between active customers and/or external markets. At the same time, new community energy systems could emerge to aggregate and coordinate the DERs and improve the potential of the consumption side to provide vital flexibility for the energy system.

Based on the above, dedicated smart grid technologies and research have emerged with the aim of upgrading existing distribution networks and enabling flexibility from community energy systems (emerging research on the topic, including relevant models, is discussed in Section 3.1). This section will provide an overview of the technical, economic, and commercial features of smart grids in the context of smart distribution networks and community energy systems, based on findings from relevant research projects and trials.

2.1 Smart distribution networks

Smart distribution networks are expected to facilitate the integration of large numbers of DERs such as electric vehicles (EVs), solar PV panels, CHP boilers, and other technologies. These DERs challenge the status quo of existing distribution networks that have not been designed to accommodate large and stochastic loads (e.g., from EVs) in combination with customer exports (e.g., from PV), or enable customers to use their new flexibility to trade energy services in different markets. If current preventive operation and restoration practices continue to be used in this smart grid context (see Section 1.3), major network investments (leading to high costs for customers) would be required to accommodate all the emerging DERs [23]. In addition, this approach would neglect the increasing customer flexibility, which is a big waste when considering the plentiful and valuable applications of consumption side flexibility [8,24]. For example, customer flexibility can be used for the active management of voltage and thermal limits that may arise during emergency conditions. This service can facilitate integration of DERs while also reducing the need for spare emergency capacity and costs associated with traditional preventive distribution network practices [19,25].

In light of the above, several UK research projects and trials have been aimed at developing and testing smart grid technologies as a means to both facilitate the integration of DER and manage customer flexibility to reduce investments in emergency capacity, such as the C_2C and Smart Street projects [26,27]. Both projects investigate the potential of customer flexibility in combination with smart network automation and reconfiguration technologies and practices to reduce the need for spare feeder capacity. As part of the C_2C project, novel automation technologies and commercial agreements were developed to test DSR in a wide range of real UK networks. The trials comprised 180 HV rings and 20 HV radial circuits, involving more than 1300 industrial and commercial entities and 300,000 domestic customers. These circuits were highly reliable, experiencing less than one interruption every 5 years, which is representative of roughly 80% of the network owned and operated by Electricity Northwest (ENWL), the DNO that proposed the C_2C method. Among these networks, 36 rings were monitored in detail to produce the required data to inform different technical and economic studies for the project. The Smart Street project considers new smart solutions in addition to the C_2C method, and is also being put through trials in ENWL networks until 2018. More specifically, the project explores combinations of smart options to enhance the automatic response of the distribution networks, such as the active use of on-load tap changers and different network configurations. These options provide conservative voltage reduction (CVR) to lower consumption from voltage-dependent loads when needed (this can be used alongside DSR) [4].

In a smart grid context (in light of C_2C and Smart Street solutions), distribution networks possess enhanced automation levels that allow rapid reconfiguration in response to contingencies while network limits can be actively managed with postcontingency DSR. As a result, alternative network configurations such as operating the feeders as closed rings could be used during normal conditions (see Fig. 6A). This ring configuration may provide technical and economic benefits in the form of improved voltage profiles and reduced power losses as well as lower energy consumption, costs, and emissions associated with CVR [2,28]. Nevertheless, these benefits are

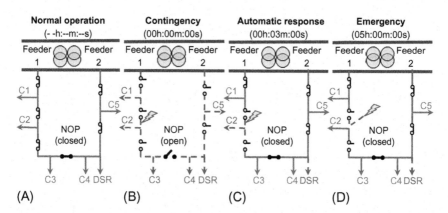

Fig. 6 Example of smart distribution network operation and restoration practices: (A) normal operation conditions, (B) actuation of protection devices after a contingency occurs, (C) automatic network reconfiguration, and (D) emergency conditions while the contingency is manually cleared.

case-specific and must be assessed on a case-by-case basis using dedicated technoeconomic analysis tools (e.g., Ref. [25]).

Assuming that the feeders are connected as rings, if a contingency were to occur in any part of the network, both feeders would be momentarily disconnected as the protection devices and now-automated NOP actuate (see Fig. 6B). This can be considered an undesired effect as, if the feeders were operated as radial networks, only the customers in the affected feeder would be disconnected. Accordingly, it can be debated whether or not the case-specific technoeconomic benefits associated with the ring configuration justify an increased customer exposure to infrequent contingencies. Regardless of the selected configuration, the improved network automation levels facilitate fast (under 3 min) restoration of more customers compared with traditional network conditions (this can be seen by comparing Figs. 4C and 6C). That is, the only remaining affected customers at this stage would be those that are directly connected to the contingency (i.e., C2 in Fig. 6C).

As in a traditional context (see Section 1.3), a crew can be sent to manually disconnect the remaining affected customers from the fault and restore their supply by either installing a mobile generator or connecting the customers to the original or a neighboring feeder (see Fig. 6D). However, in this smart context, DSR could be deployed to manage network violations that may arise after a contingency occurs and the network enters emergency conditions. These practices effectively avoid the need for investments in preventive spare network capacity. Yet, DSR would only be required if the unlikely contingency occurs during times of network stress, such as during peak time, which in the United Kingdom would be in the evening of particularly cold days during the winter season (see Fig. 2 in Section 1). Thus, DSR for distribution network capacity support is a service with low frequency and a potentially significant economic value [8]. As discussed in Section 1.4, such a service can be attractive even for traditional passive customers.

Based on the above, smart grid applications in the United Kingdom have combined the use of new technologies and commercial arrangements to enable DSR and improved network response. These applications are expected to provide attractive economic benefits (especially for DNOs and customers) as the smart distribution networks can cope with demand growth while deferring or avoiding traditional asset-based interventions (e.g., investments in feeder and substation upgrades) while still meeting or improving network security levels and reliability standards.

2.1.1 Technologies

As part of the UK network trials, selected distribution feeders were enhanced to enable improved network operation and the provision of DSR within relevant commercial and technical arrangements. Robust and low-cost remote controls were installed at the NOPs to enable automatic response as well as the option for normally closed operation. Moreover, several switches and automatic controls (e.g., 132 kV/6.6 kV/240 V switches and molded case circuit breakers) were installed throughout the networks to improve automatic network response and, potentially, isolate contingencies in smaller sections of the networks. In addition, new technologies were developed to enable the use of different network configurations. Automated and controllable low-voltage switches (called LYNXs) were developed to allow reconfiguring or meshing low-voltage networks and, from the protection perspective, fuses at the low-voltage side of the distribution transformers were replaced with smart breakers called WEEZAPs [29]. Both LYNX and WEEZAP devices can be operated remotely and can record valuable technical data (i.e., phase power, voltage, current, and so forth).

In order to enable DSR, circuit breakers fitted with automation were installed for new customers or, in the case of existing customers, available devices were upgraded with a retrofitted actuator. The devices were fitted with a remote terminal unit to provide the required communication link. The trial circuits were also enhanced with additional monitoring equipment (e.g., supervisory control and data acquisition systems). This enabled the collection of real-time network data for the analysis of real performance and to inform a series of network simulations and modeling exercises.

The automation algorithms within ENWL's *control room management system*, which is in charge of the normal operation and restoration of the networks, were updated with so-called *automatic restoration sequence algorithms* to take advantage of the enhanced automation levels, network reconfiguration capabilities, and DSR. These algorithms are based on predefined actions informed by the conditions of the network as estimated based on offline power flow estimations (using PowerOn Fusion software [30]) informed by measured demand data.

2.1.2 Commercial arrangements

New commercial agreements were developed for active customers who were willing to provide DSR. After testing several permutations of these contracts, two types of contracts (for existing and new customers) based on variations of the existing national terms of connection agreement were selected. These contracts define the conditions under which the customers are willing to provide flexibility, such as, for example, protected days, maximum number of calls per year, contract length, and other parameters. This work has led to the creation of managed contracts for the provision of DSR in the United Kingdom, which has now become BAU practices [31].

2.1.3 Network security and reliability

In the United Kingdom, distribution network security standards are based on P2/6 engineering recommendations, which are analogous to N-1 conditions but also include considerations for time and network demand [18]. Reliability levels are regulated in terms of customer interruptions (CI) and customer minutes lost (CML) for interruptions exceeding 3 min. As both security and reliability levels are based on infrequent interruptions, it was impractical to wait for occurrences in every trial network to obtain relevant security and reliability data. Therefore, network security limits and reliability levels were estimated based on simulations, which were updated in light of the five contingencies that were recorded in the trial networks throughout the duration of the C_2C project [32].

The security limits were simulated based on P2/6 considerations and AC power flows to identify the firm capacity of the network [20,25]. More specifically, contingencies in every section of the network were simulated to identify the potential network configurations during emergency conditions. AC power flows were used to identify thermal and voltage constraints and assess whether the networks were able to supply customers, considering daily load curves and emergency ratings (i.e., up to 20% overload for 2 h). The studies considered five demand growth scenarios proposed by ENWL as well as different options to upgrade the feeders and substations and deploy network automation and DSR.

Network reliability was assessed via Monte Carlo simulations while assuming, (i) real failure rates for the available trial networks (a rate of 0.05 failures/km/year was assumed when real data was unavailable [20]), (ii) the existing number of customers in each feeder (distributed across the network based on the demand at each load point), (iii) manual operation of the NOP of 1 h on average and distributed based on an exponential probability density function, (iv) no NOP automation in the existing networks, (v) mean time to failure, mean time to repair, and mean time to switch of 175.2, 5, and 1 h, respectively, and (vi) that the automated network can restore supply within 3 min after a contingency occurs. Randomly allocated faults are repeatedly simulated thought the networks to estimate average interruptions, also considering the conditions mentioned above. The results are the expected CI and CML for each network subject to different demand growth scenarios and interventions (i.e., reinforcements, enhanced automation, DSR, and so forth).

2.1.4 Economics

The economic evaluations of smart grid solutions at the distribution level, such as the use of DSR, should be consistent with existing UK regulations on distribution network assets built, namely the cost-benefit analysis (CBA) framework introduced as part of the RIIO price control [33]. This CBA framework proposed by Ofgem, herby called Ofgem's CBA framework, provides a set of mathematical tools to assess all investments in distribution assets. More specifically, as denoted by Eq. (1), investments in distribution network interventions are assessed in terms of the net present cost (NPC) criterion, which sums the discounted investment ($Investment\_costs_y$) and social ($Social\_costs_y$) costs associated with the intervention during every year throughout predefined discount rates ($d = 3\%$) and lifetime ($T = 45$ years).

$$\text{NPC} = \sum_{y=1}^{T} \frac{Investment\_costs_y + Social\_costs_y}{(1+d)^y} \tag{1}$$

It is worth noting that investment costs are attributed to all capital expenditures associated with the intervention (e.g., investment costs, annual payments, maintenance, etc.) whereas social costs are based on losses, emissions, CI, and CML, among other factors. Further details on Ofgem's CBA framework and its applications can be found in the available literature [2,19,34].

2.2   Smart community energy systems

The participation of the consumption side in the active management of the energy system is one of the key features of the smart grid paradigm [35]. As a result, significant focus has been placed on understanding and quantifying customer flexibility, particularly considering the emergence of DER. The integration of DERs at the community level provides flexibility for smart buildings and communities to improve their energy efficiency in terms of reduced economic costs and emissions. This flexibility may be attractive at the building level, but the inherent constraints of the geographical area and resources within buildings limit potential applications and benefits. Conversely, a district level perspective where buildings actively trade multiple energy flows between them (and the markets) may maximize consumption-side flexibility by exploiting diversity in technologies and energy needs [5,36]. However, properly modeling and assessing the flexibility of customers who may have access to different types of DER while also considering multienergy flows between buildings within communities and service trading between the community and the different markets is a daunting task that requires the development of new technoeconomic and business case frameworks [36,37].

To this end, as part of the ADDRESS project, new technoeconomic tools were developed to explore the value that customers, particularly small domestic and commercial customers, could accrue from DSR. It was shown that network support services, such as the DSR applications discussed above in Section 2.1, can be attractive for customers due to their high value and low frequency [8]. Conversely, most other DSR applications do not offer attractive business cases, as DSR would be called frequently (causing discomfort) and lead to low economic benefits due to the small size of customers. Nevertheless, as more and more DERs emerge, customer flexibility increases as do the relevant DSR benefits. Furthermore, as explored in the APS project, in a smart grid with significant DERs, the intelligent and coordinated operation of the different energy devices allows the provision of DER while, at the same time, meeting all energy needs of customers (i.e., customers no longer forego comfort). Accordingly, strong business cases for DSR from smart customers and, especially, communities arise in a smart grid context [38]. These applications, ideas, and lessons learned, particularly focused on improving energy efficiency, were further explored and tested in the UK context in the DIMMER project.

As part of the DIMMER project, a portfolio of tools was developed for the simulation and assessment of energy efficiency within districts as well as for the optimization of community operation to maximize benefits when partaking in different markets and mechanisms. The tools were tested on two demonstrators. The first demonstrator was based on buildings connected to the heat network that supplies the Polytechnic of Turin in Italy. The second demonstrator comprised buildings owned and managed by The University of Manchester, some of which are interconnected through shared electricity, heat, and gas networks [6]. The Manchester demonstrator is naturally a smart community multienergy system. An overview of relevant research on this topic as well as a brief description of the tools developed is presented in Section 3.2.

2.2.1 Technologies

Several technologies were required for the DIMMER field trials and studies, including multienergy technologies, data acquisition systems, and a portfolio of mathematical tools to simulate and optimize the behavior of the community energy system. The multienergy technologies under consideration were those that were already in place in the Manchester demonstrator (e.g., PV panels and gas boilers) and those that the university is exploring as a means to improve energy efficiency in the coming years (e.g., CHP boilers, PV panels, and electric heat pumps (EHP)). An example of how these technologies may interact is presented in Fig. 7.

Historical energy consumption data for all buildings throughout the University of Manchester is available with a half-hour resolution from the Coherent data system [39]. An interface was built between the DIMMER platform and the Coherent data system to acquire information in real time. The mathematical tools developed for the project included, (i) multienergy operation, simulation, and optimization models [40,41], (ii) integrated multivector district energy network models [42], (iii) an integrated district energy management system framework [24], and (iv) a CBA and multiflow mapping model [38]. These tools allow the assessment of the energy performance of community energy systems, also considering smart coordination of DERs for the minimization of energy costs and emissions, and trade of services in different markets and mechanisms. The operation of the communities takes into consideration multienergy exchanges between different buildings, which are analyzed using dedicated technical models. The business case of different energy services is evaluated from the perspective of the community as well as considering impacts to generators, DNOs, the government, and other actors (see Section 1.2.2 for a list of relevant actors).

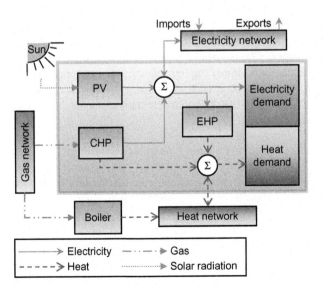

Fig. 7 Multienergy building/community system.

2.2.2 Commercial arrangements

In a traditional energy system context, passive customers only interact with retailers, which protect the customers from the dynamic prices and signals in the energy system by only exposing them to flat retail prices (see Section 1.4). As a result, customers are unable to see and react to the multiple components that constitute the customer price, such as system fees for the distribution and transmission networks, wholesale electricity prices, various taxes, capacity, the retailer margin, and so forth [38,43]. This is a reasonable approach for customers who are unable to react to these dynamic signals, but not for smart customers such as those within smart community energy systems who could use their flexible resources to benefit from these signals and relevant markets.

Allowing some of the variable components of these customer prices to be passed through to customers, which is known as a transactive energy approach [36,44], is crucial to incentivize the smart operation of communities [45]. More specifically, community energy systems could actively respond to variable signals as a means to minimize energy costs and emissions as well as to maximize revenue from participating in different markets. This smart community behavior where flexibility is used to trade services in different markets can, in principle, enhance the efficiency and security of the energy system. Accordingly, as part of the DIMMER trials, the operation of the smart community system is simulated and optimized under the consideration that different combinations of flat retail and dynamic market prices are passed down to customers.

2.2.3 Energy efficiency

The use of smart solutions and DERs at the building level are generally seen as attractive options to improve energy efficiency by reducing energy consumption. This traditional definition of energy efficiency focuses on energy savings (*Energy_savings*) that, in principle, lead to lower energy costs and emissions. That is, improving energy efficiency based on this traditional view simply implies reducing energy consumption (*Current_Energy*) compared with a baseline (*Baseline_Energy*) while considering externalities, as shown in Eq. (2). It is worth noting that externalities can be associated with weather, occupancy levels, and so forth.

$$Energy\_savings = Baseline\_Energy - Current\_Energy \pm Externalities \qquad (2)$$

The main disadvantage of this definition of energy efficiency is that it cannot consider the different dynamic costs and carbon intensities associated with different energy vectors. For example, based on the traditional energy efficiency concept, reducing electricity consumption has the same value regardless of whether this occurs when the market is saturated with highly costly and dirty fossil fuel-base energy, or clean and cheap renewable energy. This is not reasonable, as customers should not make active efforts to reduce their consumption when there is surplus renewable energy [36].

Based on the above, the traditional energy efficiency is only adequate for passive customers who cannot respond to dynamic signals. However, this concept undermines the flexibility of smart community energy systems that can actively respond to dynamic prices and carbon intensities and exploit diversity by allowing multienergy trading between buildings. Thus, in order to properly capture the flexibility inherent in smart community energy systems, new energy concepts have been proposed, such as the one denoted by Eq. (3) [5,36].

$$Energy\_Consumption = \sum_m \sum_b \sum_t Imports_{v,b,t} \times Imports\_Externalities_{v,b,t}$$
$$- \sum_m \sum_b \sum_t Exports_{v,b,t} \times Exports\_Externalities_{v,b,t} \quad (3)$$

This new concept of dynamic environmental efficiency explicitly considers multienergy (m) exchanges between buildings (b) through time (t) as well as variable externalities relevant to imports or exports. As before, these externalities can account for weather and occupancy levels, but now variable energy market conditions and the physical constraints of the integrated (electricity, heat, and gas) energy networks can also be considered. This is vital for smart community energy systems as it allows the consideration of the consumption side for the provision of flexibility to the energy system [36,46,47]. Discussion on the proposed dynamic environmental efficiency and technoeconomic models developed to implement this improved efficiency paradigm in smart multienergy communities can be found in Ref. [36].

3 Research

Smart grid research in the United Kingdom, specifically from the C$_2$C, Smart Street, ADDRESS, APS, and DIMMER projects discussed in this chapter, has focused on the trial of new smart solutions and the development of proper technoeconomic and business case models. As highlighted in the previous sections, this research can be classified into two themes: smart distribution networks and smart community energy systems. The former theme places great focus on enhancing the distribution networks to enable the use of flexibility from smart community energy systems. The latter theme centers on using community-level flexibility for the provision of different energy services to the energy system.

3.1 Smart distribution networks

As discussed in Section 1.3, distribution networks in the United Kingdom are planned and operated based on traditional preventive security standards and mainly traditional solutions (e.g., feeder and substation upgrades). This has allowed DNOs to effectively evaluate distribution network asset investments based on simple, deterministic, and rule of thumb-based planning approaches [48]. However, as more and more DERs emerge at the community level, the adequacy of existing distribution network planning approaches becomes questionable.

Emerging UK regulations (i.e., RIIO, see Section 1.2.1) are encouraging DNOs to change their current practices and propose and test new smart solutions (e.g., DSR), which could make networks more economically efficient and flexible to cope with uncertainty [25]. For example, as discussed in Section 2.1, active network management from DSR can effectively defer costly investments in assets that would normally be required to ensure that the network meets statutory limits [19]. Furthermore, in an uncertain environment (e.g., subject to uncertain uptake of EVs and PV generation), smart solutions allow DNOs to wait until demand has either (i) grown enough to justify asset investments or (ii) remained below the firm capacity of the network. This flexibility to defer or even avoid investments can be significantly valuable for DNOs [19,49].

Considering that existing DNO planning practices cannot properly address smart solutions, Ofgem has proposed a new CBA tool for distribution asset planning (i.e., Ofgem's CBA framework). However, the framework presents several drawbacks, particularly when modeling uncertainty and addressing an ever growing number of smart solutions and their combinations [2]. This has led to significant research at the University of Manchester on developing different (and improved) versions of Ofgem's CBA framework, coupled with simulation and optimization engines [6,19,25,48–51]. The different versions of the arising tool provide different means to simulate current distribution network planning and operation practices, and to optimize them in light of smart solutions.

Within the different tools developed, the simulation engines are used to replicate current practices, which undermine the valuable flexibility provided by smart solutions. These engines are particularly valuable to investigate whether new smart solutions can provide value even if the DNOs do not change their behavior. Conversely, the optimization engines can capture the value of smart solutions, especially in the face of uncertain demand growth. These engines demonstrate the value that smart solutions could bring about if DNOs were to update their practices (moving from deterministic rule of thumb to stochastic optimization approaches). These engines, some of which are provided as open source material [25], also allow the consideration of bespoke smart solutions and complex representations of uncertainty (e.g., using scenario trees and/or robust constraints as shown in Fig. 8). The tools presented here will be discussed furthered and illustrated in Section 4.1.

3.2 Smart community energy systems

In a traditional environment where customers are passive actors who cannot provide services to the energy system, community energy management is limited to reducing net energy consumption as a means to improve energy efficiency (see Section 1.4) [52]. However, this approach does not make sense in a smart grid context where community energy systems can manage their DERs to respond to costs and emissions associated with the energy mix, market prices and signals, security and reserve constraints,

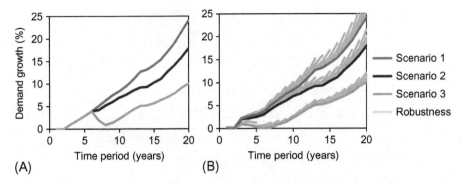

Fig. 8 Representation of uncertainty with (A) scenario trees and (B) scenarios with robust constraints.

and so forth. Accordingly, as discussed in Section 2.2.3, new concepts beyond energy efficiency must emerge. Such concepts must recognize the value of consumption-side flexibility to address dynamic energy costs and emissions brought about by the presence of renewable energy and low-carbon technologies as well as (increased) system needs for security, balancing, reserve, and so forth.

Emerging research at the University of Manchester aims to extend the energy efficiency concept to properly address the envisioned characteristics of the future smart grid, such as net zero energy building and energy positive neighborhoods, among others [52,53]. In addition, there is increasing research on the potential of smart community energy systems to provide services throughout the energy sector and, thus, improve the energy efficiency on a system level [36,54]. These services can include, but are not restricted to, energy and capacity traded in different markets [55] as well as bespoke services, for example, distribution network security and reliability support (particularly if the community can operate as a microgrid) [56,57].

In the case of the United Kingdom, where peak energy consumption is often driven by heating needs, smart community energy systems research must also recognize the role of heating and other energy vectors (e.g., gas). However, due to the complexity of the study, a portfolio of interacting simulation and optimization engines had to be developed to properly consider building level DER [40,41,58], multienergy interactions between buildings within communities [6,42,59], and trade of services between the district and the energy system [36,38]. A high-level overview of a framework that combines these models is presented in Fig. 9 [24,45]. It is worth noting that relatively few modifications would be required to apply most of these tools in the context of other countries, for instance to deal with cooling. Regardless, detailed understanding of the existing or proposed market and regulatory environments is required to update the business case assessment tool (see Fig. 10), which has to map the information, energy flow, costs, and revenue, and other parameters in light of the relevant markets and mechanisms (see Section 1.2.1). The tools presented here will be discussed further and illustrated in Section 4.2.

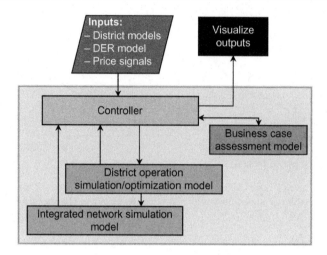

Fig. 9 Portfolio of models for community energy system modeling.

Fig. 10 Business case
assessment tool.

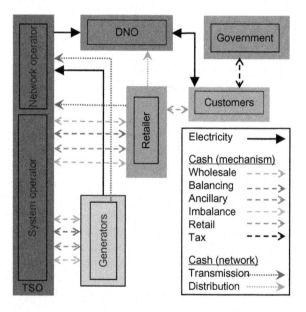

4   Case studies and field trials

This section presents applications of the smart solutions presented in previous sections
and, particularly, the tools discussed in Section 3. More specifically, studies and lessons learned from smart distribution network trials throughout the Northwest of the
United Kingdom and smart community multienergy trials in Manchester are discussed
below.

4.1 Smart distribution networks

Smart solutions based on enhanced automation levels, active network reconfiguration, and postcontingency DSR (see Section 2.1) were tested in 180 HV rings and 20 HV radial circuits owned and operated by ENWL in the Northwest of the United Kingdom. Among these systems, 36 networks where monitored and simulated in detail to inform the tools discussed in Section 3.1. The tools were used to formulate four different strategies for the implementation of smart solutions. These strategies were considered to shed light on the potential value of the smart solutions under (i) current planning practices and (ii) upgraded DNO practices based on optimizations:

- *Baseline*: This strategy is meant to reflect current BAU practices and ignore smart solutions such as improved operation and restoration alternatives from increased network automation (including NOP automation) and DSR. Accordingly, the interventions associated with this strategy are formulated with a rule-of-thumb simulation engine [48] (optimality is not guaranteed). This strategy only considers traditional feeder and substation reinforcements as options to upgrade the network in response to demand growth forecasts.
- *Smart (Radial)*: This strategy aims to emulate the impact of smart solutions under existing planning practices that have not been designed to properly address them. Based on this, the distribution networks are operated as open rings and the smart solutions are used to expedite the restoration process and actively manage thermal and voltage constraints. As in the case of the baseline, a simulation engine is used to choose the interventions to be deployed [48].
- *Smart (Ring)*: As in the previous case, this strategy aims to emulate the impact of smart solutions under existing planning practices. However, this time, the networks are operated as closed rings whenever the restoration system is enhanced. Thus, this strategy can provide economic and environmental benefits from reduced demand and power losses during normal operating conditions.
- *Optimal*: This strategy explores the benefits of smart solutions in light of improved operation and planning practices based on optimizations. That is, an optimization engine that is now available as open source material [25] is used to cater to the complex flexibility from potential combinations of traditional (i.e., feeder and substation reinforcements) and smart solutions (e.g., enhanced automation and DSR).

4.1.1 Smart distribution network applications

The consideration of smart grid solutions based on the above-mentioned strategies is illustrated here with applications to one of the 36 trial networks that have been monitored in detail as part of the C_2C project [60]. The network selected for this study is the Chamber Hall system (see Fig. 11), which is a 6.6 kV distribution network connected to a 20.96 MW substation. The two feeders selected to form the open or closed rings are the Europa House and Peel Mill No.2 feeders. The Europa House feeder is formed by 21 lines connecting 22 nodes and supplies 1950 customers (mainly urban). The Peel Mill No.2 feeder is formed by 22 lines connecting 23 nodes and supplies 1660 customers (mainly urban).

The study considers the implementation of one or combinations of the traditional and smart interventions presented in Table 1. These interventions are deployed based on different strategies (i.e., baseline, smart, and optimal) to ensure

Fig. 11 Chamber Hall
6.6 kV distribution network.

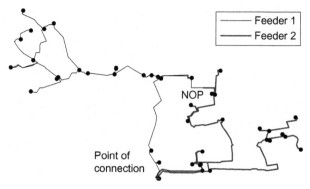

Table 1 **Interventions considered for the study**

Intervention	Description
IF1	Several sections of the feeders are reinforced. Following current DNO practices, the upgrades are made so that, after the reinforcement, the network can withstand at least an additional 5% demand growth without exceeding statutory limits.
IF2	After IF1, if firm capacity is reached again, a second set of sections of the feeders is reinforced.
IS	The substation is upgraded by installing an additional transformer.
IAL	The automation levels of the distribution network are enhanced (including automation of the NOP) to facilitate improved customer restoration. However, the network is still operated as an open ring.
IAG	The automation levels of the distribution network are enhanced (including automation of the NOP) to facilitate improved customer restoration and normal operation as a closed ring.
ID	Relevant contractual arrangements and automation and communications infrastructure are put in place to enable DSR.

that the network limits are not exceeded under the different demand growth scenarios presented in Fig. 12.

Based on studies informed with data taken from UK trials, upgrading the substation costs $338£ \times 10^3$ and takes 3 years whereas reinforcing the feeders costs $110£ \times 10^3$ and takes 2 years. The network automation levels can be enhanced at a cost of 19 k£, and up to two 0.5 kW peak DSR blocks can be contracted. For this purpose, increasing network automation levels costs 19 k£ per year whereas customer automation costs $13.5£ \times 10^3$ per block, and there are additional $6.3£ \times 10^3$ costs in billing and management fees per block. Overall, it is considered that enhancing the automation levels of the network and contracting DSR takes roughly a year. The recommended interventions and associated NPC associated with each investment strategy are presented in Tables 2 and 3, respectively.

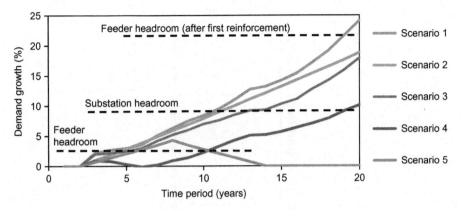

Fig. 12 Firm capacities of the Chamber Hall 6.6 kV in different scenarios.

Table 2 **Strategies to upgrade the Chamber Hall network**

Scenario	Baseline		Smart (Radial)		Smart (Ring)		Optimal	
1	IF1	3	IAL+ID	4	IAG+ID	4	IF1	3
	IS	15	IF1	14	IF1	14	IAG	15
	IF2	18	IS	42	IS	42	ID	17
							IS	42
2	IF1	4	IAL+ID	5	IAG+ID	5	IF1	4
	IS	17	IF1	15	IF1	15	IAG	10
	IF2	37	IS		IS		ID	19
3	IF1	4	IAL+ID	5	IAG+ID	5	IF1	4
	IS	17	IF1	17	IF1	17	IAG	13
	IF2	38					ID	19
4	IF	9	IAL+ID	10	IAG+ID	10	IF1	9
5	IF	5	IAL+ID	6	IAG+ID	6	IAG	1
							ID	6

The study shows that, even under current planning practices, the introduction of smart nonasset-based solutions generally leads to lower investment costs. This is mainly due to deferral or avoidance of traditional asset-based investments, particularly costly substation upgrades. The smart solutions can also bring about attractive social benefits if ring operations are allowed, regardless of the potential increase in short-term interruptions (short-term interruptions are not penalized under current regulation). However, due to inherent uncertainty in the demand growth forecasts, current practices may not properly address smart solutions and could actually lead to increased network costs, as can be seen in the application of the smart strategies on Scenario 4.

The use of smart nonasset-based solutions becomes more valuable when improved network practices, based on proper optimization tools, are considered (i.e., optimal

Table 3 Expected NPC (£ × 10³) associated with different strategies to upgrade the Chamber Hall network

Scenario	Baseline			Smart (Radial)			Smart (Ring)			Optimal		
1	320	565	884	291	566	857	291	544	835	239	529	768
2	299	542	840	246	543	789	246	522	768	199	497	696
3	299	536	834	253	539	793	253	519	772	198	496	693
4	87	501	588	147	525	672	147	512	659	87	501	588
5	102	447	548	39	494	533	39	476	515	59	442	501

strategy). As presented, optimal strategies tend to combine traditional and smart solutions as a means to minimize costs. This makes sense considering that DSR can be an attractive option to defer or avoid costly investments, but the annual DSR payment may also become pricey. Accordingly, DSR can be used to postpone costly investments until it is clear that the intervention will be required (avoiding sunk costs) while less-expensive interventions (IF1 in this case) can also be used to reduce the need for DSR, allowing the DNO to contract fewer DSR blocks. It is important to note that these smarter network practices can also identify conditions where traditional interventions are still cost-effective (i.e., Scenario 4) or when smart solutions are preferred (i.e., Scenario 5).

These results highlight that smart grid applications can bring about attractive technoeconomic benefits. However, proper tools, or at least a proper understanding of the conditions when these solutions make sense, are needed for exploiting the full potential of smart grids. Based on these premises, the optimization engine used for the aforementioned studies has been made available as open source material [25].

4.1.2 Wider network implications

A more comprehensive analysis of smart grid applications to distribution networks requires the consideration of a wide range of networks under different conditions. For this purpose, in this section, the smart solutions discussed above are addressed here in light of 36 real UK distribution networks (see Fig. 13 for the schematics of some of these networks). The following considerations were taken for the studies:

- *Costs and assumptions*: All economic criteria (e.g., discount rates, planning horizon, and price for power losses, among others) were taken from the trial networks, were recommended by the DNO, or were based on existing regulations [3]. However, these conditions may change in the future, which must be quantified. To this end, cost variations, particularly making traditional asset-based solutions more cost-effective, are explored.
- *Substation headroom*: Smart solutions are attractive options to defer or avoid costly interventions, such as substation upgrades. This conclusion is tested by varying substation headroom so that conditions arise where investments in additional transformers may not be required or may be needed in the short or long term.
- *DSR availability*: Depending on customer preferences, different levels of DSR may be available in different networks. Accordingly, the study is extended to consider conditions where only one DSR block, and up to five blocks, are available.
- *Demand scenarios*: Demand is scaled up so that line reinforcements will be triggered after a further 3% demand increase, which guarantees that at least one intervention will be required in every study. As in the previous study, demand is modeled based on the five demand growth scenarios proposed by the relevant DNO (see Fig. 12).

An overview of the average NPC values associated with the 36 networks and different studies is presented in Fig. 14. The results corroborate the initial observation that smart grid applications generally lead to cost reductions, especially if environmental, reliability, and other so-called social benefits are also internalized (e.g., Smart—Ring strategy). Nevertheless, it also becomes evident that certain conditions make

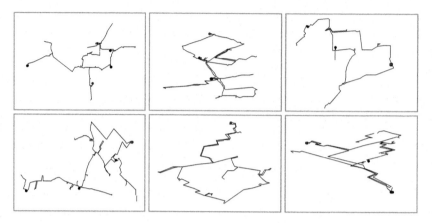

Fig. 13 Schematics of six trial distribution networks.

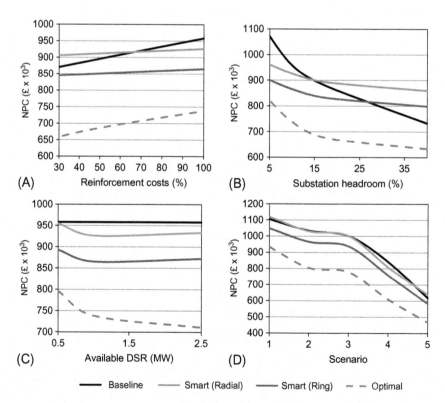

Fig. 14 Average expected NPC associated with different strategies and assumptions of (A) reinforcement costs, (B) substation headroom, (C) DSR, and (D) demand growth scenarios.

traditional asset-based solutions more convenient than smart solutions under current planning practices that are not able to effectively deploy smart solutions. This is the case for conditions where traditional asset-based solutions become cheap or costly investments are only envisioned in the far future. These effects can be seen in Fig. 14A where the baseline becomes cheaper on average than one of the smart solutions when costs of traditional solutions drop by more than 60% of their current value. This can also be seen in Fig. 14B where the baselines outperform the smart solutions in cases with high substation headroom.

The studies also highlight that the use of proper optimization tools to capture the features of smart solutions can bring about significant benefits. This is clear by comparing the performance of the optimal strategy with any other strategy in Fig. 14. Taking the analysis further, Figs. 15 and 16 provide comparisons between the baseline

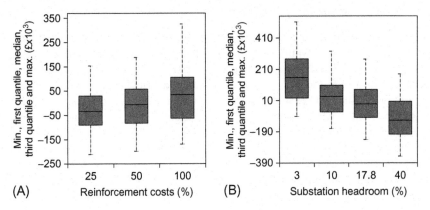

Fig. 15 NPC difference between the baseline and the Smart (Ring) strategy under different (A) reinforcement costs and (B) substation headroom conditions.

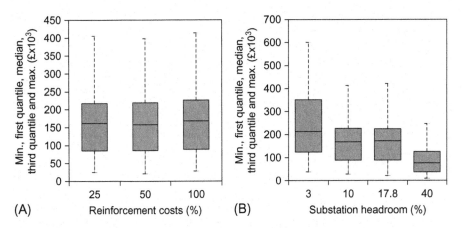

Fig. 16 NPC difference between the baseline and the optimal strategy under different (A) reinforcement costs and (B) substation headroom conditions.

and smart and optimal strategies in terms of the number of cases when one solution outperforms the other. More specifically, instead of only presenting the average values as in Fig. 14, the box plots in Figs. 15 and 16 show how often and by how much the smart or optimal strategies outperformed the baseline in terms of minimum and maximum values, first and third quantiles, and median values.

These studies provide further evidence that current planning practices could frequently cause the suboptimal use of smart solutions. This may occur even when the average value of smart strategies is greater than that of the baseline. Conversely, the use of smart solutions combined with enhanced planning practices based on optimizations resulted in economic savings (or the same costs as the baseline) in all cases considered in this study. These findings provide strong justifications for DNOs to pursue the use of improved network investment and operation planning practices.

4.2 Smart community multienergy systems

Smart grid applications to community energy systems are illustrated here with a real district at the University of Manchester. The Manchester community comprises 26 buildings owned and managed by the University. Most of the buildings are connected to the same (6.6 kV) electricity distribution, district heating, and gas networks (see Fig. 17), which makes this community a multienergy system. The energy (electricity, heat, and gas) consumption in each building is currently monitored by the Coherent data system, which provides half-hour data [39]. The community energy system is planned and operated with the aim of maximizing energy efficiency based on the novel concepts presented in Section 2.2.3. That is, the intelligent coordination of DER operation and multienergy exchanges between buildings is used to facilitate economic and environmental gain as well as the provision of services to the energy system [36].

The Manchester community was investigated in light of existing conditions as well as considering an ambitious 30% carbon reduction target set by the University of Manchester for 2020. To this end, four actuation cases were considered:

- *Reference*: This case represents the current conditions of the Manchester community. It is used as a reference to assess the performance of all other cases.
- *Conservative case*: The University invested in awareness campaigns to encourage switching off lights and computers when not in use as well as in double-glazed steel windows and waterproof roof covers (a total of $470 £ \times 10^3$). In addition, 900 kW of PV capacity and 300 kW of EHP capacity were installed throughout the university. Emissions were reduced by roughly 10% compared with the reference case.
- *Modest case*: Additional investments in energy efficiency measures throughout the university were made (a total of $627 £ \times 10^3$) whereas PV capacity increased to 2000 kW and EHP capacity increased to 1700 kW. In addition, 1900 kW of CHP capacity were installed throughout the University. Emissions were reduced by roughly 20% compared with the reference case.
- *Extreme case*: Relatively large investments (1.2 M£) in energy efficiency measures were made, coupled with significant installation of energy infrastructure. The total installed PV, EHP, and CHP capacities increased to 3500, 2600, and 2700 kW, respectively. Emissions were reduced by roughly 30% compared with the reference case.

Fig. 17 Electricity, heat, and gas networks in the Manchester demonstrator.

4.2.1 Smart community applications

The portfolio of tools discussed in Section 3.2 was used to assess the energy performance of the community based on the different cases and considering BAU and optimal operation. More specifically, in a BAU context, all controllable energy devices within the district are operated in heat-following mode, as is typically done in the United Kingdom. In a smarter context where the operation of the community is optimized, all energy infrastructure is operated with the objective of minimizing energy costs and emissions while, at the same time, partaking in the available energy markets and mechanisms (see Section 1.2). An example of the energy performance of the district under both BAU and optimal operation (considering all markets and mechanisms and prioritizing economic benefits) is presented in Table 4. A comparison of expected economic and environmental benefits compared with the reference case can be seen in Table 5.

It is important to note that, in a smart context where the operation of all energy devices can be optimized, the energy performance of the community can be flexibly customized based on customer preferences and participation in different markets. The operational flexibility of the smart community is illustrated in Fig. 18. As shown, under BAU practices, the energy performance of the community is static. This does not change much in cases where little or no controllable technologies are available, as is the case of the Reference and Conservative cases. However, as shown in the Modest and Extreme cases, the intelligent coordination of controllable devices brings about significant economic and environmental benefits. For example, in this study, it can be seen that the 30% reduction targets can only be met in the extreme case under BAU practices. However, the environmental targets can be reached while also providing attractive economic gain in the modest case when the operation of the community is optimized.

The study also shows that the community can achieve different combinations of economic and environmental benefits by changing its operation regime and without compromising customer comfort (i.e., all demand is always met). This allows valuable headroom to accommodate customer preferences and unforeseen conditions, such as higher prices and carbon intensity, than originally forecast. The latter is particularly valuable if, for example, the forecast environmental benefits from the decarbonization of electricity are not as significant as expected. In such a case, the district may still meet the environmental targets just by updating its operation and without the need for investments in additional energy infrastructures. Another example is the provision of different services such as DSR just by changing the operation of the district. This could allow the community to provide distribution network support (as presented in Section 4.1) without causing customer disutility. This makes the provision of DSR for distribution network support particularly attractive for smart community energy systems.

Assuming typical UK prices for PV, EHP, and CHP technologies of 1300, 240, and 600£/kW, respectively, the energy savings attributed to the smart community energy system would pay back relevant investments in 6 (in the conservative case) to 8 years (in the extreme case). If customers are focused on net savings and revenue, the extreme case provides the highest benefits under optimistic economic conditions (i.e., a net present value of 5 M£ considering a discount rate of 3% and a planning

Table 4 Expected energy performance of the Manchester demonstrator in the different cases

	Reference	Conservative		Modest		Extreme	
		BAU	Opt.	BAU	Opt.	BAU	Opt.
Annual costs ($£ \times 10^3$)	3102	2821	2806	2723	2437	2642	2220
Annual emissions (ktCO$_2$)	19.1	17.2	17.2	15.3	14.0	14.0	11.9
Peak electricity demand (MW)	2.91	2.91	24.3	2.85	16.4	2.84	14.1
Peak heat demand (MW)	5.65	5.65	13.03	2.67	2.75	1.56	0.90
Annual electricity consumption (MWh)	27,852	27,852	5058	22,333	3782	20,602	2702
Annual heat consumption (MWh)	18,218	18,218	2677	1290	2732	167	2627

Table 5 **Expected economic and environmental benefits associated with different cases**

	Conservative		Modest		Extreme	
	BAU (%)	Opt. (%)	BAU (%)	Opt. (%)	BAU (%)	Opt. (%)
Annual economic savings[a]	9.06	9.54	12.22	21.44	14.83	28.43
Annual carbon savings[a]	10.14	10.28	19.94	27.00	26.90	38.07
Peak heat demand reduction[a]	4.67	12.63	2.20	41.03	2.37	49.38
Peak electrical demand reduction[a]	15.50	28.48	52.80	84.88	72.51	95.07
Annual electrical consumption reduction[a]	7.84	8.10	19.82	6.21	26.03	9.82
Annual heat consumption reduction[a]	38.89	10.59	92.92	33.14	99.08	52.24

[a]Compared with the reference case.

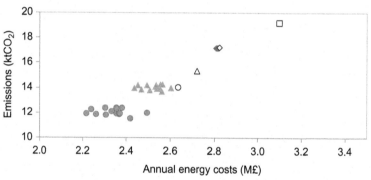

■ Reference (smart) ◆ Conservative (smart) ▲ Modest (smart) ● Extreme (smart)
□ Reference (BAU)   ◇ Conservative (BAU)   △ Modest (BAU)   ○ Extreme (BAU)

Fig. 18 Potential economic and environmental performance of the Manchester district.

horizon of 20 years). Assuming more pessimistic conditions where customers (or other actors) may demand higher returns before investing in energy technologies (e.g., considering a higher discount rate of 10%), the Modest case may become the preferred one due to its lower cost and relatively similar benefits to those in the Extreme case (see Fig. 18). The baseline case only becomes attractive under highly pessimistic conditions where technologies become too pricey and investments in energy infrastructure require significantly large expected return of investments. This provides strong evidence that smart community multienergy systems are generally economically attractive and highly profitable.

4.2.2 Wider system implications

The emergence of smart communities that, besides consuming energy also actively participate in different markets and mechanisms, can have significant impacts on the energy system and the business case of different actors (see Section 1.2). For example, as the Manchester district optimizes its operation more and more using controllable CHP boilers, the community will become less dependent on grid imports (the community will now also export energy to the different markets) while increasing gas consumption as shown in Fig. 19.

The new operation strategy of the community leads to a reduction in annual energy costs for customers. However, as shown in Fig. 20, the emergence of smart community multienergy systems has a clear effect on the business case of generators, the government, DNOs, and the TSO, who are expected to accrue less revenue. This information is critical for policy makers, who must decide whether or not each of these specific impacts are beneficial or not for the energy system and update regulations

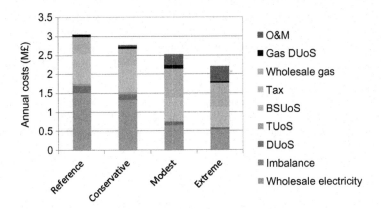

Fig. 19 District costs by component in different optimized cases.

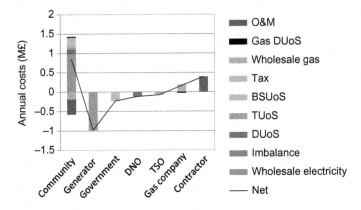

Fig. 20 District costs by component in different optimized cases.

accordingly. For example, negative impacts on generators may be deemed reasonable if they lead to decommission of inefficient generation, but DNO losses may pose an issue as distribution network upgrades may be needed to enable the smart operation of communities (see Section 2). In order to address the latter, policymakers may update regulations so that, for example, some of the economic benefits accrued by smart communities are shared with DNOs and used for the improvement of the distribution networks.

5 Conclusion

This chapter presented some of the latest UK research in smart grid applications to distribution networks and intelligent community energy systems enabled by DERs. Results from several projects carried out by the authors, involving trials in UK distribution networks and communities, were presented, including the C_2C, Smart Street, ADDRESS, APS, and DIMMER projects.

It has been demonstrated that the introduction of smart solutions can bring about attractive benefits for the energy sector, particularly when in combination with DERs. However, new tools such as the optimization and business case frameworks presented here (see Section 3) are required to effectively deploy smart solutions. More specifically:

- The use of smart solutions to enhance the planning and operation of distribution networks generally results in economic and environmental benefits. However, there may be many cases where the performance of the network may actually worsen if these solutions are deployed based on existing DNO practices.
- Improving network planning and operation practices (e.g., based on stochastic optimizations) significantly increases the benefits brought about by smart solutions as well as the conditions under which benefits arise.
- Community energy systems can provide attractive economic and environmental benefits for customers. However, without the aid of intelligent DER coordination tools, most of the flexibility provided by these communities would be wasted.
- The use of adequate tools to coordinate DERs within districts further maximizes benefits for customers while, at the same time, allows the provision of manifold valuable services to the energy system.
- The introduction of smart solutions at the community and distribution network levels has impacts on the business case of different actors throughout the value chain. Dedicated tools are vital to shed light on these impacts, which can motivate policymakers to modify the existing regulation.

References

[1] A.G. Olabi, Energy quadrilemma and the future of renewable energy, Energy 108 (2016) 1–6.
[2] E.A. Martínez Ceseña, P. Mancarella, Capacity to Customers (C2C) Development of Cost Benefit Analysis Methodology for Network Expansion Planning Considering C2C Interventions, 2014.

[3] E.A. Martínez Ceseña, P. Mancarella, Capacity to Customers (C2C) Sensitivity Analysis of the Expected Economic Value of the C_2C Method, 2015.

[4] L. Gutierrez-Lagos, L.F. Ochoa, in: CVR assessment in UK residential LV networks considering customer types, 2016 IEEE ISGT-Asia, 2016, pp. 741–746.

[5] E.A. Martínez Ceseña, N. Good, P. Mancarella, in: Energy efficiency at the building and district levels in a multi-energy context, Energycon, 2016.

[6] E.A. Martínez Ceseña, P. Mancarella, in: Distribution network support from multi-energy demand side response in smart districts, IEEE ISGT-Asia, 2016, pp. 1–6.

[7] A.L.A. Syrri, Y. Zhou, E.A. Martínez Ceseña, P. Mancarella, in: Reliability and economic implications of collabortive distributed resources, CIRED Workshop, 2016.

[8] E.A. Martínez Ceseña, N. Good, P. Mancarella, Electrical network capacity support from demand side response: techno-economic assessment of potential business cases for small commercial and residential end-users, Energy Policy 82 (2015) 222–232.

[9] M. Stewart, Future Energy Scenarios 2016, 2016.

[10] Elexon, Guidance The Electricity Trading Arrangements A Beginner's Guide, 2015.

[11] E.A. Martínez Ceseña, J. Mutale, in: Trading wind power in the UK using an imbalance penalty reduction strategy, IEEE PowerTech 2013, 2013, pp. 1–6.

[12] E.I. Batov, The distinctive features of 'smart' buildings, Procedia Eng. 111 (2015) 103–107.

[13] National Grid, Our Services, (Online), Available from: http://www2.nationalgrid.com/uk/services/, 2017.

[14] S. McGonigle, RIIO-ED1 Electricity Distribution Price Control—Overview of the Regulatory Instructions and Guidance, 2015.

[15] L. O'Brien, Price Controls Explained, 2013.

[16] Transmission Electricity Charges TNUoS Triad Data | National Grid, National Grid, 2017 (Online), Available from: http://www2.nationalgrid.com/UK/Industry-information/System-charges/Electricity-transmission/Transmission-Network-Use-of-System-Charges/Transmission-Charges-Triad-Data/.

[17] Distribution and Connection Use of System Agreement (DCUSA), 2017.

[18] A. Boardman, S. Mockford, P. Lawson, EDS 08-0119: Guidance From the Application of ENA ER P2/6 Security of Supply, 2014.

[19] E.A. Martínez Ceseña, P. Mancarella, in: Distribution network reinforcement planning considering demand response support, PSCC, 2014, pp. 1–7.

[20] A.L.A. Syrri, P. Mancarella, Reliability and risk assessment of post-contingency demand response in smart distribution networks, SEGAN 7 (2016) 1–12.

[21] J.K.W. Wong, H. Li, S.W. Wang, Intelligent building research: a review, Autom. Constr. 14 (1) (2005) 143–159.

[22] C. Langston, R. Lauge-Kristensen, Strategic Management of Built Facilities, Routledge, New York, 2011.

[23] A. Rodriguez-Calvo, R. Cossent, P. Frías, Integration of PV and EVs in unbalanced residential LV networks and implications for the smart grid and advanced metering infrastructure deployment, Int. J. Electr. Power Energy Syst. 91 (2017) 121–134.

[24] N. Good, E.A. Martínez Ceseña, X. Liu, P. Mancarella, in: A business case modelling framework for smart multi-energy districts, CIRED Workshop, 2016, pp. 1–4.

[25] E.A. Martínez Ceseña, P. Mancarella, Practical recursive algorithms and flexible open-source applications for planning of smart distribution networks with demand response, SEGAN 7 (2016) 104–116.

[26] Network Innovation | Ofgem, Ofgem, 2015 (Online), Available from: https://www.ofgem.gov.uk/network-regulation-riio-model/network-innovation.

[27] Smart Street, ENWL, 2017 (Online), Available from: http://www.enwl.co.uk/smartstreet (Accessed 06-Apr-2017).

[28] J. Broderick, Capacity to Customers (C2C) Carbon Impact Assessment Final Assessment Report, 2015.

[29] WEEZAP & LYNX, LV Meshing, Automation, & Protection, 2017.

[30] PowerON Fusion Distribution Management System, General Electric, 2017 (Online), Available from: https://www.gegridsolutions.com/UOS/catalog/poweron_fusion.htm.

[31] Managed Connections, ENWL, 2017 (Online), Available from: http://www.enwl.co.uk/our-services/connection-services/generation/managed-connections.

[32] Fault Restoration Performance on HV rings as part of the Capacity to Customers Project, 2013.

[33] Network Regulation—The RIIO Model, Ofgem, 2015 (Online), Available from: https://www.ofgem.gov.uk/network-regulation-riio-model.

[34] E.A. Martínez Ceseña, V. Turnham, P. Mancarella, Regulatory capital and social trade-offs in planning of smart distribution networks with application to demand response solutions, Electr. Power Syst. Res. 141 (2016) 63–72.

[35] A. Ipakchi, F. Albuyeh, Grid of the future, IEEE Power Energy Mag. 7 (2) (Mar. 2009) 52–62.

[36] N. Good, E.A. Martínez Ceseña, P. Mancarella, Ten questions concerning smart districts, Build. Environ. 118 (2017) 362–376.

[37] P. Mancarella, G. Andersson, J.A. Pecas-Lopes, K.R.W. Bell, in: Modelling of integrated multi-energy systems: drivers, requirements, and opportunities, PSCC, 2016, pp. 1–22.

[38] N. Good, E.A. Martínez Ceseña, L. Zhang, P. Mancarella, Techno-economic and business case assessment of low carbon technologies in distributed multi-energy systems, Appl. Energy 167 (2016) 158–172.

[39] Coherent Data Collection Server, Coherent Research, 2017 (Online), Available from: https://www.ems.estates.manchester.ac.uk/DCS/Logon.aspx?ReturnUrl=%2FDCS.

[40] E.A. Martínez Ceseña, T. Capuder, P. Mancarella, Flexible distributed multienergy generation system expansion planning under uncertainty, IEEE Trans. Smart Grid 7 (1) (2016) 348–357.

[41] T. Capuder, P. Mancarella, Techno-economic and environmental modelling and optimization of flexible distributed multi-generation options, Energy 71 (Jul. 2014) 516–533.

[42] X. Liu, P. Mancarella, Modelling, assessment and Sankey diagrams of integrated electricity-heat-gas networks in multi-vector district energy systems, Appl. Energy 167 (2016) 336–352.

[43] Ofgem, Supply Market Indicator, 2014.

[44] R. Masiello, J.R. Aguero, Sharing the ride of power: understanding transactive energy in the ecosystem of energy economics, IEEE Power Energy Mag. 14 (3) (2016) 70–78.

[45] A. Monti, D. Pesch, K.A. Ellis, P. Mancarella, Energy Positive Neighborhoods and Smart Energy Districts: Methods, Tools and Experiences from the Field, Elsevier Academic Press, Massachusetts, 2016.

[46] N. Good, K.A. Ellis, P. Mancarella, Review and classification of barriers and enablers of demand response in the smart grid, Renew. Sustain. Energy Rev. 72 (2016) 57–72.

[47] D. Six, E.A. Martínez Ceseña, C. Madina, K. Kessels, N. Good, P. Mancarella, Techno-economic analysis of demand response, in: A. Losi, P. Mancarella, A. Vicino (Eds.), Integration of Demand Response into the Electricity Chain: Challenges, Opportunities, and Smart Grid Solutions, Wiley-ISTE, London, 2015, p. 296.

[48] E.A. Martínez Ceseña, P. Mancarella, in: Distribution network capacity increase via the use of demand response during emergency conditions: a cost benefit analysis framework for techno-economic appraisal, CIRED Workshop, 2014.

[49] E.A. Martínez Ceseña, P. Mancarella, in: Economic assessment of distribution network reinforcement deferral through post-contingency demand response, IEEE ISGT-Europe, 2014, pp. 1–6.

[50] A.L.A. Syrri, Y.Z. Yutian Zhou, E.A. Martínez Ceseña, J. Mutale, P. Mancarella, in: Reliability and economic implications of collaborative distributed resources, CIRED Workshop, 2016.

[51] E.A. Martinez Cesena, P. Mancarella, in: Distribution network capacity support from flexible smart multi-energy districts, 2016 IEEE T&D-LA, 2016, pp. 1–6.

[52] R. Ruparathna, K. Hewage, R. Sadiq, Improving the energy efficiency of the existing building stock: a critical review of commercial and institutional buildings, Renew. Sust. Energ. Rev. 53 (2016) 1032–1045.

[53] I. Sartori, A. Napolitano, K. Voss, Net zero energy buildings: a consistent definition framework, Energy Build. 48 (2012) 220–232.

[54] A.M. Carreiro, H.M. Jorge, C.H. Antunes, Energy management systems aggregators: a literature survey, Renew. Sust. Energ. Rev. 73 (Jun. 2017) 1160–1172.

[55] N. Good, E.A. Martínez Ceseña, L. Zhang, P. Mancarella, Techno-economic and business case assessment of low carbon technologies in distributed multi-energy systems, Appl. Energy 167 (2016) 158–172.

[56] A.L. Syrri, E.A. Martínez Ceseña, P. Mancarella, in: Contribution of microgrids to distribution network reliability, IEEE PowerTech, 2015, pp. 1–6.

[57] N. Good, E.A. Martinez Cesena, P. Mancarella, in: Mapping multi-form flows in smart multi-energy districts to facilitate new business cases, Sustainable Places, 2014. (Online), Available from: http://docplayer.net/10793122-Mapping-multi-form-flows-in-smart-multi-energy-districts-to-facilitate-new-business-cases.html.

[58] N. Good, L. Zhang, A. Navarro-Espinosa, P. Mancarella, High resolution modelling of multi-energy domestic demand profiles, Appl. Energy 137 (2015) 193–210.

[59] E.A. Martínez Ceseña, F.R. Dávalos, Evaluation of investments in electricity infrastructure using real options and multiobjective formulation, IEEE Lat. Am. Trans. 9 (5) (2011) 767–773.

[60] Capacity to Customers (C₂C) project, Electricity Northwest (Online), Available from: http://www.enwl.co.uk/c2c.

Impact of smart meter implementation on saving electricity in distribution networks in Romania

9

Gheorghe Grigoras
Gheorghe Asachi Technical University of Iasi, Iasi, Romania

Abbreviations

ANRE	National Energy Regulatory Authority
DNO	distribution network operator
EC	European Commission
EU	European Union
LV	low voltage
MEPS	minimum energy performance standards
RAB	regulated asset base

1 Overview: Romania and the European situation

1.1 Introduction

The installation of smart meters can be considered a discreet revolution that is already underway in many countries around the world, including Romania. Smart meters must ensure the bidirectional communication between consumer/prosumer and the distribution networks operator (DNO) while promoting energy efficiency services. Thus, on the one hand, the customers can get accurate information online regarding electricity consumption and production, which allows them to take measures to become more energy efficient. On the other hand, DNOs provide accurate data on the consumption pattern, enabling them to manage and better plan their investments in networks. DNOs will ensure, in accordance with the result of mandates M/441/EC [1] and M/490/EC [2] in the field of smart metering and smart grids, that consistent and more cost-effective developments in the fields of communications and information will establish a more effective management of the network [3].

It all began in 2009 when the European Parliament approved the third energy package of the European Union (EU) (Directive 2009/72/EC), a legislative initiative aimed at principally improving the competition in the electricity and gas markets. In Romania, the early stages for the implementation of smart metering began that same year. Initially, several pilot projects that focused on the installation of automatic

Application of Smart Grid Technologies. https://doi.org/10.1016/B978-0-12-803128-5.00009-X

Table 1 **Estimation of the smart meter market in Romania [4]**

Consumer category	Consumer number	Smart meters installed (pc.)	Smart meters installed (%)
Public consumers (larger nonresidential customers)	20,000	9700	48.5
Small and medium nonresidential customers	600,000	–	0
Residential consumers	8,380,000	34,300	0.4
Total	9,000,000	44,000	0.48

meter reading—with only 15,000 customers that implemented advanced metering management—were ongoing. For these stages, during 2009–12, the information is presented as a whole in Table 1 [4]. The residential consumers completely dominate, constituting 99% of all electricity customers in Romania. The share of nonresidential consumers (such as industry and small and medium enterprises) is approximately 1%. A very small number of the smart meters was until now installed at residential customers, and hence a high potential for the implementation of smart metering system can be identified [5].

At the end of 2012, the Romanian National Energy Regulatory Authority (ANRE) published a study carried out by the Global Management Consulting Firm—A.T. Kearney, at the request of the European Bank for Reconstruction and Development (EBRD). The study focused on the feasibility of introducing and using the smart meters in the markets for electricity, natural gas, and heat from Romania. Based on this study and others made by some DNOs, the smart electricity distribution has become a priority of ANRE, which, starting from Law No. 123/2012 on electricity and gas, issued Order No. 91/2013 that provides as a national target the implementation of smart metering systems for electricity to approximately 80% of the final consumer number by 2020 [6]. In accordance with Romanian law on electricity and natural gases, the same study assessed the implementation of smart metering systems from the viewpoint of long-term costs and benefits to the market, profitability, and feasible implementation deadlines [7]. The reduction of electricity losses, lower operating costs, and utilities could lead also to increase the benefits. Other important aspects were about relevant elements of the cost-benefit analysis and an effective approach to install the smart meters [5]. The feasibility study, including the results of the cost-benefit analysis, indicated a positive net present value. This led to the conclusion that the implementation of smart metering in the electricity sector has the potential to be a profitable investment due to the benefits from reducing electricity losses in electric networks and decreasing operating costs under certain conditions and assumptions.

Because the results of the cost-benefit analysis were positive, ANRE has proposed a plan to implement the smart metering system, taking into account the best practices, latest technologies, appropriate standards, and the importance of developing the

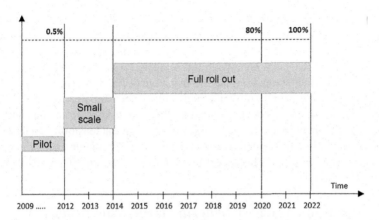

Fig. 1 Smart metering in Romania: implementation time frame in three steps [8].

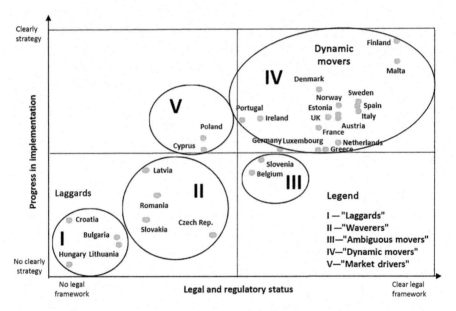

Fig. 2 Status of implementation and regulation in the European countries [9].

electricity and gas markets. A suggestive time frame for the implementation of smart metering in Romania is presented in Fig. 1 [8].

1.2   Romania versus European countries

The monitoring done since 2011 by the European Commission (EC) [9] showed that progress has been strongest in countries with a significant impulse in the regulatory field (Fig. 2).

It can be observed that many so-called "dynamic movers" took into account very ambitious rollout plans. Furthermore, the minimum requirements of smart meters have been clarified in most of these countries. On the other side, countries that are classified as "market drivers" had only moderate progress. In general, it should also be said that in many countries there is still a lot of skepticism among consumers and private companies. Romania belongs in the category "waverers" [9].

Along with 16 other states from the EU that have decided in favor on a large scale (>80%) for the implementation of electricity smart metering after cost-benefit studies, Romania had an official decision pending. In these studies, Romania and Poland were specifically mentioned, emphasizing that although the feasibility analyses indicated positive results, the formal decisions regarding the beginning of some national projects have been delayed. This aspect was highlighted by the EC in 2014, and it is shown in Table 2 [10].

Table 2 Status of electricity metering large-scale rollout in the EU [10]

Country	Wide-scale roll-out[a]	CBA conducted	Outcome of the CBA[a]
Austria (AT)	Yes	Yes	Positive
Denmark (DK)	Yes	Yes	Positive
Estonia (EE)	Yes	Yes	Positive
Finland (FI)	Yes	Yes	Positive
France (FR)	Yes	Yes	Positive
Greece (EL)	Yes	Yes	Positive
Ireland (IE)	Yes	Yes	Positive
Luxembourg (LU)	Yes	Yes	Positive
The Netherlands (NL)	Yes	Yes	Positive
Sweden (SE)	Yes	Yes	Positive
Great Britain (GB)	Yes	Yes	Positive
Poland (PL)	Yes—Official decision pending	Yes	Positive
Romania (RO)	Yes—Official decision pending	Yes	Positive
Malta (MT)	Yes	No	NA
Spain (ES)	Yes	No	NA
Italy (IT)	Yes	NA	NA
Czech Republic (CZ)	No	Yes	Negative
Lithuania (LT)	No	Yes	Negative
Belgium (BE)	No	Yes	Negative/inconclusive
Portugal (PT)	No	Yes	Inconclusive
Germany (DE)	Selective	Yes	Negative
Latvia (LV)	Selective	Yes	Negative
Slovak Republic (SK)	Selective	Yes	Negative
Cyprus (CY)	No decision	In progress	NA
Hungary (HU)	No decision	In progress	NA
Slovenia	No decision	In progress	NA
Bulgaria	No decision	NA	NA

[a]At least 80% of customers by 2020.

Different indicators resulted from the study made for Romania [5,6]. These indicators reflected the total costs per the information collected, adjusted to reflect the best electricity-specific activities. Thus, the total costs per categories were estimated as 648.2 million euros, and the average costs per metering point per categories (based on a 15-year modeling period) were 77.35 euros. These costs are some of the smallest from the EU. A comparison between costs from different countries is presented in Figs. 3 and 4 [11]. The signification of abbreviations for countries is given in Table 2.

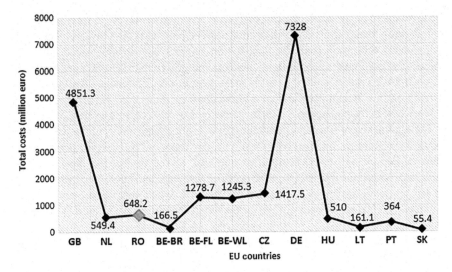

Fig. 3 The total costs (in millions of euros)—15-year modeling period.

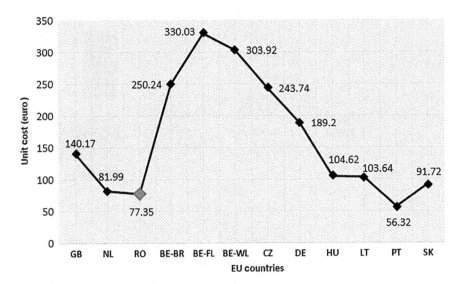

Fig. 4 Smart Meter Unit Cost (euros)—15-year modeling period.

However, the analysis based on the recent pilot projects has led to the conclusion that, in Romania, the average cost per metering point increased to €122. The components of this cost are represented by the following: smart meters (€75), communications (€10), information technology (€4), distribution (€25), and others undefined costs (€8) [11].

1.3 Configuration and key functionalities of the smart metering system in Romania

The assessment of the model corresponds to a "middleware layer" design. It contains the following main components: data concentrators and balancing meters installed on each electric substation from the distribution network and meters placed with customers. The data communication occurs through power line communication (PLC) wiring from the smart meters to concentrators and through various communication channels from concentrators to a central database.

Most DNOs have adopted a similar configuration, as depicted in Fig. 5.

The establishment (through regulatory tools and mechanisms) and the use of open standard communication protocols played a very important role in the connecting of smart meters to the data concentrators and then to the central database. Such an approach avoided the massive investment in equipment that is not interoperable, cannot log, or generates errors in the data transmission when purchased from different suppliers [5]. The key functionalities of the smart metering system used in Romania are indicated in Fig. 6.

The adopted solution for the smart metering system in Romania has a low cost with minimum functionality. However, the benefits could decrease due to the minimum functions.

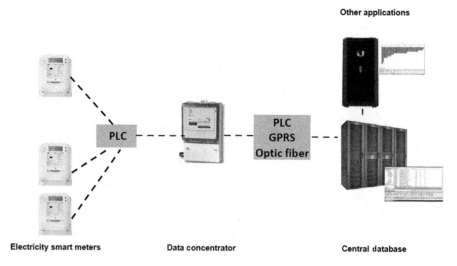

Fig. 5 The most common configuration for smart metering systems from Romania.

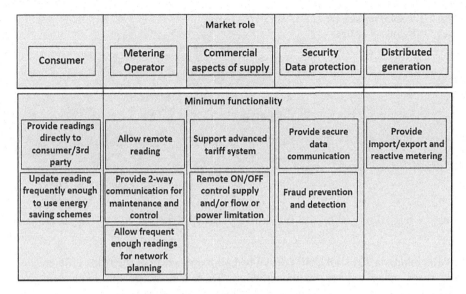

Fig. 6 Key functionalities of the smart metering system in Romania.

2   The current status of smart metering in Romania

2.1   General aspects

In Romania, there are eight DNOs that serve more than 9 million customers [7]:

1. CEZ Distributie (CEZ)—15.1% electricity distributed nationwide.
2. ENEL Distributie Banat (ENEL-B)—9.8% electricity distributed nationwide.
3. ENEL 1 Distributie Dobrogea (ENEL-D)—8.4% electricity distributed nationwide.
4. ENEL Distributie Muntenia (ENEL-M)—16.2% electricity distributed nationwide.
5. E.ON Moldova Distribuţie (E.ON) (in January 2017, changed its name to DELGAZ GRID)—10.3% electricity distributed nationwide.
6. ELECTRICA Distributie Transilvania Sud (Electrica-TS)—13.3% electricity distributed nationwide.
7. ELECTRICA Distributie Transilvania Nord (Electrica-TN)—11.6% electricity distributed nationwide.
8. ELECTRICA Distributie Muntenia Nord (Electrica-MN)—16.2% electricity distributed nationwide.

The share of electricity distribution nationwide corresponds to the year 2015, when the total electricity distributed was approximately 42.4 TWh.

Each company is responsible for the exclusive distribution of electricity within its licensed region, based on a concession agreement with the Romanian state acting through the Ministry of Energy (see Fig. 7).

Fig. 7 Licensed regions for
each DNO from Romania.

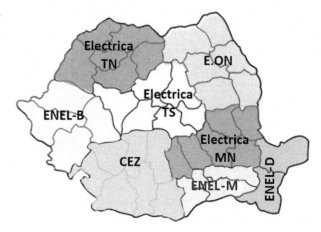

Beginning in 2013, all DNOs developed pilot projects to evaluate the efficiency of the smart meter implementation program. Thus, a homogeneous implementation (pilot projects in the regions covered by DNOs) is given by [6]:

- The installation of smart meters for consumers from the urban/rural areas served by a DNO for each demonstration project.
- The permanent monitoring of electricity consumption by consumers (online) to enable them to benefit from the positive impact of energy efficiency measures to be implemented as well as the immediate impact of pricing policies to be applied on the behavior of nonresidential electricity customers.
- The recorded data that must be sufficient to analyze and optimize the functioning of distribution facilities and equipment in order to increase the efficiency of networks. These data must allow the rapid identification of action points.

The expected results of the pilot projects were the following [6]:

- Installation of smart meters enables DNOs to reach their target of approximately 80% implementation of final customers by 2020, as required by Law No. 123/2012 [12].
- For selected areas, the electricity consumption will decrease by 10% due to the smart management at the consumer and the distribution levels.
- To support the model that DNOs use for the implementation of smart metering.
- Savings of around 300 MWh/year/project (based on the annual average electrical consumption of urban families in Romania).

Through these projects, the technical possibilities for achieving the solution of smart metering and the customers' benefits in the new system are investigated. On the other hand, an analysis related to data protection in the system of remote reading and the protection of communication lines is deepened.

Based on the experience acquired in 2013, by the end of the next year, ANRE had issued an order regarding smart metering, called the "Order on the implementation of smart metering in electricity" [13], which provides, in addition to a range of general functionalities of smart metering systems, general data on the implementation plan and a description of the pilot projects.

According to the regulatory stipulations on the deployment of smart metering systems, DNOs.

submitted proposals to ANRE for the realization of some pilot projects in 2015. Those results were to provide needed information to establish the conditions and tools on the developing of a schedule and a national plan for the implementation of smart metering.

ANRE approved 18 pilot projects worth 69,639,770 lei (approximately 15.48 million euros) in 2015 (see Table 3 [5]). For these projects, a set of performance indicators was defined for smart metering systems so that progress could be monitored during the implementation process and a period thereafter. The performance indicators relate to the following issues: the status of implementation, the structure of intelligent metering systems, the economic effects, the quality, and the security of information. These indicators will be applied to all these pilot projects so as to verify the achievement of proposed objectives.

The implementation areas were defined and the characteristic data and information have been set in terms of the technical, economic, sociodemographic, and qualitative aspects. The processing of these data is aimed at prioritizing a hierarchy of the areas for each DNO through a multicriteria analysis to assess the potential implementation of smart metering systems and the preparation of the national plan.

The analysis report on the pilot projects for each DNO given by ANRE in August 2016 [14] showed the following important aspects:

- Postimplementation results on the benefits in question are not relevant to all pilot projects.
- Cost-benefit analyses did not allow a comparative study based on the obtained results.
- The initial assumptions are different, depending on the business strategies, with the specific aims and focus of each DNO.

Table 3 **The pilot projects approved by ANRE in 2015**

DNO	Number of Pilot projects	Number of customers	Total cost (million euro)
ENELDistributie Banat	3	9961	1.17
ENEL Distributie Dobrogea	4	10,000	1.10
ENELDistributie Muntenia	1	11,392	1.48
CEZ Distributie	2	20,150	3.57
EON Distributie	2	23,237	1.85
ELECTRICA Transilvania Sud	2	23,047	5.09
ELECTRICA Transilvania Nord	2	5335	0.9
ELECTRICA Muntenia Nord	2	2143	0.32
Total	18	105,265	15.48

• Cost-benefit analyses are positive for ENEL Distributie, E.ON Distributie, and ELEC-TRICA Transilvania Sud but negative for CEZ Distributie.

The result was the need for ANRE to impose a detailed model of cost-benefit analysis or the performance of a cost-benefit analysis for all DNOs, possibly through an impartial external consultant in order to avoid allegations of nontransparency and lack of objectivity [14].

However, the final targets vary from a DNO to another as follows: ENEL target 80% by 2020 and ELECTRICA and CEZ <50% while E.ON targets more than 50% [15].

2.2 The implementation of smart metering by DNOs

2.2.1 ENEL

ENEL provides services to 2.8 million customers in three key regions of the country: Muntenia Sud (MS) (including the Bucharest zone), Banat (B), and Dobrogea (D), representing one-third of the electricity distribution market in Romania. Since 2014, the company has carried out an extensive program of investments to modernize the infrastructure in Banat, Dobrogea, and Muntenia [16,17]. In Table 4, the pilot projects and the number of smart meters in the three branches are given. It can observed that more than 30,000 smart meters were installed.

From 2016, the pilot projects were selected in both urban and rural areas. The information is given per total, residential, and nonresidential customers in Table 5 [16].

The total smart meters installed in pilot projects from 2014 to 2016 are indicated in Fig. 8 per each county served by the three subsidiaries of ENEL.

Table 4 **The number of smart meters installed from 2014 to 2015 [16]**

Zone	Residential customers	Nonresidential customers	Total
ENEL Muntenia			
Bucuresti	10,520	872	11,392
ENEL Banat			
Arad County	3004	99	3103
Hunedoara County	2485	228	2713
Timis County	3813	332	4145
ENEL Dobrogea			
Constanta County	1951	694	2645
Tulcea County	3014	342	3356
Calarasi County	3339	156	3495
Calarasi County	477	27	504
Total ENEL			
All zones	28,603	2750	31,353

Table 5 **The number of smart meters installed in 2016**

Zone	Area	Number of places	Total
ENEL Muntenia			
Bucuresti	Urban	>3	22,294
Ilfov County	Rural	3	14,256
	Urban	1	9329
Giurgiu County	Rural	9	7219
ENEL Banat			
Timis and Caras Severin Counties	Rural	7	2741
	Urban	2	11,742
Hunedoara County	Rural	7	4145
Arad County	Rural	1	1001
	Urban	1	6662
ENEL Dobrogea			
Constanta and Tulcea Counties	Urban	6	11,383
	Rural	>6	5708
Calarasi County	Rural	3	4030
	Urban	2	6831
Total ENEL			
All zones	Rural	>36	39,100
	Urban	>15	68,241
	Total	>51	107,341

Fig. 8 The total smart meters installed in the pilot projects of ENEL from 2014 to 2016.

The number of smart meters installed until now is low in relation to the target assumed by the national program (see Fig. 1), but ENEL considers that the number will increase up to 80% of total customers until 2020.

2.2.2 CEZ

CEZ has been present on the Romanian market since 2005, ensuring the electricity supply to seven counties in the southern area. The company provides services to 1.427 million customers. The modernization of the metering system represents an ambitious investment by CEZ, which carried out pilot projects in four areas of Dolj County in the period 2013–14 [18]. The pilot projects were selected from both urban and rural areas (see Table 6).

In 2015, the implementation of smart metering continued, using the pilot projects from the same county (Dolj). The number of smart meters is separately given for residential and nonresidential customers (see Table 7 [18]).

The investments continued in 2016 with Olt County, but their value is lower (see Table 8 [18]).

The total smart meters installed in the pilot projects of CEZ from 2013 to 2016 are indicated in Fig. 9.

Table 6 The number of smart meters installed between 2013 and 2014

Zone	Area	Number of places	Total
Dolj County	Urban	2	12,164
	Rural	2	1033
Total		4	13,197

Table 7 The number of smart meters installed in 2015

Zone	Area	Number of places	Residential consumers	Nonresidential customers	Total
Dolj County	Urban	1	17,620	1389	19,009
	Rural	1	1039	102	1141
Total		2	18,659	1491	20,150

Table 8 The number of smart meters installed in 2016

Zone	Area	Number of places	Residential consumers	Nonresidential customers	Total
Olt County	Rural	1	54	9	63
Total		1	54	9	63

Fig. 9 The total smart meters installed in the pilot projects of CEZ from 2013 to 2016.

2.2.3 ELECTRICA

ELECTRICA consists of three electricity distribution subsidiaries (see Fig. 7), namely:

- ELECTRICA Muntenia Nord ensures the electricity supply to six counties from the southeast part of Romania.
- ELECTRICA Transilvania Nord ensures the electricity supply to six counties from the north and northwest parts of Romania.
- ELECTRICA Transilvania Sud ensures the electricity supply to six counties from the central part of Romania.

The total number of customers supplied is approximately 3.6 million [19].

Information about the implementation of smart metering is poor. The data are based on presentations from various scientific events and the declarations of managers in the Romanian press.

The program for the implementation of smart meters began in 2012 within some pilot programs. The implementation stage and the adopted solutions at the level of 2012 are presented in Table 9 [19].

In 2012, subsidiaries reported that the degree of implementation for smart meters was the following [11]: ELECTRICA Transilvania Nord—1%, ELECTRICA Transilvania Sud—3%, and ELECTRICA Muntenia Nord—5%.

Currently, the plan is to install 250,000 smart meters in the three distribution subsidiaries; however, only 200,000 had been installed by 2016 through ongoing pilot projects [20].

Table 9 **The number of smart meters installed until 2012 [19]**

Technical solutions	Area	Number of pilot projects	Customer type	Total
PLC communications	Urban	6	Residential customers	9215
	Rural	2	Residential customers	
	Urban	1	Nonresidential customers	
GSM/GPRS communication	Urban	15	Residential customers	45,660
Optical fiber and wire/ GSM communications	Urban	18	Nonresidential customers	9726
Total		32		64,601

2.2.4 E.ON (DELGAZ grid)

E.ON provides services to approximately 1.4 million customers in six counties from the northeastern part of Romania. The company has carried out an extensive program of investment to modernize the infrastructure since 2012. From 2010 to 2012, E.ON has implemented a "load curves collecting" system that allowed the testing of technology and the functionality of smart metering. The main objective of the project was to develop specific load profiles for the different categories of customers in order to facilitate the opening of the retail market for small users [21].

By the end of 2015, E.ON aims to install a total of 226,000 smart meters within an investment program of 30 million dedicated to the development of such networks. By the end of 2012, E.ON installed about 14,300 smart meters. In 2013, another 35,000 smart meters [22] were installed, followed by 147,000 units in 2015. In 2016, plans called for the installation of an additional 55,000 smart meters. The investment for this purpose from 2014 to 2016 was about 20 million euros [23]. The evolution of the implementation program from 2012 to 2016 is given in Fig. 10.

3 Impact of smart metering implementation on saving electricity in distribution networks in Romania

3.1 General aspects

Smart metering provides more accurate billing of the electricity used based on real consumption and decision making on the tariff for the flattening of the load curve and reduction of the invoice. Additional reasons for the implementation of smart metering in the electricity sector are the benefits to the entire society, which may result from the installation of smart grid solutions (smart grids) based on a smart metering infrastructure [24]. Also, significant reductions in electricity consumption and CO_2 emissions can be obtained by providing information about consumption on central

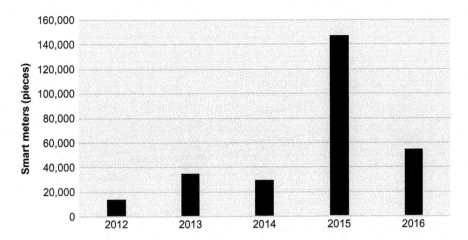

Fig. 10 The evolution of implementation program by E.ON from 2012 to 2016.

Internet portals, which consumers can access directly or at electronic devices located at the consumption point. In addition, the "demand response" solutions can help to reduce the peak load using information from the smart meters. The level of expenditure, the losses in the electricity distribution networks, and the volume of investments are supervised indicators in order to improve the quality of the distribution service due to the implementation of smart metering in Romania.

Thus, smart metering will have a direct impact on:

- The need to increase energy efficiency through a greater transparency of metering information as well as encouraging the consumers to change their consumption behavior.
- The pressure of regulatory authorities for a decrease of the cost by reducing electricity losses and the costs of meter reading and through a better identification of the required investments.
- The education of consumers to reduce the consumption at peak load.
- The environmental protection (reducing the peak load will lead to a decrease in electricity production and the use of power plants with high emissions of carbon dioxide).
- The security of the electricity supply by introducing the flexible infrastructure and minimizing the number of power plants while increasing the share of electricity production from renewable sources.

From the point of view of DNOs, smart metering could follow and improve the state parameters and operation conditions of distribution networks, contributing to increasing energy efficiency. This could be done by reducing their own technological consumption (technical and commercial), the interruptions in power supply, the number of incidents, the voltage dips, the expenditure level, the investments, the operational costs with meter reading, etc.

In Fig. 11, an evolution of electricity losses assumed by each DNO beginning with 2014 is presented [24].

These values will be obtained based on the investments of DNOs in proper distribution networks aiming for modernization measures (replacement of aging equipment and installations) and measures for the implementation of smart grid technologies (distribution automation, distributed generation, smart metering, etc.).

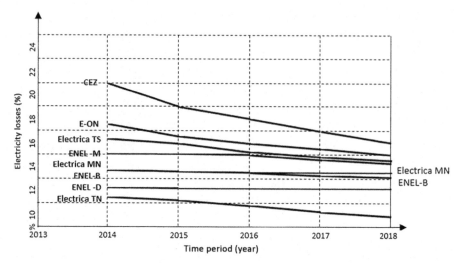

Fig. 11 The evolution of electricity losses (LV level) for all DNOs.

Fig. 12 shows the estimated electricity saved from 2014 to 2020 from distribution networks set out in the national action plan to achieve the target assumed by Romania and the compliance with requirements from Directive 2012/27/EU [24].

The targets are obtained only if the above technical measures are adopted. The implementation of smart metering influences the operation of distribution networks. Thus, the optimal decision making based on information from the smart meters can lead to higher electricity savings.

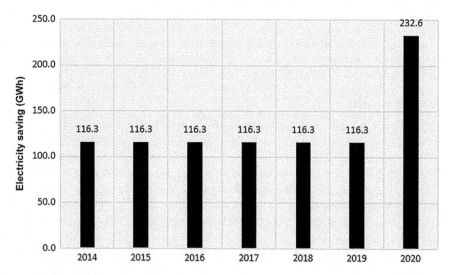

Fig. 12 The estimated electricity savings from 2014 to 2020 from distribution networks set out in the Romanian national action plan.

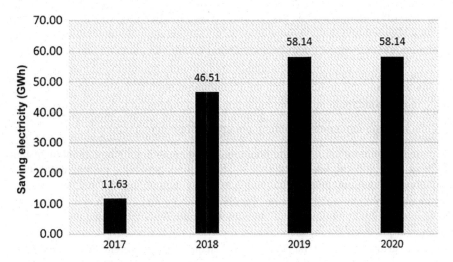

Fig. 13 The estimated electricity savings from the implementation of smart metering in Romania from 2017 to 2020.

For the implementation of smart metering in Romania in the period 2017–20, the estimated electricity savings are presented in Fig. 13 [24]. These savings lead to a decrease in nontechnical losses.

In terms of pollutant emissions, about 0.437 kg of CO_2 are emitted per kWh generated. Thus, the decrease of pollutant emissions (for the data from Fig. 13) can be seen in Fig. 14. It can be observed that total pollutant emissions decrease with approximately 76,221 ton of CO_2.

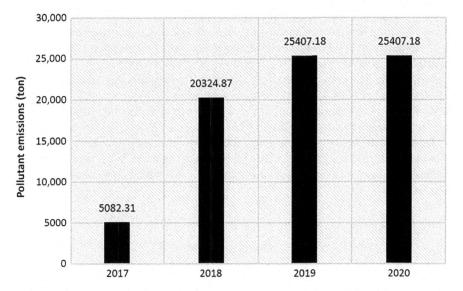

Fig. 14 The estimated decreases of pollutant emissions from 2017 to 2020 due to the implementation of smart metering in Romania.

In the following paragraph, the results obtained in three case studies will be presented to demonstrate the positive impact of implementation of smart metering on the operation and planning of distribution networks.

3.2 Case studies

3.2.1 Phase balancing using smart metering

Due to the distribution of single-phase loads on phases and their random instant demand values in the low voltage (LV) distribution networks, current unbalances can occur. These unbalances increase electricity losses and the risk of capacity constraint violation while deteriorating the power quality. The disadvantages can be eliminated by an optimal phase allocation of customers to balance the LV feeder using data provided by smart meters.

The case study was conducted on an LV electric distribution network included in a pilot project of smart metering program from Romania [25]. The structure of this network is presented in Fig. 15 [25].

The circles signify poles at which consumers are connected via service cables. The total numbers of poles from the network is 198. The line length between two poles is 40 m, as specified in Romanian standards. The LV electric distribution network has three feeders with the primary characteristics presented in Table 10.

The model for an analyzed distribution network was created using a professional package. The conductor properties were entered into the model using the conductor specification data sheets. The lengths for the trunks of LV distribution feeders (conductors between poles) and service cables to each customer were determined via a physical onsite inspection.

Each consumer is connected to a distribution pole via the service cable. The initial phase allocation represents the real situation corresponding to the connection of each customer to the network. The consumer types and their consumption categories are indicated in Table 11.

All customers from the network are integrated into the smart metering system. The individual load profiles for each consumer were taken from that system. The steady-state calculations were made for each hour from the days of the study period (1 week) and the results are presented in Table 12.

An important aspect of the analysis relates to the degree of current unbalance of each LV electric distribution feeder from the analyzed substation zone. This was evaluated using the unbalance factor calculated by DNOs from Romania with relation [25]:

$$CUF = \frac{1}{3}\left[\left(\frac{i_R}{i_{av}}\right)^2 + \left(\frac{i_S}{i_{av}}\right)^2 + \left(\frac{i_T}{i_{av}}\right)^2\right] \tag{1}$$

where: i_{av} is the average value of the current.

The value of CUF should not exceed 1.10 r.u.

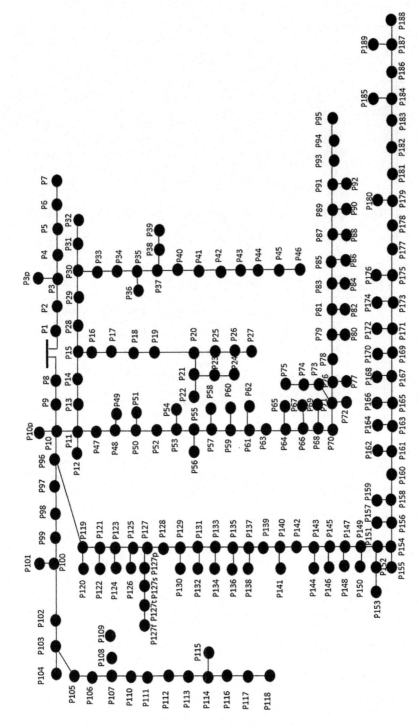

Fig. 15 The pilot distribution network [25].

Table 10 **The primary characteristics of LV feeders**

No. feeder	Total length (m)	Customer number	Conductor type	Conductor size (phase+neutral) (mm²)	Length of section (m)
1	280	13	Classical	$1 \times 50 + 50$	280
2	3520	163	Stranded	$3 \times 35 + 35$	120
			Classical	$3 \times 50 + 50$	3760
3	3880	167	Stranded	$3 \times 50 + 50$	120
			Classical	$3 \times 50 + 50$	2080
			Classical	$3 \times 35 + 35$	960
			Classical	$1 \times 25 + 25$	280
			Classical	$1 \times 16 + 25$	80
Total	7680	343	Classical		7440
			Stranded		240

Table 11 **The characteristics of customers**

Consumer type	No.	Phase			Consumption category (kWh/year)				
		R	S	T	0–400	400–1000	1000–2500	2500–3500	>3500
1-Phase	335	83	152	100	150	108	65	5	7
3-Phase	8	–	–	–	5	2	0	0	1

Table 12 **Energy losses for the study period (1 week)**

No.	Day	Load (kWh)	ΔW (kWh)	ΔW (%)	Input energy (kWh)
1	TU	1064.3	67.5	6.0	1131.8
2	WE	960.4	52.2	5.2	1012.6
3	TH	941.4	54.7	5.5	996.1
4	FR	887.2	45.8	4.9	933.0
5	SA	984.2	60.5	5.8	1044.7
6	SU	828.2	40.1	4.6	868.3
7	MO	912.3	49.9	5.2	962.2

Because feeder 1 is very short and supplies only 13 consumers, this was removed from the analysis. For feeders 2 and 3, the values obtained were 1.25 r.u. and 1.09 r.u., respectively. It can be observed that, for the second feeder, the value of CUF exceeded with 15% the admissible maximum value. This aspect can be highlighted by the loading of phases on the first section of the feeder, presented in Fig. 16.

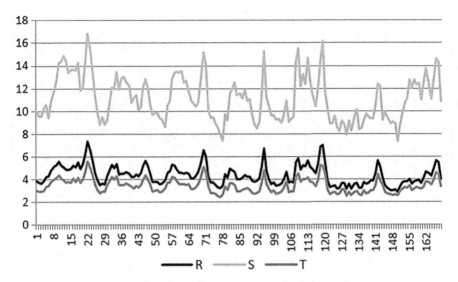

Fig. 16 The loading of phases on the first section of the second feeder (kW).

For this feeder, it proceeded at the reallocation of customers on phases such that the degree of current unbalance would be below the admissible maximum value (1.1 r.u.). The obtained results are indicated in Table 13.

These achievements have resulted in a significant reduction of the degree of unbalance (CUF = 1.016 r.u.). This translates into loads much closer on phases, as evidenced by the graphs from Fig. 17.

With the new data (see Table 13), the steady-state calculations have led to the results presented in Table 14.

The analysis of the results presented in Table 14 highlights a significant reduction of weekly electricity losses in the pilot zone with 1.7%, from 5.3% to 3.6%. The variation of hourly power losses from the studied period for both cases (unbalanced and balanced) is presented in Fig. 18, when a decrease of the power losses can be observed. In addition, the final situation of customer allocation led to improving the voltage level on the phase S (highly loaded phase) and at a voltage balance in the final nodes of can observed distribution feeder. The variation of voltages in Node 95 (both cases) is shown in Figs. 19 and 20.

An extension of the analysis at 1 year revealed that the electricity savings can be by 8.7 MWh/year. This amount of electricity savings was assessed considering the forecasting coefficients for load-in function by each season [25]. This value could be obtained if data from the smart meters are used by DNOs for the purpose of an optimal operation and planning of the network.

3.2.2   Reduction of commercial losses based on smart metering

One of the biggest benefits corresponding to the implementation of smart metering is decreasing the electricity losses and especially commercial losses. The commercial

Table 13 **Customers' reallocation on phases**

No.	Pole	Initial phase	Allocated phase	No.	Pole	Initial phase	Allocated phase
1	20	R	S	32	48	S	R
2	20	R	S	33	48	S	T
3	23	R	S	34	49	S	R
4	26	T	S	35	54	S	T
5	27	T	S	36	54	S	R
6	34	T	R	37	54	S	R
7	34	T	R	38	54	S	T
8	34	T	R	39	55	S	T
9	37	T	R	40	65	S	R
10	37	T	R	41	66	S	R
11	38	T	R	42	66	S	T
12	39	T	R	43	67	S	R
13	39	T	R	44	68	S	T
14	39	T	R	45	68	S	T
15	39	S	R	46	70	S	R
16	40	T	S	47	72	S	T
17	40	T	S	48	75	S	T
18	40	T	S	49	76	S	R
19	41	T	S	50	78	S	T
20	42	T	S	51	79	S	R
21	43	T	S	52	80	S	T
22	43	T	S	53	82	S	T
23	44	T	S	54	83	R	T
24	45	T	S	55	84	S	R
25	45	T	S	56	87	S	T
26	45	T	S	57	88	S	R
27	45	T	S	58	89	S	T
28	46	T	S	59	90	S	R
29	46	T	S	60	93	S	T
30	47	R	S	61	95	S	T
31	47	S	R	61	95	S	T

electricity losses result from the errors and consumptions of measuring equipment as well as from electricity theft. In many implementation zones, the installation of smart meters is essential to restore or change components from networks in order to have good results and proper monitoring. Also, the electric substations are equipped with balanced smart meters, which measure the total losses based on the difference between the measured electricity and the sum of all measured electricity by the smart meters from customers. Due to these balanced smart meters, the nontechnical (commercial) losses can be easily evaluated based on the difference between total losses and

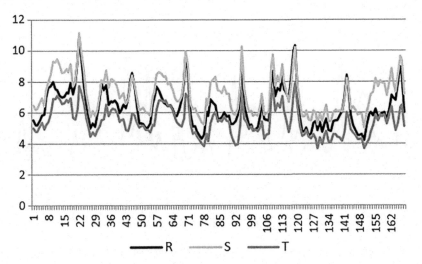

Fig. 17 The loading of phases on the first section of feeder 2, after consumers' reallocation (kW).

Table 14 **Comparative situation of the energy losses between the initial and final customer allocation**

		Initial case (unbalanced)			Final case (balanced)		
Day	Total load (kWh)	Input energy (kWh)	ΔW (kWh)	ΔW (%)	Input energy (kWh)	ΔW (kWh)	ΔW (%)
MA	1064.3	1131.8	67.5	6	1108.5	44.1	4
MI	960.4	1012.6	52.2	5.2	995.6	35.1	3.5
JO	941.4	996.1	54.7	5.5	977.3	35.8	3.7
VI	887.2	933	45.8	4.9	918	30.6	3.3
SA	984.2	1044.7	60.5	5.8	1023.3	39	3.8
DU	828.2	868.3	40.1	4.6	855.8	27.5	3.2
LU	912.3	962.2	49.9	5.2	945.5	33	3.5
Total	6578	6948.7	370.7	5.3	6824	245.1	3.6

technical losses. The technical losses can be calculated exactly because the smart meters can be connected to a central database provided with dedicated software that can determine the technical losses in a very short time. Thus, the technical and commercial losses can be evaluated more accurately and very fast.

In Predescu and Vornicu [26], taking into account these aspects, the results obtained after the implementation of smart metering in some pilot networks from Romania were presented. The conclusions were that the commercial losses reached all areas, but in some, the reduction is significant (see Fig. 21).

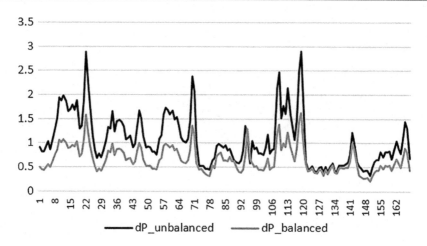

Fig. 18 Comparison between the hourly power losses in unbalanced and balanced cases (kW).

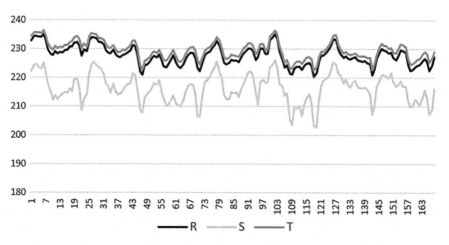

Fig. 19 The voltages in Node 95 in the unbalanced case (V).

The electricity savings for each area are shown in Fig. 22. The total electricity savings obtained from reducing the commercial losses was approximately 0.548 GWh.

3.2.3 Smart metering-based decision making for replacement of distribution transformers

The energy efficiency of distribution networks can be increased adopting some measures focused on the reduction of electricity losses. Thus, the investments in new electricity generation units and its transmission and distribution to consumers will decrease.

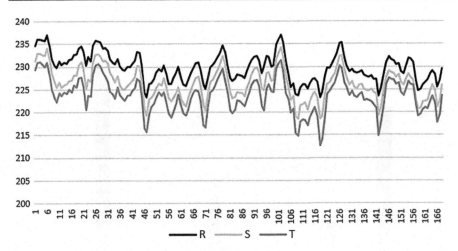

Fig. 20 The voltages in Node 95 in the balanced case (V).

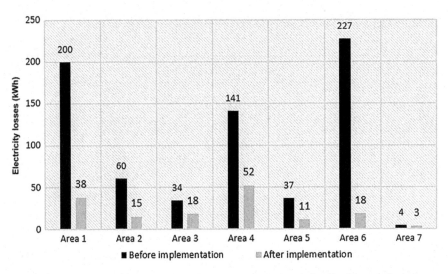

Fig. 21 Comparison between the commercial losses before and after implementation of smart metering for each analyzed area.

The distribution networks from Romania as well as other European countries are characterized by a vast majority of components (lines, transformers, switching equipment, etc.) at the end of their lifetime. These components have been in operation since the 1960s. Fig. 23 presents two profiles: the first indicates the investments made in electric distribution networks, and the second suggests the time frame for investments in the replacement of aging equipment, at the limit of the lifetime [27].

An analysis of the situation presented in Fig. 23 leads to the conclusion that in the forthcoming period, the vast majority of components from the electric distribution

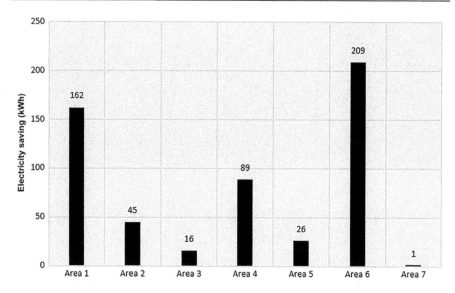

Fig. 22 The electricity savings obtained from reducing the commercial losses.

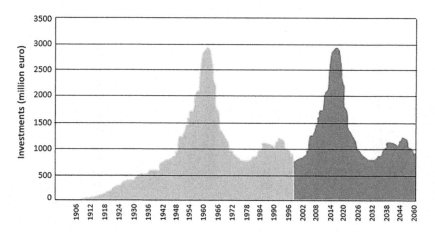

Fig. 23 The typical profiles of investments made in distribution network—*light gray* (£ million) and suggested time frame for aging equipment replacement—*dark gray* [27].

networks will be replaced. It can be observed that the height of this modernization is reached during the period 2016–20. In this context, the DNOs must propose and implement strategies that lead to improved energy efficiency. The eight DNOs from Romania have permanent multiannual investment programs that take into account the energy efficiency. Among the general objectives of the development/modernization strategies envisaged by DNOs can be mentioned [28–30]:

- Increasing the security of the electricity supply to all consumers.
- Reducing electricity losses.
- Increasing the degree of security in operations and function safety.

- Decreasing the amount of damage in electric installations and the interruption duration of consumers.
- Ensuring the quality parameters of electricity supplied in accordance with ANRE.
- Reducing the amount of undelivered electricity to consumers as a result of unplanned outages.
- Reducing the maintenance and repair costs.

The measures that can be taken to achieve all objectives of development/modernization are as follows:

- Modernization of the transformer substations.
- Increasing the voltage level from 6 kV up to 20 kV.
- Development of distribution automation by installing the reclosers and remotely controlled switches.
- Integrating the SCADA system in all electric substations.
- Introduction of the smart metering.
- Installing the automatic voltage regulators in all electric substations.
- Introduction of the asset management system.

The achievement of all objectives is directed toward better consumer satisfaction from the viewpoint of power quality, sustainable energy development, and high economic performance.

In this paragraph, only the modernization of MV/LV electric substations based on the replacement of transformers was treated. Thus, the results obtained in a pilot project will be presented. The distribution transformers with older minimum energy performance standards (MEPS) were replaced with efficient transformers (European performance standard EN 50464). Decision making was based on the data provided by the balanced smart meters installed in MV/LV electric substations. For example, the weekly loading profile of a transformer (provided by the balanced smart meter) is presented in Fig. 24.

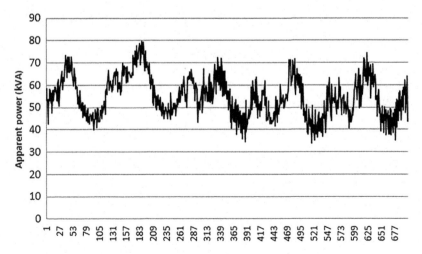

Fig. 24 Weekly loading profile of a transformer provided by the smart meter.

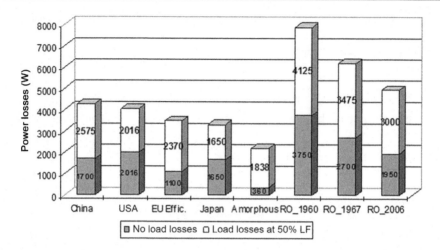

Fig. 25 The international MEPS from the world and Romania (distribution transformer with rated power $Sn = 1000\,kVA$ at a 50% loading).

Table 15 Minimum energy performance standards for transformers

MEPS		RO_1967	RO_1972	RO_2006	Total
Number	(pieces)	12	32	7	51
	(%)	24	63	14	100%

The aim is obtaining the technical and economic benefits from the replacement of all overload and underload old transformers. The old transformers are considered the transformers with the following MEPS: RO_1960, RO_1967, and RO_2006 (see Fig. 25 [31]).

The analysis revealed 51 distribution transformers (20/0.4 kV) that meet the above criteria, as you can see in Table 15. The information about the loading of transformers was obtained from the load profiles recorded by the balanced smart meters installed in each electric distribution substation.

In the initial stage, the loading factor had values between 3% and 160%. The final rated power of transformers has been flattened, now situated in the range 40%–80%. This aspect is reflected by the loading factor represented in Fig. 26. Also, Fig. 27 shows the initial versus final rated power.

As to the annual energy losses, they decreased by 330.9 MWh (about 63%). A comparison between the annual electricity losses in the initial and final stages is presented in Fig. 28.

Further, the specific electricity savings (MWh/year/transformer) were calculated. The obtained value for this indicator is 6.5 MWh/year/transformer. Finally, an assessment of the total annual cost of electricity losses was calculated, as you can see in Fig. 29, and the obtained benefit was by 26,483 euro/year.

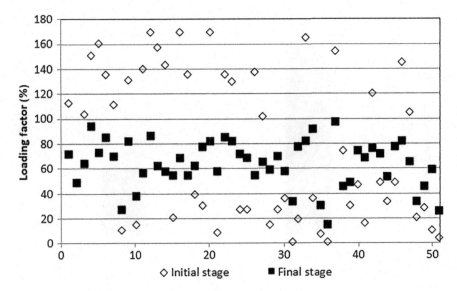

Fig. 26 The initial and final loading factor of transformers.

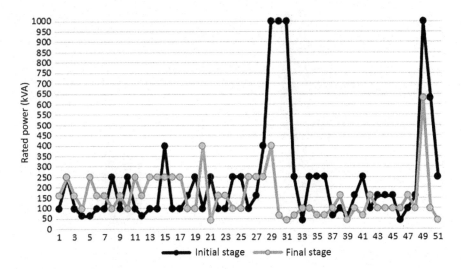

Fig. 27 The initial versus final rated powers of transformers.

Due to the tightening of the regulatory framework, the decisions of DNOs for the replacement process must be based on detailed technical and economic analysis, having as input the data provided by the smart metering system. These decisions should lead to:

- Increasing the electricity savings.
- Reducing the maintenance and operation costs.
- Reducing the security risks of people and goods.
- Reducing the risk of penalty arising from noncompliance with the quality conditions of the distribution service imposed by the performance standards.

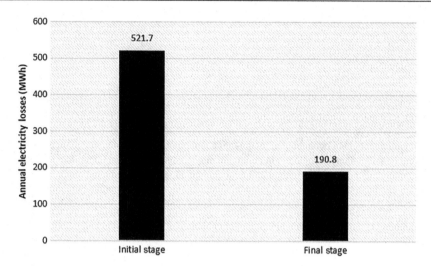

Fig. 28 Total annual electricity losses for all transformers.

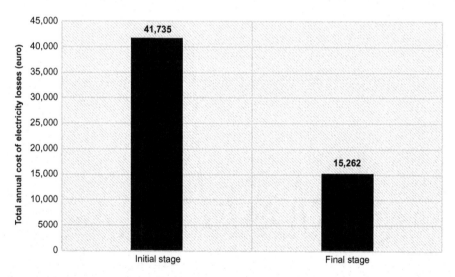

Fig. 29 The total annual cost of electricity losses.

4 Conclusions and future trends

The most applied model by DNOs from Romania to the implementation of smart metering from 2012 to 2016 was an exponential model that involved the installation of a smaller number of smart meters in the early years (to allow electricity distribution companies to adjust the parameters, to plan and learn from the experience of the first years) followed by a balanced implementation for the coming years.

ANRE proposed beginning with 2014 to implement pilot projects and evaluate the specific aspects of the distribution network in order to establish the conditions for the final installation of intelligent metering systems by each operator, targeting the informational campaign for customers and logistics activities. The pilot projects represent part of the national program for the implementation of smart metering and are included in the investment plans of DNOs [32]. These projects take into account urban and rural areas with high electricity losses, areas that require the modernizing of electric networks, and urban and rural areas in good technical condition and small electricity losses.

The investments corresponding to the implementation of smart metering are recognized in the regulated asset base (RAB) by ANRE and paid the regulated rate of return, which also applies to other investments of DNOs. The implementation of smart metering systems can be made through accessing some funding sources or grants, including from European funds, these being considered investments of DNOs.

This funding will not establish the beginning corresponding to the implementation of the smart metering system and the values of financed investments will not be included in the RAB calculation, according to ANRE. The document stipulates that, to the request of the customer/DNO paying a fee, the smart metering systems should include advanced pricing structures, records of the consumption depending on the period, and a remote financial control without being passed through the computer system (for respecting confidentiality) [32].

The establishment of tariffs is made by the supplier with the data only available to the customer. Load curve can be an important element in the correct setting of the tariff for consumers. The smart meters are designed to help the consumers save energy as well as force the DNOs to ensure a better quality of service.

The results of pilot projects for smart metering showed that the installation of smart metering makes economic sense only in urban areas [16]. There are many DNOs that believe the smart meter helps just to know consumption in real time, to reduce thefts and commercial losses. Thus, it is very important for DNOs to explain to customers the benefits for them: the real time readings of consumption, ensuring the ideal power consumption curve, or balancing the power consumption at the level of each household [33].

At this moment, ANRE is in the stage of cost-benefit assessment on eight regions where the smart meters were implemented. The analysis of the implementation stage at the end of 2015 revealed the following [7]:

- The postimplementation results concerning the expected benefits are not relevant to all pilot projects due to the very short time period from the completion of the implementation up to the reporting date of the achievements (lack of a relevant period for monitoring); it does not provide indicative premises for decision making on a rollout.
- Cost-benefit analyses submitted by DNOs did not allow a comparative analysis of the results. These were achieved on different analysis models following the business strategy of DNO.
- The results of cost-benefit analyses are not positive for all DNOs.

The preliminary conclusions of cost-benefit analysis are not favorable in achieving the proposed level. ANRE will launch in 2017 the national proposal for smart metering,

Table 16 **The situation regarding the implementation of smart metering for 2016–20**

DNO	Number of zones for implementation	Number of smart meters to customers	Number of smart meters in electric substations	Number of smart meters for replacement in 2016–20
CEZ	2606	1,398,919	13,268	1,202,115
ENEL Banat	448	866,366	7697	360,022
ENEL Dobrogea	344	626,627	5498	301,874
ENEL Muntenia	254	1,201,804	6805	569,298
E.ON (DELGAZ GRID)	2552	1,154,729	9857	947,300
ELECTRICA Transilvania Sud	1782	1,078,383	6059	627,864
ELECTRICA Transilvania Nord	1963	1,224,969	8719	576,320
ELECTRICA Muntenia Nord	1414	1,224,969	10,218	730,766
Total	11,363	8,798,299	68,103	5,315,559

but it is already clear that Romania cannot reach a smart metering degree of approximately 80% from the electricity consumption by 2020 as forced by European regulations. One of the explanations is linked to the fact that there are eight DNOs, four entities legally, each with its own design, cost, and specifications.

The situation from 2016 to 2020 is presented in Table 16 [7].

It can be observed that the number of smart meters that should installed from 2016 to 2020 represents approximately 57% from the total.

References

[1] European Commission, Directorate-General for Energy, Standardization Mandate to CEN, CENELEC, and ETSI in the field of measuring instruments for the developing of an open architecture for utility meters involving communication protocols enabling interoperability, M 441/EN, (online). Available from: https://ec.europa.eu/energy/sites/ener/files/documents/2011_03_01_mandate_m490_en.pdf, 2009.

[2] European Commission, Directorate-General for Energy, Standardization Mandate to European Standardisation Organisations (ESOs) to support European Smart Grid

deployment, M/490 EN, (online). Available from: https://ec.europa.eu/energy/sites/ener/files/documents/2011_03_01_mandate_m490_en.pdf, 2016.

[3] Electrica Distributie, Energy Effificency (in Romanian), (online). Available from: https://www.electrica.ro/activitatile-grupului/distributie/eficienta-energetica.

[4] M. Anastasiu, A. Milandru, A. Vintea, Roadmap for smart metering services to the final consumer for Romania (in Romanian), (online). Available from: http://www.smartregions.net, 2012.

[5] A.T. Kearney, Smart metering in Romania, Final Report, 2012.

[6] Ministry of European Funds, Large scale infrastructure operational program 2014–2020. Improving energy efficiency through the implementation of intelligent smart metering systems to low-voltage electrical networks, (online). Available from: http://www.jaspersnetwork.org/download/attachments/18743427/6.%20JASPERS%20NP%20Smart%20Grids%2025%20March%202015%20-%20Romania%20smart%20metering.pdf?version=1&modificationDate=1427307467000&api=v2.

[7] ANRE, Energy Efficiency Department, Report on progress towards in achieving the national objectives regarding the energy efficiency (in Romanian), April 2016.

[8] ENEL Distribution Romania, Meter project in Romania, Sustainable Energy, 24–28 June 2013.

[9] R. Hierzinger, M. Albu, H. van Elburg, A. Scott, A. Łazicki, L. Penttinen, F. Puente, H. Sæle, European smart metering landscape report (IEE), (online). Available from: https://ec.europa.eu/energy/intelligent/projects/sites/ieeprojects/files/projects/documents/smartregions_landscape_report_2012_update_may_2013.pdf, 2013.

[10] J. Žádník, Cost benefit studies (CBA) for smart metering roll-out in the EU, overview and way forward, smart grids in practice, September 29 2015.

[11] Institute of Communication & Computer Systems of the National Technical University of Athens ICCS – NTUA for European Commission, Study on cost benefit analysis of in EU Member States Smart Metering Systems, Final Report, 25 June 2015.

[12] The National Energy Regulatory Authority from Romania, Law no. 123/2012 on electricity and natural gas, (online). Available from: www.anre.ro, 2012.

[13] The National Energy Regulatory Authority from Romania, Order on the implementation of smart metering in electricity, (online). Available from: www.anre.ro, 2014.

[14] The National Energy Regulatory Authority from Romania, National Report 2015, August 2016, www.anre.ro.

[15] V. Janoušek, J. Macek, F. Müller, Advanced meter management. From EU legislation to real aplications in CEZ, ŠpindlerĤv Mlýn, Czech Republic, 13–14 April 2016.

[16] ENEL Distributie, Modernization Programme (in Romanian), (online). Available from: http://eneldistributie.ro/ro-RO/Pagini/Programe-de-modernizare.aspx.

[17] Capital Magazine (in Romanian), (online). Available from: http://www.capital.ro/enel-lanseaza-contorul-inteligent-care-sunt-beneficiile-pentru-clienti.html.

[18] CEZ Distributie, Smart metering system (in Romanian), (online). Available from: http://www.cez.ro/ro/proiecte/sistemul-de-masurare-inteligenta-a-energiei-electrice/proiecte-pilot-2013-2014.html.

[19] D. Federenciuc, Concerns of Electrica Distribution for the best practical in implementation of pilot projects for smart grid/smart metering (in Romanian), FOREN 2012, Neptun, Romania, 2012.

[20] Economica.net (in Romanian), (online). Available from: http://www.economica.net/sute-de-mii-de-contoare-inteligente-se-monteaza–incepand-cu-acest-an--in-transilvania–nordul-munteniei-si- oltenia_114101.html.

[21] M.L. Cazacu, G. Cosarca, Smart metering systems to E.ON Moldova Distribution (in Romanian), Energetica Magazine 3 (2014) 111–113.

[22] Energynomics (in Romanian), (online). Available from: http://www.energynomics.ro/ro/distribuitorii-vor-%20instala-inca-100-000-de-contoare-inteligente-pana-la-sfarsitul-anului/.

[23] E.ON Romania (in Romanian), (online). Available from: http://www.eon-romania.ro/cps/rde/xchg/SID-E5F1D520-D3EF31A0/eon-romania/hs.xsl/3514_MTQ4MzA2NTYwMzE.htm.

[24] National Action Plan for Energy Efficiency (in Romanian), (online). Available from: http://www.escorom.ro/images/Planul%20national%20de%20actiune%20in%20domeniul%20eficientei%20energetice-2020.pdf.

[25] G. Grigoras, M. Gavrilas, M. Cazacu, G. Cosarca, Assessing energy losses in unbalanced low voltage electric distribution networks, Energetica Magazine 2 (2016) 53–59.

[26] L. Predescu, D. Vornicu, Better oversight and management of electricity use with smart meters, CIRED Workshop, Helsinki, 14–15 June 2016, Paper 0165.

[27] Northcote-Green J. and Speiermann, M., Third generation monitoring system provides a fundamental component of the smart grid and next generation power distribution networks. In: 20th International Conference and Exhibition on Electricity Distribution – Part 1, CIRED 2009, June 8–11, 2009.

[28] C.E.Z. Distributie, The general objectives of the development strategy of distribution installations (in Romanian), (online). Available from: http://www.cez.ro.

[29] Electrica Distributie, The general objectives of the development strategy of distribution installations (in Romanian), (online). Available from: http://www.electrica.ro.

[30] EON Romania, The general objectives of the development strategy of distribution installations (in Romanian), (online). Available from: http://www.eon-energie.ro.

[31] G. Grigoras, M. Gavrilas, S. Buliga, C. Strugaru, Solutions regarding the efficient use of transformers in electric distribution networks (in Romanian), Energetica 63 (4) (2015) 145–149.

[32] Econet-Romania, Smart metering. Smart meters to most of electricity consumers (in Romanian), (online). Available from: http://www.econet-romania.com/ro/market/474/smart-metering-contoare-inteligente-pentru-majoritatea-consumatorilor-de-energie.html.

[33] Energynomics, Smart metering is feasible economically only to urban areas (in Romanian), (online). Available from: http://www.energynomics.ro/ro/anre-contorizarea-inteligenta-este-fezabila-economic-numai-la-oras.

Smart grid digitalization in Germany by standardized advanced metering infrastructure and green button

10

Jürgen Meister, Norman Ihle, Sebastian Lehnhoff, Mathias Uslar
OFFIS—Institute for Information Technology, Oldenburg, Germany

Acronyms

3D EMT	a special kind of AEE that collects all data relevant for energy billing (e.g., from the smart meter gateway and the supplier) and provides this data to the customer
AEE	authorized external entity
AMI	advanced metering infrastructure
BDEW	German Association of Energy and Water Industries
BMWi	Federal Ministry for Economic Affairs and Energy
BNetzA	Bundesnetzagentur (Federal Network Agency), the highest regulation authority in Germany
BSI	Bundesamt für Sicherheit in der Informationstechnik, the German Federal Office for Information Security
CLS	controllable local system
DER	distributed energy resource
DMD	Download My Data, a process to receive Green Button data
DR	demand response
DSO	distribution system operator
EEG	Erneuerbare-Energien-Gesetz (Energy Sources Act)
Green Button	an American data format for access of usage information by customers
HAN	home area network
IEC	International Electrotechnical Commission
IF_3D_CON	the interface between the 3D EMT and the customer
IF_GW_CON	the interface between the smart meter gateway and the customer
IF_GW_WAN	the interface between the smart meter gateway an the wide area network
IP	Internet protocol
ISO	International Organization for Standardization
LMN	local metrological network
NAESB	North American Energy Standards Board
OBIS	object identification system defines identification codes for metering data
PTB	Physikalisch-Technische Bundesanstalt, the national metrology institute in Germany

Application of Smart Grid Technologies. https://doi.org/10.1016/B978-0-12-803128-5.00010-6

PV	photovoltaic
RTU	remote terminal unit
SCADA	supervisory control and data acquisition
SINTEG	Smart Energy Showcases—Digital Agenda for the Energy Transition, a funding program initiated by the German government
SMGW	smart meter gateway
SMGWA	smart meter gateway administrator
SM-PKI	smart metering-public key infrastructure
TAF	Tarifanwendungsfall, a type of tariff
TLS	transport layer security, a cryptographic protocol for secure Internet communication
TSO	transmission system operator
USEF	Universal Smart Energy Framework
VPP	virtual power plant
WAN	wide area network
XML	extensible markup language, a data format for document encoding

1 Applications in Germany

Germany has achieved a significant share in renewables due to guaranteed feed-in tariffs for renewables and a good state of grid infrastructure. A further increase in renewables demands market-oriented tariffs and more smartness in power grids. Therefore, German regulation and the grid operators need to address the following smart grid applications:

- Provision of secondary and tertiary reserve by virtual power plants (VPPs) composed of distributed energy resource (DER), for example, wind or photovoltaic (PV) power plants.
- Feed-in management as direct control of the DER by grid operators in case of power grid congestion.
- Demand response (DR) management by energy suppliers, VPP operators, or other kinds of aggregators.
- Application for consumer attitude change to a lower CO_2 footprint or for shifting energy consumption in times with oversupply by renewables.

Secondary reserve, tertiary reserve by DER, and feed-in management are already established applications. Despite a few established markets and regulations, DR management and applications for consumer attitude change are still a moving target. In Germany up until now, they have mainly been addressed in R&D projects only, for example, in decentralized energy management systems [1], VPPs [2], or for the purpose of providing decentralized ancillary services from renewable generation units [3]. Soon, new applications will arise in this area that will be put into practice. These applications will require better availability of detailed metering data and secure communication with DER and consumers.

In this chapter, at first a short overview of the aforementioned smart grid applications in Germany is presented. After this, the new bill called the "Digitalization of the Energy Transition" and the technological implications of this bill for smart grid applications in Germany will be introduced.

1.1 Role and applications of the transmission system operator

German transmission system operators (TSOs) provide two main applications for the integration of renewable energy sources into the transmission grid: the market for control reserve (regelleistung.net) and feed-in management for renewables.

The German TSOs are responsible for maintaining the balance between electricity generation and consumption within their control areas at all times. For the execution of this task, the TSOs can engage different types of control reserve (primary control reserve, secondary control reserve, and tertiary control reserve). The types of control reserve differ according to the principle of activation and their activation speed. Close cooperation between the TSOs contributes to keeping the total requirements for control reserve as low as possible [4]. Therefore, the control reserve may be provided not only by conventional power plants but also by VPPs composed of renewable power generation units, electricity storage units, and flexible loads. In the last few years this led to a decrease of market prices for tertiary and secondary control reserve.

Communication between TSOs, VPPs, and DERs is usually implemented mostly by using the telecontrol standard IEC 60870-5. In case of an activation of a secondary control reserve or a tertiary control reserve, the supervisory control and data acquisition (SCADA) system of a TSO sends an activation signal to a VPP SCADA by the serial protocol IEC 60870-5-101 and gets the meter data about power delivery on the same telecontrol channel. The VPP SCADA communicates with pooled DERs, often using the IP-based protocol IEC 60870-5-104.

The Federal Network Agency (BNetzA) provides a guideline on feed-in management by grid operators under the Renewable Energy Sources Act (EEG). This guideline regulates priorities for the reduction of energy production by conventional and renewable power plants in case of an impending congestion for power grid areas. Briefly, the energy production by renewable power plants has a much higher priority than the feed-in by conventional power plants. Furthermore, the EEG provides guidance on compensation for renewable power plant operators in case of production throttling by a grid operator.

Feed-in management applications of TSOs and distribution system operators (DSOs) apply ripple control systems or IEC 60870-5-104 based automation systems for communication of reduction signals to DERs in a case of feed-in management activation. Historically, most German DSOs implement ripple control systems for the support of their own feed-in management or the feed-in management by TSOs. A ripple control system implements a radio-based unidirectional communication from DSO SCADA to all DERs in an electricity grid. The main drawbacks of this technology are missing feedback about the success of a requested DER deactivation and roughly defined communication groups, which do not always match with the area of impending congestion. In practice, grid operators often deactivate more DERs than required for prevention of congestion in a grid area. Therefore, new installed DERs have to provide an IEC 60870-5-104-based bidirectional remote terminal unit (RTU) [5]. This technology allows a tightly focused feed-in management for congestion-relevant subareas in the grid and therefore helps to increase the overall amount of renewable energy in the power grids.

In summary, regulatory, market, and technical rules for the control reserve market and feed-in management by TSOs in Germany are well known and broadly applied.

1.2 Role and applications of the distribution system operator

DSOs implement feed-in management in the same manner and with similar technologies as TSOs. The usage of ripple control systems for feed-in management in distribution grids has the same shortcomings as in transmission grids. Implementation of bidirectional communication with small DERs, for example, household PVs, based on "classic" RTUs from substation automation is often too expensive. Therefore, there is a demand for a new approach for bidirectional communication between small DERs and feed-in management systems.

New applications for flexibility management in distribution grids are, at the moment, a matter of discussion and research in context of the position paper "Smart Grid Traffic Light Concept" of the German Association of Energy and Water Industries [6]. The proposed traffic light concept describes a model for market participants and network operators to interact with one another in the future. The BDEW model uses the logic of a traffic light as an analogy. In the green phase, the energy market participants may act without restrictions. In the red phase, there is a critical congestion and the grid operator overrules market participants by feed-in management. The BDEW position paper defines a new yellow intermediate phase. The yellow phase is entered if a potential bottleneck in a network segment of a distribution grid has been identified. In the yellow phase, the DSO calls upon the flexibility offered by DER operators or their aggregators in that network segment in order to prevent a red phase situation. The discussion about the possible flexibility products for DSOs is still a work in progress, nationally and EU-wide [7]. The German government initiated the funding program "Smart Energy Showcases—Digital Agenda for the Energy Transition" (SINTEG) that aims to develop and demonstrate new approaches to safeguarding secure grid operations with high shares of intermittent power generation on the basis of wind and solar energy in five model regions. The initiated projects in this program address, among other things, the issue of flexibility products for the yellow phase in distribution grids in order to secure an increased share of renewable energy in the energy mix.

The feed-in management applications and those within the SINTEG projects developed market-based flexibility applications for DSOs require detailed knowledge about the grid state and reliable bidirectional communication to the DERs also in distribution grids.

1.3 Role and applications of consumers and decentralized producers in electrical grids

In many situations, it is beneficial if a consumer is able to adapt his consumption behavior. One idea is that he is rewarded if he shifts his power demand in times with high production or even an overproduction from renewable energy sources. To reach this goal, smart metering solutions with real time visualization for the customers are proposed as well as flexible tariffs that adapt to the situation in the grid. Furthermore, there is a need for bidirectional communication to customers with flexible and controllable demand in order to implement DR applications for the yellow phase.

All DERs with an electrical connection point to medium-voltage grids and few of the DERs with a connection point to low-voltage grids have to provide bidirectional communication to feed-in management of DSOs. Due to decreasing subsidies for renewable energy production, DER operators will have to cooperate with VPP operators and other aggregators in the long term in order to sell the produced energy. Therefore, there will be an additional demand for affordable bidirectional communication and metered data from producers and consumers in real time. This demand is addressed by the new bill called "Digitalization of the Energy Transition" [8].

1.4 Distribution grid digitalization as enabler for energy transition

As was seen in the examples of application above, the next step is the energy system transition demands, a more active integration of consumers and DERs in distribution grids and markets. Due to high hardware costs and engineering efforts, grid automation technologies such as the existing IEC 60870-5-104 RTUs are not widely applicable in Germany; therefore a different approach is proposed. The main issue for integrating more renewables into power grids is the digitalization of the distribution grid. The German government addressed this issue in 2016 by a new bill called "Digitalization of the Energy Transition" [8]. The bill will boost digitalization by implementing a Germany-wide advanced metering infrastructure (AMI), including the staggered nationwide rollout of smart meters. On the one side, the planned AMI will provide utilities, customers, and other authorized external entities (AEEs) with metering data in order to better manage the power grids or optimize energy usage. On the other hand, the AMI will implement a secure communication backbone to control decentralized energy production as well as end user consumption.

In the following sections, the basic concepts, features, and technology of the planed Germany-wide AMI will be introduced; it will be explained how the AMI can be combined with the American Green Button Standard for better meter data management; and describe how the AMI will be used for bidirectional communication with DERs and controllable loads in an inexpensive and highly secure manner.

2 Smart grid features

Smart grid applications for transmission grids and the high- and medium-voltage level in transmission grids can and are already implemented by using existing power automation technologies [9]. These technologies were originally developed for substation automation and have their origins in the late 1980s and early 1990s. IEC 60870-5-101 and IEC 60870-5-104 based RTUs and other automation components have a few drawbacks. In practice, interface implementations are too expensive to be an economical feasible alternative for small DERs and loads on the medium- and low-voltage level in distribution grids. However, most significant are the shortcomings in cybersecurity issues and the lack of built-in plug and play support. TSOs and DSOs use the "air gap" approach to achieve required security levels. System engineers often place the automation technology "as is" in dedicated communication networks. Grid operators manage and often own these networks. This security approach is not transferable to low-voltage level and small DERs because of the need to use public networks and strict cost restrictions.

The bill "Digitalization of the Energy Transition" especially addresses, in addition to infrastructure regulation, security issues and cost restrictions. The Germany-wide roll out of the AMI shall provide DSOs, consumers, energy suppliers, aggregators, and authorized others with a set of basic services related to metering and communication issues. Smart metering-related services of AMI are:

- Collection and storage of metering data according to German measurement and instrumentation regulations of the National Metrology Institute of Germany (PTB, Physikalisch-Technische Bundesanstalt).
- Providing secure and privacy protection of metering data to customers, energy suppliers, grid operators, and other AEEs.
- Innovative tariff options.

A specific feature of the German AMI is the use of a so-called smart meter gateway (SMGW) as a secured communication device at the customer site. The SMGW collects the data from one or more attached smart meters, stores them, and even applies tariff rules for them before communicating the result to AEEs. The SMGW has three security zones: local metrological network (LMN), home area network (HAN), and wide area network (WAN) (see Section 3.1). Each security zone has dedicated functions, for example, smart meters are connected to the SMGW via the LMN. For communication purposes with AEEs (via the WAN interface), the SMGW has to be equipped with a hardware security module that is responsible for encryption and authorization tasks. The SMGW, together with a smart meter, is in Germany often referred to as an "intelligent metering system" while a digital meter (smart meter) is referred to as a "modern metering device."

Potentially, the most underestimated basic service of the AMI is the possibility of establishing a secure communication channel to controllable local systems (CLS) at the customer site. In the future, grid operators could use this communication channel of a SMGW for feed-in applications in low-voltage grids. Furthermore, the CLS channel may be used for additional applications in smart grids or other smart city usage areas, for example, assisted living for old or handicapped persons. The CLS channel fulfills the actual security requirements of the German Federal Office for Information Security (BSI), which is the national cybersecurity authority and a regulatory authority of the German government.

In the following, some details of the planned AMI in Germany will be described. First, basic concepts, roles, and components of the German AMI as defined in Refs. [8,10,11] are described. After that, some use cases based on the AMI are presented. These use cases will be the ones implemented first after the rollout for the Germany-wide AMI.

2.1 Basic concepts, roles, and components of German AMI

Regulations for the German AMI differentiate few key roles and components. Fig. 1 illustrates these roles and components with its dependencies.

The role of "smart meter operator" is responsible for installation and operation of the metering infrastructure on customer sites. Persons or companies with the role "customer" are the owner of metering data. Customers may provide the role "authorized external entities" (AEEs) with metering data for billing purposes or efficiency consulting projects. Furthermore, customers can entrust AEE with the usage of the CLS communication channel for various purposes.

The system configuration at the customer's site includes a digital smart meter, a SMGW, and an optional device, the "controllable local system" (CLS). The SMGW is the actual communication device that collects data from one or more smart meters

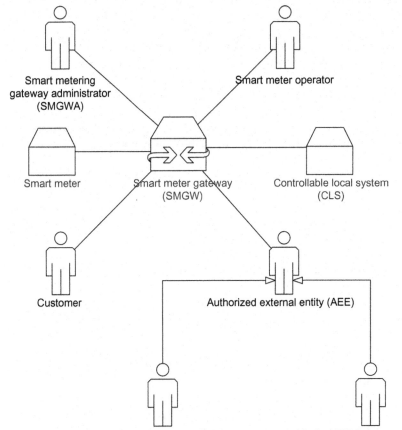

Fig. 1 Actors and components of German AMI.

and sends it to authorized receivers. The SMGW is also responsible for the establishing of secure communication channels for further purposes, for example, the control of local CLS devices. All these components must fulfill the security and privacy requirements of the BSI [10]. The smart meter and SMGW must provide a metrological certificate of the PTB [11].

The role "smart meter gateway administrator" (SMGWA) is a kind of a sovereign role. SMGWA is an independent authority that checks authentication and authorization credentials of an AEE and can deny their requests for metering data or for communication with CLS devices on customer's sites.

Regulations for smart metering differentiate between passive and active AEE. The passive AEE only receives metering data from the SMGW whereas the active AEE can interact with CLS devices on customer's sites, for example, for implementation of the feed-in management use case. Furthermore, there are regulated and non-regulated AEEs. Smart meter operators are obliged to support regulated AEEs (customer, energy supplier, and grid operator). The government has defined a price limit for the regulated services in Ref. [8]. Additionally, smart meter operators can charge nonregulated AEEs with usual market prices for nonregulated services in order to generate additional revenues with smart metering infrastructures. We will describe various use cases for typical German AEEs later in more detail.

The distinctive feature of German AMI is the "don't call me, I'll call you back" principle of the SMGW. This means that an AEE cannot initiate communication directly with an SMGW. The SMGW always initiates the communication between the AEE and SMGW. The BSI chose this procedure in order to minimize cybersecurity risks in communication with smart meters and DERs. The UML sequence diagram in Fig. 2 describes in detail the procedure for establishing communication between the SMGW and AEE in case of a special need.

An AEE asks SMGWA for metered values or for a communication request to a dedicated SMGW. SMGWA authenticates the AEE and checks its rights. If the authorization checks of the AEE are successful, then SMGWA sends the request notification to the SMGW. In case of a data request, SMGW sends metering data in the COSEM [12] format to the AEE. In case of a communication request, SMGW opens a secured communication channel between the CLS device and the AEE. The communication channel can last up to 24h before it shuts down.

Besides lower security risks, this communication paradigm reduces dependencies from a centralized communication server because different AEEs may establish communication to an SMGW without a communication server in the middle. Besides the described sequence, the SMGW can be configured using a so-called communication profile to send meter data regularly to authorized AEEs. The communication profile is provided to the SMGW by the SMGWA. This way the SMGWA doesn't have to be contacted for every communication case.

In consequence, the regulations of the BSI [10] and the Bundestag [8] lead to a system architecture described in Fig. 3. On each customer site, we have an SMGW and at least one smart meter. Optionally, we may have one or more CLS devices on a customer site specializing in different domains, for example, a CLS device for

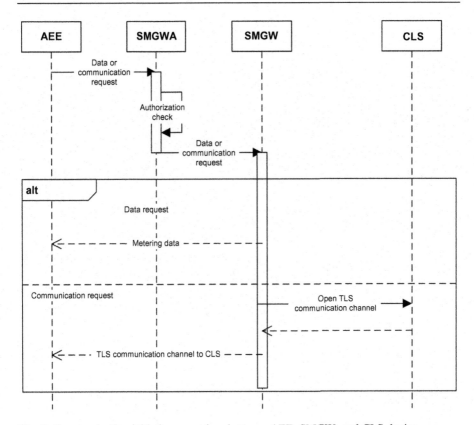

Fig. 2 Communication initiation procedure between AEE, SMGW, and CLS device.

supporting feed-in management by a home PV-system or a CLS device for remote health monitoring. In order to control a local CLS device, each AEE can establish a dedicated communication channel to the customer SMGW, if it is authorized by law or by a customer and fulfills the security requirements of the BSI.

2.2 AMI-based use cases for the German Energiewende

In Fig. 4, you see an overview of use cases for the German Energiewende that are supported by the AMI. Regulated services are mandatory. Smart meter operators must support these use cases. Nonregulated services are optional and provide a chance to make additional revenues for smart meter operators. Use cases for nonregulated services described in this chapter are only showcase examples. There are a potential for various other use cases in the context of smart home applications. In the following, the use cases will be described in detail.

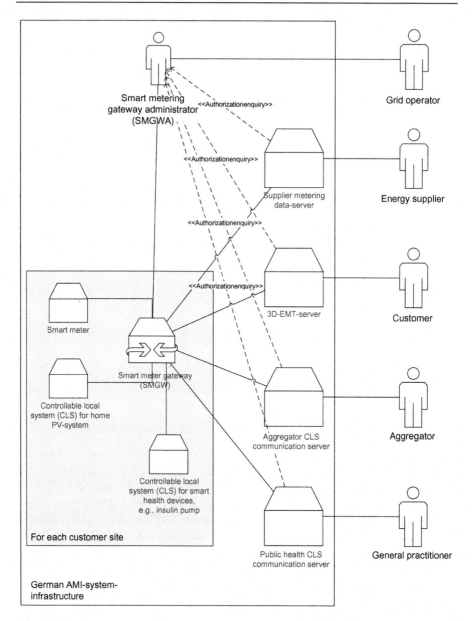

Fig. 3 System architecture of the German AMI.

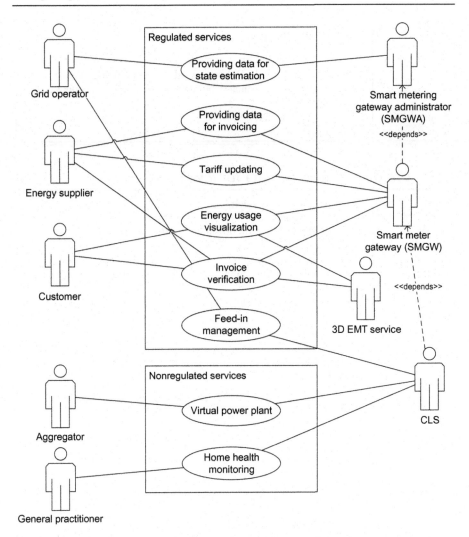

Fig. 4 Overview of AMI-based use cases for the German Energiewende.

2.2.1 Use case "providing data for state estimation"

Name	Providing data for state estimation
Scope	Grid operators should be supported with metered values of electrical power usage from customer sites.
Objective	• Increasing the usage rates of renewable energy. • Support of fine-grained feed-in management in case of power grid congestion.

Continued

Name	Providing data for state estimation
Narrative	The grid operator subscribes to metered values at SMGWA for his grid region. SMGWA acquires measured values from SMGWs, then transforms the data into anonymous data and sends the anonymous data to the grid operator.
Actors	Grid operator, SMGWA, SMGW
Prioritization	Regulated, mandatory

2.2.2   Use case "providing data for invoicing"

Name	Providing data for invoicing
Scope	Energy suppliers (and grid operators) must be supported with metered values and auxiliary billing data from smart meters on customer sites in order to prepare monthly bills.
Objective	• Supporting near time billing, at least on a monthly basis (instead of the usual annual billing). • Enabling innovative electricity tariffs in order to motivate customers for energy savings and/or demand shifts into times with a surplus of renewable energy.
Narrative	Energy supplier (or grid operator) applies for a subscription of metered values and auxiliary data for his customers at SMGWA. If the energy supplier is eligible, then the SMGWA configures mechanisms in the SMGW for automated data delivery. In the case of lost data, the energy supplier may request an ad-hoc data delivery for a specific time range.
Actors	Energy supplier, grid operator, SMGW, SMGWA
Prioritization	Regulated, mandatory

2.2.3   Use case "tariff updating"

Name	Tariff updating
Scope	The energy supplier may arrange various tariff configurations with customers. Most types of these tariffs can be configured to be applied directly in the SMGW. This way the SMGW can directly rate the metered values for the configured tariff or provide auxiliary billing data, for example, a timestamp when contracted demand peak lasts longer then 30 min.
Objective	• Enabling innovative electricity tariffs in order to motivate customers for energy savings and/or demand shifts.
Narrative	Energy supplier negotiates with the customer a tariff and a customer-specific tariff configuration. After that, the energy supplier sends the new tariff configuration to the SMGWA. The SMGWA deploys this configuration into the SMGW. After (usually time-based) activation of the new tariff configuration, the SMGW sends precalculated energy

Name	Tariff updating
	consumption data for configured tariff-stages to the energy supplier regularly.
Actors	Energy supplier, SMGW, SMGWA
Prioritization	Regulated, optional

Regulatory specifications differentiate the following types of tariffs:

- TAF 1 is a data-minimizing tariff configuration; only meter readings or sums about energy consumption will be sent to the energy supplier. This tariff is comparable to a manual readout of metering data, such as the state of the praxis in Germany now.
- TAF 2 is a time-of-use tariff configuration; the price for the customer depends on configured daytime ranges.
- TAF 3 is a load-based tariff configuration; the price depends on the metered loads.
- TAF 4 is a consumption-based tariff configuration; the price depends on consumed energy in a specific time range, for example, the price may significantly increase if the customer uses more energy in a month than contracted.
- TAF 5 is an event-based tariff configuration; the price depends on external events. The supplier may send external events by using the CLS communication channel to the SMGW.
- TAF 7 is based on the metered load profile data; different than other tariffs, suppliers calculate energy quantities for various tariff stages in their data centers. In TAF 1 to TAF 5, these calculations are done by SMGWs.

There are some other auxiliary TAF options in SMGW without relevance to price calculations for the customers.

2.2.4 Use case "energy usage visualization"

Name	Energy usage visualization
Scope	One of the main goals of smart metering in Germany is the rise of customer energy awareness. Therefore there is a need for near-time visualization of energy usage in households and companies.
Objective	• Customer is sensitive to his energy use by a simple visualization of the energy usage data. • Minimizing of energy usage in order to reduce the CO_2 footprint in private households and industry.
Narrative	Customer gets metered values about his energy usage from the 3D EMT and uses an app to visualize this data. The 3D EMT Service acquires the metered values from the SMGW and provides this data to the customer or his app in a standardized data format.
Actors	Customer, 3D EMT Service, SMGW, SMGWA
Prioritization	Regulated, optional

2.2.5 Use case "invoice verification"

Name	Invoice verification
Scope	Due to complex tariff options in deregulated markets and digital metering, customers need tool support for verification of the energy supplier's bills. This tool support has to take in account the requirements for PTB as German authority for measurement and instrumentation regulations and must be independent from energy suppliers.
Objective	• Simple and energy supplier independent process and tools for invoice verifications by the customer.
Narrative	Customer requests metered values for invoice verification from the 3D EMT portal. The 3D EMT Service collects the required data from the SMGW, the SMGWA, and others. The 3D EMT service provides the metered values to the customers, including a digital signature that was generated from the SMGW. In case of TAF 7, the 3D EMT service additionally gets verified tariff calculation rules from the energy supplier and sends this information to the customer, too. The customer uses the received data with the free app Transparenz-Software for invoice verification. In the first step, the app checks, based on digital signatures, whether metered values were manipulated. In the second step, the app visualizes data according to relevant tariff stages ergonomically, so that the customer can simply compare invoiced data with original data.
Actors	Customer, 3D EMT Service, SMGW, SMGWA, Energy supplier
Prioritization	Regulated, mandatory

2.2.6 Use case "feed-in management"

Name	Feed-in management
Scope	In the case of grid congestion, the grid operator can reduce the energy production of a DER in congestion-related subareas of the power grid.
Objective	• Highly selective feed-in management for fine-grained subareas in low- and middle-voltage power grids requires bidirectional communication with the controllable local system of the DER.
Narrative	In the case of grid congestion, the grid operator opens a communication channel to the CLS of the DER in the affected subarea of the grid. After the connection is established, the grid operator sends reduction signals to CLSs and monitors the reductions.
Actors	Grid Operator, CLS, SMGW, SMGWA.
Prioritization	Regulated, mandatory (especially for DERs in low-voltage grids)

2.2.7 Use case "virtual power plant"

Name	Virtual power plant
Scope	For direct marketing of electricity from a renewable DER, the DER owners usually have to mandate an independent aggregator with reselling. The aggregator integrates DERs in pools of his virtual power plants and sells the aggregated produced energy or the aggregated flexibilities in energy production to different markets.
Objective	• Reducing subsidies for renewable DERs by increasing direct marketing of renewables by aggregators.
Narrative	The aggregator opens a communication channel in order to send schedules to a DER or a signal for starting or stopping energy delivery. Aggregators usually choose automation protocols. They also define data models and communication procedures for CLS because CLS devices must align with the virtual power plant solution of the aggregator.
Actors	Aggregator, CLS, SMGW, SMGWA
Prioritization	Nonregulated, optional

2.2.8 Use case "home health monitoring"

Name	Home health monitoring
Scope	There are usage scenarios besides the energy sector that can benefit from the high security and privacy requirements of the German AMI. One of these usage scenarios may be the public health sector.
Objective	• Prevention responsibilities include reducing security risks for the Internet of Things applications in the public health sector.
Narrative	A general practitioner authenticates himself with his digital ID card at the SMGWA. After authorization checks, the SMGWA requests the SMGW to open a communication channel between the software system of the general practitioner and a CLS for health monitoring linked to the SMGW. Then the general practitioner may remotely monitor and configure for example, an insulin pump for his patient.
Actors	General practitioner, CLS (for health monitoring), SMGW, SMGWA
Prioritization	Nonregulated, optional

This use case "home health monitoring" is, to our knowledge, not yet on the rollout road maps of the AMI stakeholders in Germany. It is only an example for the market potential of the German AMI.

3 Technology

The focus of the AMI in Germany is cybersecurity, privacy, trustworthiness, and the extensibility for feature requirements in energy and other sectors. In the following subsections, we first give a short overview of the system and security architecture of the AMI as predefined by BSI. Second, we introduce a German adaption of the Green Button format for providing trustworthy data to customers.

3.1 System and security architecture of the German AMI

The system architecture of the AMI differintiates three security zones (cf. Fig. 5):

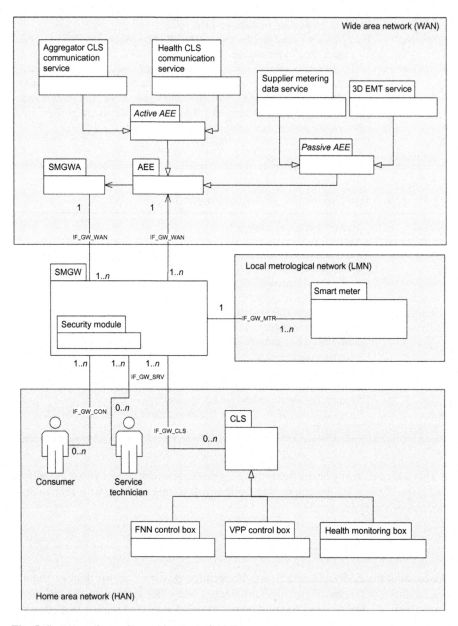

Fig. 5 System and security architecture of AMI.

- The LMN contains digital metering devices. Devices in this zone communicate only with a dedicated SMGW.
- The HAN hosts security-certified devices. These may be visualization devices for customers, PCs of service technicians, or CLS devices. CLS devices can communicate to physical devices such as a PV on the roof over secured automation protocols. But neither CLS device nor linked devices are allowed to communicate with other networks, for example, the public Internet.
- The WAN hosts the mandatory SMGWA service and services for different kinds of AEEs. Communication between an SMGW and an AEE service is always initiated by the SMGW after security approval by the SGMWA. The provider of the AEE services must have the ISO 27001 [13] certification because this service may have communication channels to unsecure networks.

This system architecture design reduces high cybersecurity risks because of strict security zones and mandatory authorization of AEEs by the SMGWA. Furthermore, the SMGWA can force a disconnection between the SMGW and an AEE in case of security threats. All SMGWAs in Germany will jointly maintain black lists of untrustworthy AEEs.

The SMGW implements collection and processing of data at the content level as well as the establishment of trusted channels to connected devices and AEEs [14]. All metered values that are prepared to be sent to an AEE or connected devices in the HAN are encrypted and integrity-protected at the content level. Furthermore, the SMGW always enforces the use of encrypted, integrity-protected and mutually authenticated channels for all external communication to AEEs and connected devices. This concept of two-layer encryption facilitates scenarios in which the SMGWA is not the final recipient of metered values from the smart meters. In this way, for example, the SMGWA receives meter data that it forwards to a different party, the real recipient of the data. Here, the SMGWA is the endpoint of the trusted communication channel but cannot read the encrypted meter data addressed to the third party. System administration messages and data exchanged between SMGW, SMGWA, and service components of AEEs are also encrypted and integrity-protected using the cryptographic message syntax of the BSI.

The BSI technical directive "smart metering-public key infrastructure (SM-PKI) for Smart Meter Gateways" [15] specifies requirements for security certificates and how they should be used in context of the German AMI. SM-PKI distinguishes between certificate authorities and certificate users. Certificate authorities release and manage security certificates. Security certificates are X.509 certificates according to requirements in Ref. [16]. All security certificates have a limited validity period. User's certificates are only valid for 2 years. Other certificates may be valid up to 5 years.

Security certificates guarantee secure communication and data exchange between the components and actors of the AMI on various levels of communication. For this purpose, the SM-PKI differentiates three kinds of security certificates:

- *Transport layer security (TLS) certificates*: AMI components use TLS certificates for implementation of encrypted communication channels between each other.
- *Encryption certificates*: SMGWs use encryption certificates for encryption of meter data before sending the data to an AEE. Thus, it ensures a very high level of privacy because only the receiver with the correct certificate can read the meter data.

- *Signature certificates*: A SMGW signs all meter data on arrival with its own signature certificate so that manipulations of metered data by third parties in the SMGW memory or outside the SMGW can be uncovered by signature checks.

Certificate authorities can invalidate security certificate before the end of their validity. A certificate authority publishes invalid certificates in a certificate blacklist. All components and actors in the AMI must always check the validity of the public certificates of their communication counterparts before establishing a connection channel or encrypting the metered data.

In summary, the strict differentiation of network zones LMN, HAN, WAN, and the hard security requirements of the SM-PKI ensure that only trustful AEEs can receive metered data or communicate with CLS devices over the AMI. In case that an AEE or another AMI actor is compromised, for example, by hackers, certificate authorities can isolate the compromised AEE by adding its security certificates to the blacklist. Furthermore, the SMGWA can instantly separate open communication channels of the compromised AEE to SMGWs and CLS devices.

3.2   Data privacy compliant metered data delivery

The main issue in the design of the German AMI is the high requirements in data protection and privacy. The system and security architectures provide interfaces and technologies for the implementation of data protection. The primary use cases "Providing Data (to grid operators) for State Estimation" and "Providing Data (to energy suppliers) for Invoicing" can be implemented with concepts and technologies defined in Ref. [10]. However, these BSI specifications do not exactly describe the delivery of the metered data to customers (respectively consumers). BSI specifications define only a communication channel from SMGW to consumers in HANs, such as the IF_GW_CON communication channel in Fig. 5. But the data model and the protocol for IF_GW_CON are not specified. This implies that data visualization solutions for customers are vendor-specific. It can be acceptable for the use case "energy usage visualization" but it does not work for the use case "invoice verification." Here, a solution with open interfaces and data formats is needed that supports both. Where the IF_GW_CON lacks standardization, the German National Metrology Institute PTB defined a large set of requirements to enable the customer to verify the bill. They defined that the customer not only has the right to receive data from the SMGW in a digital way, but also that the tariff rules that were applied in order to make the energy bill comprehensible. The requirements also define the metered data to be secured against manipulation using an additional digital signature that includes a hash function over the actual metered value, its status, and the time stamp associated with it.

Since the IF_GW_CON was not fully specified, the idea was to use the IF_GW_WAN interface and an independent instance of an AEE called 3D-EMT to collect the necessary data for the customer and to provide it in a standardized manner using the so-called IF_3D_CON interface. The 3D-EMT serves as an interface between the highly secure AMI with its encrypted machine-to-machine communication and the customer who might not have a digital certificate to provide identification and encryption. It also serves as a bridge between the highly regulated web service-based communication infrastructure that was built up for high automation and security and transforms the data into one file that can be handled by the customer. Fig. 6 shows the system architecture where the 3D-EMT plays a major role as the data collector and processor. The customer logs himself on to the portal of the 3D-EMT to find the meter data of his smart meter together with additional information in one file. The file can be downloaded securely by standard Internet encryption technologies, for example, those used for Internet banking applications.

The data format that is used to provide the data to the customer is defined by the German standard application rule IF_3D_CON [17]. This application rule adapts and extends the American Green Button format to German requirements. Green Button is based on the standard NAESB REQ.21—Energy Services Provider Interface [18] and is already applied widely in providing meter data and additional information to customers in the United States. The format is XML-based and uses the Atom Link Model for a hierarchical structure of the data. Over time, two application variants have emerged: "Download My Data" and "Connect My Data." Download My Data (DMD) describes the original use case: the data is provided to the customer as one downloadable file. This file can then be provided to other applications for evaluation or can be viewed in the web browser. Connect My Data (CMD) extends this use case by enabling the customer to authorize third-party applications and service providers to

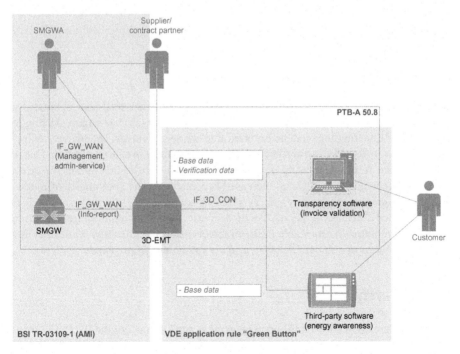

Fig. 6 System environment for customer data provision.

download the data file on behalf of the customer. Respective authorization mechanisms and processes are provided.

In Germany, the DMD concept is applied in a first step. Green Button provides extension points in the format definition that were used to extend the format to the German requirements. The idea was to keep the German version as compatible as possible to the original definition. So-called "transparency software" will be provided to the customer that can import the Green Button data file and check the digital signatures of the metered data. Only if the metered data is identified as unaltered will the data be presented to the customer. In the application rule that describes the format, the data is structured into two different parts: "base data" and "verification data." The base data describes the actual meter data for a point of delivery. This relies in most parts on the original Green Button format and was extended to provide additional information on the meter and the meter data that is required in Germany. One example of additional information is the OBIS-figure (based on IEC 62056-61) that is used to identify the kind of metered values. The base data also includes the digital signature for the meter data and the respective certification data of the SMGW. Since the American Green Button format does not include any information on tariffs and invoicing rules, this information had to be completely added into the German version. Therefore, the verification data provides possibilities to describe a variety of tariff rules such as time of use or amount of use rules, for example. For purposes besides invoice validation, for example, energy consumption visualization or energy awareness applications, the base data is sufficient. Fig. 7 shows the process and the different instances for downloading the data to the customer's PC as well as verifying the meter data using the transparency software.

4 Economics

Until now, German energy transition has been driven by guaranteed feed-in tariffs for renewable energy sources. However, the German government has begun shifting to more competitive funding models for renewable energy resources. Established renewables such as wind parks can and should directly participate in energy markets. Due to the actual and future decreasing of guaranteed feed-in tariffs, this option for direct marketing of renewables finds an increased acceptance in the community of renewable energy source operators.

The planed Germany-wide AMI will decrease communication and engineering costs for integration of renewable DERs in direct marketing of energy while enabling the direct control of smaller DERs by grid operators and aggregators.

5 Research

Although there are well-established management applications for the flexibility management of renewable power generation in the transmission grid, the day-to-day matching of flexible demands at customer sites with fluctuating power generation by renewables is still an ongoing challenge and a matter of research projects.

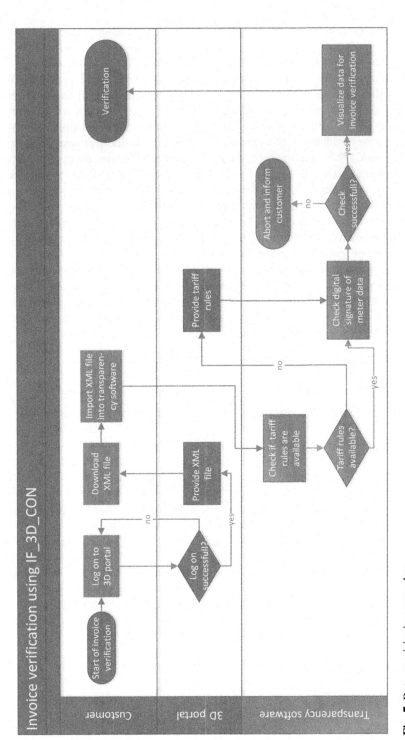

Fig. 7 Data provisioning procedure.

Discussions between industrial organizations and regulation authorities on this topic are ongoing.

New applications for flexibility management in distribution grids are a matter of discussion and research in context of the position paper "Smart Grid Traffic Light Concept" of the German Association of Energy and Water Industries [6]. In the amber phase, the DSO calls upon the flexibility offered by DER operators or their aggregators in that network segment in order to prevent a red phase situation. The discussion about the possible flexibility products for DSOs is a work in progress, nationally and EU-wide [7]. The German government initiated the funding program "Smart Energy Showcases—Digital Agenda for the Energy Transition" (SINTEG) with the aim of developing and demonstrating in model regions new approaches to safeguarding secure grid operation with high shares of intermittent power generation on the basis of wind and solar energy. The initiated SINTEG projects address the issue of flexibility products for the amber phase in distribution grids in order to secure an increased share of renewables in the energy mix.

The ongoing and planned research concentrates often on different aspects of the flexibilization of power demand, the automation of DR management, and the discovery of new market mechanisms for flexibility trading on demand and generation sides. The discussed solutions often require a secure, trustable, reliable, and inexpensive AMI for implementation, such as the German AMI.

6 Market

Guaranteed feed-in tariffs for renewable energy sources are continuously declining for new installed DERs over the last few years. Additionally, the German government is now replacing the mechanism of nationwide identical feed-in tariffs by project-specific feed-in tariffs. The project specific feed-in tariffs for, for example, a new wind farm, will be defined by a tendering procedure. The wind farm developer with the lowest demanded feed-in tariff will win a public bid and have to deliver the renewable energy by a lower tariff then other competitors. This shift in government regulations implies a new demand for direct marketing for renewable energy providers.

Whereas the German energy exchange in Leipzig and control reserve market of the transmission operators adapted their mechanisms in the last 10 years to support the trading of renewable energy, new innovative trading mechanisms are now under discussion for distribution grids in local markets, by providing the production and demand flexibilities to aggregators or to distribution grid operators. Furthermore, a few German utilities are experimenting with peer-to-peer markets in low-level grids.

7 Future developments

The German AMI provides sophisticated concepts for meter data management and solutions. The build-up of the nation-wide AMI will take several years, so there might be only a few further developments regarding the AMI concepts during the build-up phase. One of them may be a deeper adoption of the Green Button standard for Germany. The substandard CMD could provide improved metered data services for energy consumer companies. With the implementation of CMD, consumer companies will be able to automatically redirect metered data to their energy management subcontractors for different purposes, for example, regularly providing the obligated ISO 50001 energy reports to government agencies.

In the medium to long term, the broader usage for secured CLS communication channels for Internet of Thing applications outside the energy sector appears to be a promising perspective. Here an integration of future wireless 5G technologies with smart grid gateways (SMGW) would be needed. More importantly, the opening of the registration, authentication, and authorization processes of the smart grid gateway administrators (SMGWA) to other industries than just energy could be crucial for the success and acceptance of a nationwide AMI.

8 Case studies and field trials

The German government initiated the funding program "Smart Energy Showcases—Digital Agenda for the Energy Transition" (SINTEG). SINTEG aims to develop and demonstrate in model regions new approaches to safeguarding secure grid operation with high shares of intermittent power generation based on renewable energy source. The five projects in this program that were started in 2017 address, among other things, the issue of flexible products for the amber phase in distribution grids in order to secure an increases share of renewables in the energy mix and big field trials of the German AMI. In the middle of these projects, it is planned to revise the regulations and standards of the AMI in order to adapt the technical guidelines, market communication processes, and perhaps the regulation, according to experiences in field trials. After these presumably small revisions, a mass rollout of smart meters will be strengthened.

9 Summary and conclusion

In the previous chapters, the smart grid activities in Germany were outlined with a focus on the smart metering infrastructure that is currently built up. In Germany, large customers will be equipped with intelligent metering systems first, before customers with less power consumption will get the systems. The smart metering infrastructure will enable grid operators and suppliers to implement new use cases that have not been possible before. As the smart grid infrastructure is highly regulated, the roles and concepts of the relevant components within the infrastructure are well defined and the communication channels specified as well as the underlying security requirements. One main advantage of the designed overall system is that even players from outside the energy industry can use the infrastructure. With the CLS-channel, the regulator offers so-called "authorized external entities" a quite flexible way to use the infrastructure for new use cases. We introduced in detail some of the use cases foreseen to utilize the infrastructure. From a smart grid point of view, the most important ones are "feed-in management," "providing data for state estimation," and "virtual power plant" as they might support grid operators in their daily work keeping the grid stable. More applications regarding smart home applications and smart city concepts are possible but still have to be developed. They might show the full potential of the support that the smart metering infrastructure might offer to the smart grid.

The security architecture behind the infrastructure is complex with a high focus on data security and data privacy. It builds up some obstacles for the infrastructure to be used by third parties but helps the system to be accepted by customers. Since metered data may be tariffed in the SMGW, the "BundesDisplay" project was initialized to define a standardized interface between the smart meter infrastructure and the customer. It enables the customer to request the billing data for the energy consumption directly from his gateway device. The data can be used for visualization of the power consumption or can be provided to third-party software for further analysis.

It can be concluded that an advanced infrastructure such as the smart metering infrastructure as part of the smart grid offers new possibilities for controlling and influencing a large number of customers. High security requirements and extensive regulation have been developed in order to raise the acceptance of the customers for the system. Once widely installed, the infrastructure might be used to unleash the flexibility that is within the customer's energy consumption and therefore might support grid operators keeping the frequency within a smart grid with a high number of decentralized energy sources.

References

[1] S.L. Christian Hinrichs, in: A decentralized heuristic for multiple-choice combinatorial optimization problems, Operations Research Proceedings, 2014, pp. 297–302.

[2] H.F. Wedde, S. Lehnhoff, in: Dezentrale vernetzte Energiebewirtschaftung (DEZENT) im Netz der Zukunft, Wirtschaftsinformatik 49 (2007) 361–369.

[3] S. Lehnhoff, T. Klingenberg, in: Distributed coalitions for reliable and stable provision of frequency response reserve, Proceedings of the 2013 IEEE International Workshop on Intelligent Energy Systems (IWIES), 2013, pp. 11–18.

[4] Regelleistung.net, Regelleistung.net, Von, 2017, https://www.regelleistung.net/ext/static/market-information abgerufen.

[5] Netze BW GmbH, Technische Richtlinie Erzeugungsanlagen am Mittelspannungsnetz der Netze BW, Netze BW GmbH, 2013.

[6] BDEW German Association of Energy and Water Industries, Smart Grid Traffic Light Concept—Design of the Amber Phase, BDEW, Berlin, 2015.

[7] USEF Foundation, Universal Smart Energy Framework Explained, USEF Foundation, Arnhem, Netherlands, 2015.

[8] Bundestag, Gesetz zur Digitalisierung der Energiewende, Bundesanzeiger, Bonn, 2016. www.bundesanzeiger-verlag.de.

[9] A. Claassen, S. Rohjans, Application of the OPC UA for the Smart Grid, Proceedings of the 2011 2nd IEEE PES International Conference and Exhibition on Innovative Smart Grid Technologies (ISGT Europe), 2011, pp. 1–8.

[10] BSI, Technische Richtlinie BSI TR-03109—Anforderungen an die Funktionalität, Interoperabilität und Sicherheit an Smart Metering Komponenten, BSI, Bonn, 2015.

[11] PTB, PTB-A 50.8—Anforderungen an Smart Meter Gateway, Physikalisch-technische Bundesanstalt, 2014.

[12] IEC 62056-1-0:2014, Electricity Metering Data Exchange—The DLMS/COSEM Suite—Part 1-0: Smart Metering Standardisation Framework, IEC, Switzerland, 2014.

[13] ISO, DIN ISO/IEC 27001, Information Technology—Security Techniques—Information Security Management Systems—Requirements, Beuth Verlag, 2015.

[14] BSI, SMGW-PP, in: Protection Profile of the Gateway of a Smart Metering System (Smart Meter PP), BSI, Bonn, 2014.

[15] BSI, BSI TR-03109-4, in: Smart Metering PKI—Public Key Infrastruktur für Smart Meter Gateways, BSI, Bonn, 2016.

[16] D. Cooper, S. Santesson, IETF RFC 5280, in: Internet X.509 Public Key Infrastructures—Certificates and Certificate Revocation List (CRL) Profiles, 2008.

[17] VDE, VDE-AR-E 2418-6 Schnittstelle zur Anzeige von Smart Meter Gateway Daten für den Letztverbraucher (IF_3D_CON), VDE Verlag, Berlin, 2017.

[18] NAESB, REQ.21 – Energy Services Provider Interface, 2017, https://www.naesb.org/ESPI_Standards.asp.

Analysis of the future power systems's ability to enable sustainable energy—Using the case system of Smart Grid Gotland

11

Carl J. Wallnerström, Lina Bertling Tjernberg
KTH Royal Institute of Technology, Stockholm, Sweden

Acronyms

AMI	advanced metering infrastructure
DLR	dynamic line rating
DR	dynamic rating
DSO	distribution system operator
HVDC	high voltage direct current
PMU	phasor measurements unit
SCADA	supervisory control and data acquisition
SMHI	The Swedish Meteorological and Hydrological Institute
SR	static rating
VSC	voltage source converter (also known as HVDC light)

1 Introduction

1.1 Background

The electric power system is being modernized to enable a sustainable energy system. New developments include the possibilities and challenges with generation, delivery, and usage of electricity as an integrated part of the energy system. This involves new forms of electricity usage, for example, for transportation and demand response and the updating of existing electricity infrastructure. For electricity generation, the trend is toward new large-scale developments, such as offshore wind farms, as well as small-scale developments such as rooftop solar energy. At the same time, the digitalization of society is creating new opportunities for control and automation as well as new business models and energy-related services. The overall trend for technology development is new possibilities for measurement and control. An example is phasor measurements units (PMUs), generally located in the transmission network, which provide measurements of voltage and current up to 30–20 times per second. Another example is smart meters placed with the end consumer, which enables the integration of private small-scale electricity production from solar cells or energy storage from

Application of Smart Grid Technologies. https://doi.org/10.1016/B978-0-12-803128-5.00011-8

electric vehicles and general distributed control of energy use. Europe has been at the forefront of several of these deployments, especially in the areas of managing large penetrations of renewable sources of energy, advanced metering infrastructure (AMI), and advanced information technology [1].

The main challenges for the current development of the European power system can be summarized as follows (resulting from the EU project ERA-Net):

- Enabling an increased flexibility of the power system to cope with the growing share of intermittent, variable, and decentralized renewable generation as well as managing the complex interactions.
- Increase network capacity to support increased generation and transmission resulting from renewables and in support of the internal energy market.
- Provide information, services, market architecture, and privacy guarantees to support open markets for energy products and services while facilitating the active participation of customers.

Solutions for this modernized power system are rapidly evolving and have been analyzed from a number of perspectives [2,3], often from a data communication, monitoring, and control perspective such as [4]. In addition, studies of capacity-increasing benefits [5] and different solutions to facilitate the integration of intermittent distributed generation [6] have been conducted, such as demand response [7,8], energy storage [9], and dynamic rating (DR) [10,11] or looking at reliability aspects of this [12]. Comprehensive studies have been performed on the relationship between weather parameters and power system properties [13].

This chapter presents an approach in analyzing various solutions for a future sustainable energy system using the case system of the Swedish demonstration project of Smart Grid Gotland.

1.2 Smart grid demonstration project at the Swedish island Gotland

The island Gotland in Sweden has been used for exploring the new technologies for a sustainable energy system. Recently, the Swedish government expressed visions to further investigate the expansion of generation supply from renewable energy resources instead of an extension with an additional capacity connection with mainland Sweden.

Gotland is an isolated power system by geography. It is small enough to get a good overview but still large enough to illustrate and analyze comprehensive challenges and synergies [14]. There are already major challenges and opportunities of renewable distributed generation as electricity from wind power has approached the ceiling of what the system can handle with traditional technology and existing infrastructure. Gotland's subtransmission system can partly be described as a transmission system in miniature because Gotland is isolated from the rest of the Nordic power system transmission system.

Gotland is connected to the mainland with a high voltage direct current (HVDC) link [8]. There are both export and import situations for the HVDC link. At the island, there is a double onshore VSC link (HVDC Light) to connect large amounts of electricity, today typically from wind power from the south parts of Gotland, with the area around Visby. Visby is the largest city on Gotland having the highest electricity consumption. The mainland HVDC link connection outgoes from Ygne (just south of Visby). This 96 km long HVDC Link was awarded an IEEE Milestone in 2017 for being the worlds' first commercial HVDC transmission link using the first submarine HCDC cable [15]. Fig. 1 illustrates the case system of Smart Grid Gotland.

In 2011 the total installed wind turbine capacity was 170 MW. From a smart grid analysis perspective, it is of interest that the power system with current technology and infrastructure is expected to handle 195 MW, that is that installed wind power today is close to its upper calculated limit. The electricity production in 2011 was 340 GWh, representing 38% of Gotland's electricity consumption. However, both electricity consumption and electricity generation from wind turbines are unevenly spread over the year, which means that it can have both high import and export peaks. In the long run, there will be capacity problems in both ways of the HVDC link to the mainland; there are plans to increase its capacity. Furthermore, Gotland has a relatively high amount of electricity consumption by industrial customers [8]. Another difference with an average Swedish area is a relatively large amount of summer visitors, which could possibly partly compensate for the industry's low season from a power consumption perspective.

As new technologies and solutions involve unknown risks and opportunities, it is valuable to complement theoretical research and commercial development with large-scale smart grid demonstration sites. Ref. [16] provides examples for such projects in Europe. There have been three large national smart grid demonstration projects in Sweden, involving different actors and testing different aspects of a smart grid: the Royal Seaport, Hyllie, and Smart Grid Gotland. Smart Grid Gotland is a development and demonstration project to illustrate and investigate possibilities of modernizing an existing power system to handle more renewable energy while maintaining or improving power quality [17,18]. The project started in 2011 and is currently being evaluated. Smart Grid Gotland consists of nine subprojects: (1) Market tests, (2) Integration of wind power, (3) Power quality with distributed generation, (4) Market installations, (5) Smart meters, (6) Smart secondary substations and rural networks, (7) Communication technology, (8) Energy storage, and (9) Smart SCADA. For more detailed information about Smart Grid Gotland, see, for example, Refs. [8,17,18].

1.3 Initial data analyses used as input when developing the method

Based on the initial studies presented in this section, different models for calculating the average, maximum, and minimum values of electricity consumption and production as a function of weather parameters have been developed, presented, and used in later sections of this chapter.

The hourly data of electric consumption and wind power generation were obtained for the period October 2011 to September 2012 from the DSO of Gotland [8]. The Swedish Meteorological and Hydrological Institute (SMHI) is a governmental body that has made both historical and real-time data freely available online [19]. Hourly weather data from 1970 to 2003 at Visby Airport has been retrieved from SMHI.

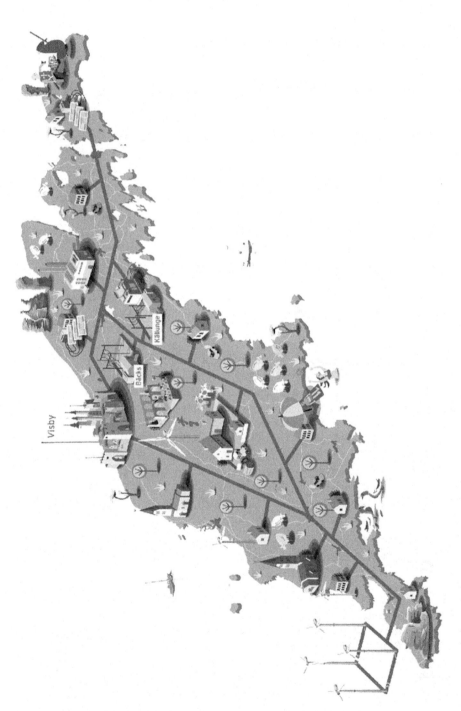

Fig. 1 Overview of the case system of Smart Grid Gotland.

In addition, weather data have been gathered for a shorter time period from Huborg, another weather station at Gotland. Long data series are important to capture improbable risks and to achieve high statistical validation when different scenarios are studied. Visby Airport was chosen because it has a long data series, a high consumer density, and it's close to the mainland link. However, Huborg is located on the southern tip where a considerable part of Gotland's wind power capacity is located.

Some smart grid solutions are directly affected by weather parameters [11]. At the same time, the capacity requirements of components often indirectly depend on the weather in different ways. Hence it is valuable to investigate the weather dependency of power system utilization. Heating and air conditioning are examples of human behavior that depend on outdoor temperature and that affect electricity consumption; this provides strong correlation between temperature and consumption [20]. The tendency of an increased amount of intermittent distributed generation gives situations where the power line congestion depends on the weather [21]. An example is the dependency between wind power production and wind speed, a weather parameter that also affects the dynamic capacity of overhead lines [22].

Table 1 provides correlations between wind power generation, electricity consumption, wind speed, and outdoor temperature. Note that electric generation and consumption refer to the entirety of Gotland while the weather parameters are only from the Visby Airport. When handling transfer limits to areas with both electricity generation and consumption, these can partly cancel each other out in some cases [21]. Fig. 2 shows the average value of electricity consumption, wind power generation, and net imports as a function of temperature. The wind power generation dependency can be explained by high pressure and low wind speed when it is hot or cold. Higher wind speeds in average occur during spring and fall.

The electricity consumption's temperature dependency is low when the temperature is low or high, but in midrange it is almost linear. That could be explained by maximal heating at about $-10°C$ and that most households do not heat above 15°C. These tendencies have also been observed in other studies; see, for example, Fig. 3 [23]. When it comes to electricity consumption, it is unlikely that it will reach >80% of the peak value if the temperature is >0°C and in the summer it rarely reaches >60%. The import needs are highest when it is cold.

Because Gotland has a high share of industrial customers, a hypothesis is that the correlation between temperature and electricity consumption is lower compared with an average Swedish power distribution system. The correlation in a study of another Swedish power system [23] was "−0.90" compared to "−0.71" for Gotland. Fig. 3 shows all hourly measurements of temperature vs. electricity consumption, including

Table 1 Correlation between input data in this study

	Generation	Consumption	Temperature
Consumption	0.22		
Temperature	−0.05	−0.71	
Wind speed	0.84	0.22	−0.04

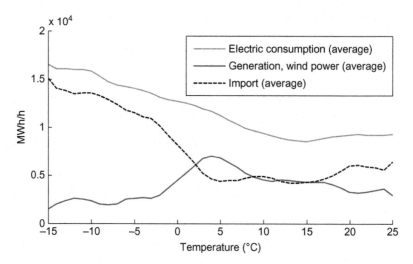

Fig. 2 Average electricity consumption, wind power generation, and import as a function of outdoor temperature.

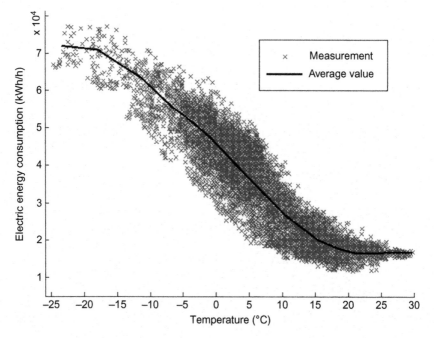

Fig. 3 Electricity consumption as a function of temperature.

the average trend in the other power distribution system. Compared to Gotland, there are many similarities. The conclusion is that the general model developed based on Gotland data provides a statistical relationship whose shape is similar to other power systems. This means that results using these dependences give results that are generally applicable.

Fig. 4 depicts the average electricity consumption and wind power generation as a function of the wind speed. Wind power generation strongly depends on the wind speed. The generation at 0 m/s is explained by the fact that the wind measurement is situated at one place and the height difference while wind power generation is an average for the whole island. Another interesting observation is that wind power generation increases with wind speed up to ~10 m/s and after ~13 m/s decreases. Note that it can be significantly much harder winds that hit the rotor blades than at the weather station, which explains that tendency already after 13 m/s.

Table 2 exemplifies the average, maximum, and minimum electricity consumption and wind power generation if the year is divided into four seasons. The minimum vale of wind power generation is 0% despite the season and is omitted in the table.

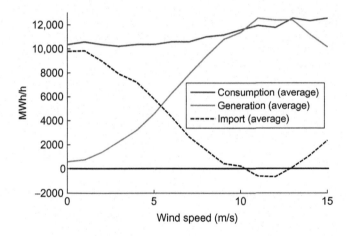

Fig. 4 Average electricity consumption, wind power generation, and import as a function of wind speed.

Table 2 Maximum and minimum values based on seasons

Seasons[a]	Electric energy consumtion[b]			Wind power generation[c]	
	Average (%)	Max (%)	Min (%)	Average (%)	Max (%)
Spring	53	81	7	31	100
Summer	44	59	19	23	97
Fall	51	72	7	35	98
Winter	68	100	45	43	98

[a]Spring = March-May; Summer = June-August; fall = September-November; Winter = December-February.
[b]The annual high = 100%, in this example equivalent to 195,900 kWh/h.
[c]The annual high = 100%, in this example equivalent to 151,500 kWh/h.

2 Proposed method

2.1 Introduction

The proposed method is general and can study a power system part of any size and voltage level as well as with flexible composition of generation and electricity consumption. The algorithm is illustrated in Fig. 6. The model contains of calculation modules, which are flexible to separately develop and improve:

- Electricity consumption, see Section 2.2.
- Wind power generation, see Section 2.3.
- Solar power generation, see Section 2.4.
- Dynamic rating, see Section 2.5.
- Energy storage, see Section 2.6

The basic idea is illustrated in Fig. 5: a local system area (C) is connected with the rest of the power system (A) by a set of components with limited transmission capacity (B).

- *Block A* represents the rest of the power system. From this area it is assumed that desired electrical energy either can be exported to or imported from.
- *Block B* symbolizes a link between A and C where the transmission capacity is limited and may correspond to a set of several components. The transmission capacity can both be modeled as equal or different in both directions. This can in reality correspond to limited transfer by various factors. For example, if generation from solar power gives voltage problems, it may cause different transfer restrictions on imports versus exports as opposed to thermal restriction.
- *Block C* symbolizes a system area flexibility to define. One or several of the following categories is connected: (a) electricity consumption, (b) wind power, (c) solar power, and/or (d) energy storage. The sizes of these in relation to each other are free to define.

The method focuses on the worst case, for example, high electricity consumption and low generation given the weather if import transfer limitations are evaluated and the opposite for export limitations. The calculations are deterministic, apart from the energy storage module that has a mix of deterministic and stochastic parts; the latter is explained and justified in Section 2.5. The approach is to go through a huge database of hourly weather conditions. This will give a good statistical basis with real measured historical data and consequently decrease the need for using hypothetical assumptions.

Fig. 6 shows the algorithm of the developed method:

I. Matlab receives input data from an Excel file. No upper limit for the number of records; the more input, the more reliable results. For each hour: month $(1-12)$, time $(1-24)$, temperature ($°$C), and wind speed (m/s).

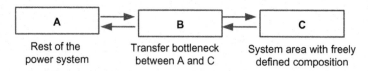

Fig. 5 Illustration of the overall concept.

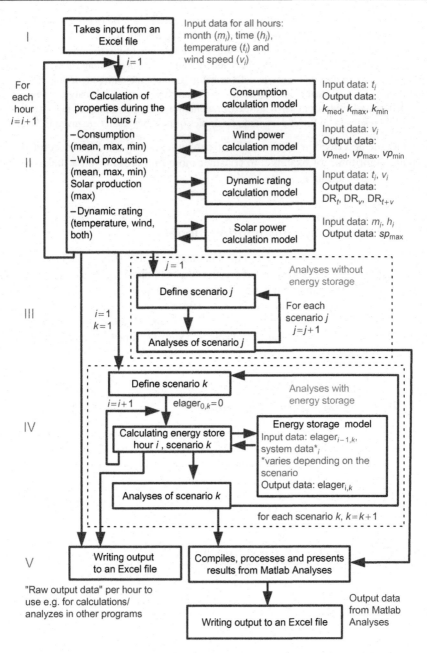

Fig. 6 Flow chart of the overall algorithm.

II. Hourly calculations: Calculate various properties for each hour by sending inputs to modules and getting back results. The modules are described in Sections 2.2–2.6.

III. Analyses are performed by using the large amounts of raw results that were produced during *step B*. By defining system characteristics and analysis questions, different analysis results are produced. These are defined before running the entire algorithm.

IV. Energy storage calculations: Just as in *step C*, system properties and analyses questions are defined and results from *step B* are used. In addition, a module for energy storage is contacted.

V. Write results to an Excel document: Examples of possible results are given in Section 3.

Each module described in Sections 2.2–2.6 can be separately modified or refined from the main algorithm, for example, to consider more properties or base these on other input data than used here.

2.2 Electric energy consumption model

Input data consists of the outdoor temperature and the calculation model returns maximum, minimum, and mean values of electricity consumption. Estimated averages for different temperature ranges are based on the initial analysis in Section 1.3. The unit is procent of its peak value while maximum and minimum values are two standard deviations from the mean (normal distribution is assumed) and validated against real input data. The model is presented in Table 3. Extreme values are important because worst-case scenarios are analyzed.

2.3 Wind power generation model

Input data consists of the wind speed and the calculation model returns maximum, minimum, and average values of wind power generation. The model is based on analyses presented in Section 1.3 and is presented in Table 4. Wind power generation is in general significantly less predictable compared to electric consumption, which is reflected by a larger range between extreme values. Note, low wind speeds can give generation. That is explained in Section 1.3.

Table 3 **Electric energy consumption model**

Temperature (°C)	Share (%)	Electric energy consumption (% of annual max)		
		Min	**Average**	**Max**
<-10	1.46	67	83	100
-10 to -5	2.15	54	74	94
-5 to 0	10.14	49	67	85
0–5	26.05	43	61	78
5–10	24.06	34	51	69
>10	36.15	32	45	58

Table 4 **Electric energy wind power production model**

Wind speed (m/s)	Share (%)	Electric energy production (% of annual max)		
		Min	Average	Max
<1.5	7.9	0	5	20
1.5–2.5	12.2	0	9	29
2.5–3.5	12.5	0	15	43
3.5–4.5	16.2	0	21	54
4.5–5.5	11.5	0	30	68
5.5–6.5	12.1	0	41	84
6.5–7.5	9.5	0	52	96
7.5–8.5	5.9	20	62	100
8.5–9.5	5.6	32	71	100
9.5–15.5	6.8	44	79	100
>15.5	0.1	0	42	100

2.4 Solar power generation model

Input data consists of month (1–12) and hour (1–24) and the model returns the maximum value of solar power generation in % of annual maximum production. The model is illustrated by Table 5 and calculates the peak production at perfect conditions, based on how high the sun is over the horizon in Visby [24]. This is justified because the model focuses on the worst case and that the minimum value is assumed to be zero.

2.5 Dynamic rating introduction, assumptions, and model used

Overhead line DR is used. It is possible to replace this module with other DR models or to expand it with more power component models. Wind power generation has a correlation with the weather in a positive way [12], that is, high production correlates with high dynamic transfer capacity. Furthermore, in Sweden there are positive correlations between low temperature and high energy consumption [21]. A challenge is that the measured wind speed often is optimistic to use because the wind speed is neither the same through the entire line nor perpendicular. One way to handle this is to introduce a scale parameter that scales down the available wind speed input data to have a margin.

IEEE standard 738 [22] provides a formula for calculating the maximum current allowed [A] of bare OH conductors. Based on this, a DR model has been developed and is proposed in Ref. [11]. The method is simplified to be implemented in the daily operation and is also evaluated by comparing it with the more detailed standard [11]. Evaluations performed conclude that its results only differ minimally from more complex models. The proposed calculation model calculates the dynamic line rating capacity (DLR_x) as a function of static line capacity, wind speed, and ambient temperature. DLR_x has no unit and indicates how many times the transfer capacity

Table 5 Electric energy production model from solar power

	1	2	3	4	5	6	7	8	9	10	11	12
≤2	0%	0%	0%	0%	0%	0%	0%	0%	0%	0%	0%	0%
3	0%	0%	0%	0%	5%	11%	7%	0%	0%	0%	0%	0%
4	0%	0%	0%	4%	18%	24%	20%	9%	0%	0%	0%	0%
5	0%	0%	0%	18%	32%	38%	34%	23%	9%	0%	0%	0%
6	0%	0%	13%	32%	47%	53%	48%	38%	24%	8%	0%	0%
7	0%	8%	26%	46%	61%	67%	63%	52%	37%	20%	5%	0%
8	6%	20%	38%	59%	74%	80%	76%	65%	49%	31%	15%	5%
9	15%	28%	47%	68%	85%	91%	87%	75%	58%	39%	22%	12%
10	19%	34%	53%	74%	91%	98%	95%	81%	63%	43%	25%	17%
11	21%	36%	55%	76%	92%	100%	97%	83%	64%	43%	25%	17%
12	19%	34%	53%	72%	88%	95%	93%	80%	60%	39%	22%	14%
13	14%	29%	46%	64%	79%	86%	84%	72%	52%	31%	15%	8%
14	5%	20%	36%	53%	67%	74%	73%	61%	41%	21%	5%	0%
15	0%	9%	24%	40%	53%	60%	59%	47%	28%	9%	0%	0%
16	0%	0%	11%	26%	39%	46%	44%	33%	15%	0%	0%	0%
17	0%	0%	0%	12%	23%	31%	30%	18%	0%	0%	0%	0%
18	0%	0%	0%	0%	11%	18%	16%	4%	0%	0%	0%	0%
19	0%	0%	0%	0%	0%	6%	4%	0%	0%	0%	0%	0%
≥20	0%	0%	0%	0%	0%	0%	0%	0%	0%	0%	0%	0%

is compared to static line rating. If, for example, the static line rating is 3 MW and DLR_x is 2, the new DR is equal to 6 MW.

$$\mathrm{DLR}_t \approx \frac{\sqrt{(t_{max} - t_i)}}{\sqrt{(t_{max} - t_{SLR})}} = c1*(t_{max} - t_i)^{0.5} \tag{1}$$

Eq. (1) calculates the DLR_x, taking temperature into consideration (DLR_t). t_{max} is the maximal line temperature allowed, t_{SLR} is the worst-case outdoor temperature that static rating (SR) is based on, t_i is the outdoor temperature that is input to the calculation, and $c1$ is a constant defined in Ref. [11].

$$\mathrm{DLR}_v = \max \begin{cases} 1 \\ c2*v_i^{0,26} \\ c3*v_i^{0,30} \end{cases} \tag{2}$$

Eq. (2) calculates the DLR_x, taking the wind speed into consideration (DLR_v) where v_i is the wind speed. For information on how to calculate $c2$ and $c3$, see Ref. [11]. The contribution of wind speed and outdoor temperature can be approximately assumed to be independent of each other [11]. Therefore, taking both wind and temperature into consideration can be calculated by multiplying the respective contribution (Eqs. (2), (3)) to each other:

$$\mathrm{DLR}_{t+v} = \mathrm{DLR}_v * \mathrm{DLR}_t \tag{3}$$

2.6 Energy storage algorithm

The energy storage status regarding previous hour $i - 1$ ($EStorage_{i-1}$), actual transfer limitation, electric generation, and consumption data is used as input data to this algorithm. The output data is the energy storage status of actual hour i ($EStorage_i$). The analyses are technique-neutral and produce specifications of requirements for energy storage. $EStorage_i = 0$ does not necessarily mean empty energy storage, but its "normal level" (is explained later). Required size and the normal level are determined after $EStorage_i$ has been calculated for all hours.

Unlike other models proposed, this calculation model uses both deterministic and stochastic parts. The current status of the energy storage is based on both current hour and earlier hours. The probability of a worst-case situation during a specific hour is not negligible, but having that situation constantly during many hours or even several days is unlikely. Such an assumption would put unrealistic demands on the energy storage. However, it is important that here we note the precautionary principle and calculate pessimistically. A compromise is that a value between the mean and the worst-case extreme value is randomized (uniform distribution). Because the energy storage can be used to both handle import and export restrictions, two different pessimistic values are calculated see Eqs. (4)–(9).

$$EG_1 = [\max solar\,gen] * \text{Table 5} + [\max wind\,gen]$$
$$*U(averer\,\text{Table 4, max Table 4}) \tag{4}$$

$$EG_2 = [\max wind\,gen] * U(\min\,\text{Table 4, } aver\,\text{Table 4}) \tag{5}$$

$$EC_1 = [\max e\,cons] * U(aver\,\text{Table 3, max Table 3}) \tag{6}$$

$$EC_2 = [\max e\,cons] * U(\min\,\text{Table 3, } aver\,\text{Table 3}) \tag{7}$$

where EG_1 and EG_2 are upper and lower estimates of electricity generation, respectively, and EC_1 and EC_2 are upper and lower estimates of electricity consumption, respectively.

$$High\ import\ level\ stochastic\ estimation = EC_1 - EG_2 \tag{8}$$

$$High\ export\ level\ stochastic\ estimation = EG_1 - EC_2 \tag{9}$$

If the import is higher than the import transfer capacity, the energy storage needs to deliver electrical energy to temporarily compensate for transfer limitation and hence be fully charged to be as prepared as possible. If, however, the export is higher than the export transfer capacity, the energy storage needs to store electrical energy to temporarily compensate for transfer limitation and hence be empty to be as prepared as possible. If no transfer limitations are reached, the energy storage is assumed to return to its normal level as quickly as possible with considered transfer constraints, trying to return to its normal level.

The algorithm is illustrated in Fig. 7. $EStorage_i$ has the unit "hour." When $EStorage_i$ is multiplied by a transfer limit that has the unit power (e.g., MW), it gives

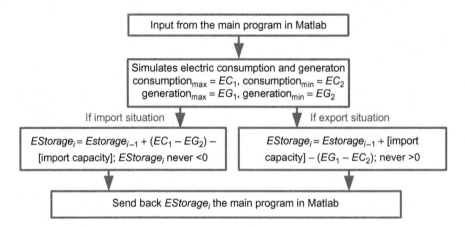

Fig. 7 Algorithm of energy storage calculations.

the unit of energy (e.g., MWh) (see Eq. 10). Power requirements are calculated by taking the highest absolute value of the difference between $EStorage_i$ and $EStorage_{i-1}$ in both directions.

$$[Energy\ storage\ size\ needed] = [transfer\ limit]*(\mathrm{Max}\,(EStorage_i)$$
$$-\mathrm{Min}\,(EStorage_i)) \tag{10}$$

$$[Normal\ level] = \frac{-\mathrm{Min}\,(EStorage_i)}{[Energy\ storage\ size\ needed]} \tag{11}$$

Calculation example: Assume that $EStorage_i$ varies between -2.5 and $+10.0$ during all hours analyzed and that the power transfer limitation is 1 MW (import). Then an energy storage of 12.5 MWh is needed ($=10-(-2.5)$) and the normal level is 20% charged $\left(\frac{-(-2.5)}{10-(-2.5)}\right)$, i.e., 2.5 MWh.

3 Case study results

3.1 Dynamic rating and utilizing correlations

Table 6 exemplifies results from DR analyses where the potential of taking advantage of identified weather correlations has been investigated. The figure of 100% corresponds to the maximum effect the overhead line can handle with a classic SR.

The potential benefits of using synergies between electricity consumption and generation have been analyzed, that is, how these factors cancel each other out in a mixed system part. Increased average electricity consumption provides the ability to handle more installed generation. Even when using pessimistic assumptions, each MW of peak power using traditional assumptions makes an additional 0.32 MW power possible. The opposite, that is, increasing electric consumption if intermittent

Table 6 **Transfer capacity of dynamic rating compared with static**

	Static rating = 100%, max of:		
DR that consider:	**Consumption (%)**	**Wind power (%)**	**Solar power (%)**
Temperature	134.4	101.9	103.2
Wind speed	100.0	219.2	100.0
Both	134.4	224.1	112.3

Table 7 **Analyses of nodes with both Electric production and consumption**

		(%) With different rating approaches[c]			
Allowed amount of[a]:	**Combined with[b]:**	**SR**	**DR1**	**DR2**	**DR3**
Wind power	Consumption	132.0	133.9	251.6	257.4
Solar power	Consumption	132.0	135.5	132.0	144.3
Consumption	Wind power	Unchanged	134.4	Unchanged	134.4
Consumption	Solar power	Unchanged	147.1	Unchanged	147.1

[a]Compared with either only electric production or consumption combined with static rating. That base case is set to 100%.
[b]It is assumed that the annual maximum of consumption and production in the node is equal in the base case and only one of them is increased.
[c]SR, static rating; DR1, dynamic rating considering temperature; DR2, DR considering only wind speed; DR2, DR considering both.

generation increases is, however, not possible because the electricity generation sometimes is 0. With a variety of smart solutions where electricity consumption (e.g., demand response) and/or electricity generation can be controlled, the benefits of utilizing such synergies will probably increase more significantly. Table 7 shows results analyzing transfer limitation to mixed system parts. These analyses capture synergies of both the DR and taking advantage of cancellation at the same time. These two effects sometimes becomes stronger compared with just adding one of these two solutions.

3.2 Energy storage

Energy storage specifications as a function of a variable parameter to handle transfer limitations regarding different scenarios are presented in Table 8.

Fig. 8 illustrates a general comparison of how large energy storage is required for the three categories analyzed. Solar power generation is more advantageous compared with electricity consumption and wind power generation. The significant difference can be explained by the fact that the peak generation from solar power only lasts a few hours during summer days, making time to restore the energy storage to its normal

Table 8 **List of energy storage requirements**

Max^a (%)	103			107.5			112		
	C	**W**	**S**	**C**	**W**	**S**	**C**	**W**	**S**
Size^b	8.0	9.2	3.9	34.41	37.7	17.1	179.3	128.1	36.5
Usage^c	0.1	2.0	1.0	0.39	6.3	2.4	1.2	12.4	4.5
Power^d	3.0	3.0	3.0	7.5	7.5	7.5	12.0	12.0	12.0

^{a}Maximum electric consumption or production where 100% is without energy storage. *C*, consumption; *W*, wind power; *S*, solar power.
^{b}Energy storage size needed in relation to system size. If, for example, 100% = 200 MW, then size = "10" means 20 MWh and "15" means 30 MWh.
^{c}Share (%) of time the energy storage is in use.
^{d}How fast it must be capable of charging/delivering electric energy.

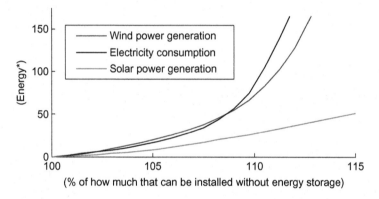

Fig. 8 Comparing energy storage size needed.

level before the next peak. For the other two, there are longer periods with peak values with no time restoration. It would, however, probably be a more favorable outcome for using energy storage if the model considers electricity patterns between different times of the day.

Another aspect is that energy storage is unused for most of the year; analyses of possible usages for other purposes during these periods are addressed in Section 3.3. Fig. 9 compares the utilization levels. Energy storage has the lowest utilization while managing electricity consumption.

Analysis results of energy storage to manage a mixed system part that has periods of both import and export transfer limitations are provided in Table 9. Electricity consumption with solar power generation is exemplified. Note that energy storage size needed as a function of electricity consumption is low compared to the system part with only electricity consumption. That indicates the positive synergies of using both energy storage and taking advantage of cancellation effects.

Fig. 10 exemplifies how the energy storage level may vary over time regarding a node with 106.5% electricity consumption and 150% solar power generation, where the energy storage is about 42% charged in standby. The analysis extends

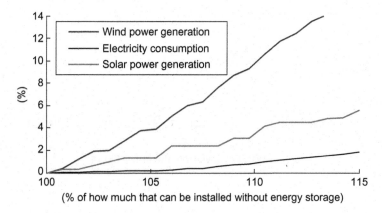

Fig. 9 Comparing energy storage utilization levels.

Table 9 Energy storage requirements in a mixed system part

Consumption[a]:	105				100	102.5	105.5	108.5
Solar production[a]:	132	137	143	149	150			
Size[b]	16.4	20.3	33.0	50.7	43.8	46.2	56.0	85.7
Normal[c]	100	81	50	32	0	12	34	60
Usage[d]	0.2	0.3	1.0	2.3	2.9	2.8	2.6	2.7
Power1[e]	0.0	3.4	9.4	15.4	18.0	17.2	16.2	15.3
Power2[e]	5.0	5.0	5.0	5.0	0.0	2.5	5.5	8.5

[a]Maximum electric consumption or production where 100% is without energy storage and without considering positive synergy effects.
[b]Energy storage size needed in relation to system size. If for example 100% = 200 MW, then size = "10" means 20 MWh and "15" means 30 MWh.
[c]How much percentage charged in standby when not in use.
[d]Share (%) of time the energy storage is in use.
[e]How fast it must be capable of charging(Power1)/delivering(Power2) electric energy.

over 40 years, 2 of these years are illustrated in the upper part of the figure. The energy storage during these two exemplified years never reaches its highest level. Clear seasonal variations can be discerned. Three days are showed in the lower part to get a detailed picture of how the energy storage may vary during a short period. These days are selected at the time when the energy storage was lowest.

3.3 Discussion and analyses of further energy storage utilization

Energy storage analysis results presented in Section 3.2 indicate that energy storage is unused most of the time. The possibility of using energy storage for another application, such as improved reliability, is best regarding scenarios where energy storage is used to manage electricity consumption as it is fully charged during normal operation. Furthermore, this scenario has the lowest utilization.

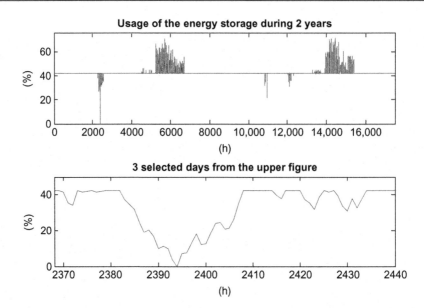

Fig. 10 Example of how energy storage that handles both high electric generation and consumption is used.

Table 10 shows the percentage of time the energy storage is unused, the size, and how long outages theoretically can be handled at low and high load. Energy storage to handle peak generation needs to be empty during normal operations. Thus, these are not available when there is a transfer limitation risk but can, however, be used during low risk periods. This risk can be predicted by weather forecasts, season, and time of day. Solar power generation can reach higher than 90% of its max generation only ~3% of the time and >80% during ~6%. Another positive aspect is that the summer months have few outages compared to the rest of the year in Sweden [20].

Table 10 The possibility of energy storage primarily used to handle overloading also handling outages when available

Max electric consumption allowed (%)[a]	Not in use (%)	Size[b]	Length possible to mitigate[c]	
			High load	**Low load**
101	99.97	2	1.2	3.6
105	99.80	15	9	27
110	99.10	85	51	153
115	98.13	378	227	680

[a]Maximum electric consumption where 100% is without energy storage.
[b]Energy storage size needed in relation to maximum power. If for example 100% = 200 MW, then size = "10" means 20 MWh and "15" means 30 MWh.
[c]Minutes full energy storage can compensate for an outage.

4 Conclusions

A method has been developed to analyze different smart grid solutions from a system perspective. The method includes integration of wind and solar power, electricity consumption, DR, and energy storage. An essential part is to find synergies that essentially consider the inclusion of weather parameters. Initial data analyses results were used for method development but can also be useful as reference material in related studies.

Results from this study confirm that the dynamic line rating increases the possibility of utilizing power systems more effectively. Taking advantage of the fact that electric generation and consumption can cancel each other out has been analyzed. Energy storage is analyzed from a technique-neutral perspective and results are presented as requirement lists of sizes, power (in both directions), and charging levels in normal operations. Size requirements to manage electric consumption and wind power generation are increasing exponentially rather than linearly as a function of the overload that needs to be handled. When more than one of the transfer capacity-increasing solutions is implemented at the same time in an analysis, the positive effect often is significantly higher compared with adding results from an analysis where these are analyzed separately.

Analysis results show that energy storage will be unused most of the time. To use them the rest of the time to increase reliability instead of standing unused has also been investigated.

References

[1] L. Bertling Tjernberg, in: The smart grid experience in Europe, IEEE Smart Grid News-Letter, 2014. http://smartgrid.ieee.org/august-2014/1132-the-smart-grid-experience-in-europ%C3%A9.

[2] P. van den Oosterkamp, P. Koutstaal, A. van der Welle, J. de Joode, J. Lenstra, K. van Hussen, R. Haffner, The Role of DSOs in a Smart Grid Environment, European Commission, DG ENER, Amsterdam/Rotterdam, 2014.

[3] European Commission Joint Research Centre (JRC), Smart Grid Projects Outlook 2014, Available from: http://ses.jrc.ec.europa.eu/sites/ses.jrc.ec.europa.eu/files/u24/2014/report/ld-na-26609-en-n_smart_grid_projects_outlook_2014_-_online.pdf. Accessed 12 December 2014.

[4] D. Niyato, Q. Dong, P. Wang, E. Hossain, Optimizations of power consumption and supply in the smart grid: analysis of the impact of data communication reliability, IEEE Trans. Smart Grid 4 (1) (2013) 21–35.

[5] L. Ochoa, L. Cradden, G. Harrison, in: Demonstrating the capacity benefits of dynamic ratings in smarter distribution networks, Innov. Smart Grid Technol. (ISGT), Gaithersburg, MD, USA, 2010.

[6] Grid Integration of Large-Capacity Renewable Energy Sources and Use of Large-Capacity Electrical Energy Storage, IEC, Geneva, Switzerland, 2012.

[7] T. Broeer, J. Fuller, F. Tuffner, D. Chassin, N. Djilali, Modeling framework and validation of a smart grid and demand response system for wind power integration, Appl. Energy 113 (2014) 199–207.

[8] D. Brodén, Analysis of Demand Response Solutions for Congestion Management in Distribution Networks, KTH Royal Instiurute of Technology, Stockholm, 2013 (Master Thesis).

[9] Electrical Energy Storage, IEC, Geneva, Switzerland, 2011.

[10] M. Lachman, P. Griffin, W. Walter, A. Wilson, Real-time dynamic loading and thermal diagnostic of power transformers, IEEE Trans. Power Delivery 18 (2003) 142–148.

[11] C.J. Wallnerström, Y. Huang, L. Söder, Impact from dynamic line rating on wind power integration, IEEE Trans. Smart Grid 6 (1) (2015) 343–350.

[12] A. Leite, C. Borges, D. Falcao, Probabilistic wind farms generation model for reliability studies applied to Brazilian sites, IEEE Trans. Power Syst. 21 (4) (2006) 1493–1501.

[13] L. Hernández, C. Baladrón, J.M. Aguiar, L. Calavia, C. Belén, A. Sánchez-Esguevillas, D.J. Cook, D. Chinarro, J. Gómez, A study of the relationship between weather variables and electric power demand inside a smart grid/smart world framework, Sensors 12 (9) (2012) 11571–11591.

[14] C.J. Wallnerström, P. Hilber, J. Gadea Travi, in: Implementation and evaluation of commonly used risk analysis methods applied to a regional power distribution system, International Conference on Electricity Distribution (CIRED), Stockholm, 2013.

[15] IEEE Milestone, Gotland High Voltage Direct Current Link, 1954, Approved 2017. Available from: http://ethw.org/Milestones:Gotland_High_Voltage_Direct_Current_Link.

[16] K. Chatziioannou, J. Guštinčič, L. Bertling Tjernberg, On Experience of Smart Grid Projects in Europe and the Swedish Demonstration Projects, Chalmers, Göteborg, 2013.

[17] Smart Grid Gotland, Available from: http://www.smartgridgotland.com. Accessed 17 June 2014.

[18] GEAB, Vattenfall, ABB, KTH, Smart Grid Gotland Pre-Study, 2011.

[19] Swedish Meteorological and Hydrological Institute (SMHI), Available from: http://www.smhi.se/en. Accessed 22 December 2014.

[20] C.J. Wallnerström, P. Hilber, Vulnerability analysis of power distribution systems for cost-effective resource allocation, IEEE Trans. Power Syst. 27 (1) (2012) 224–232.

[21] C.J. Wallnerström, L. Bertling Tjernberg, P. Hilber, S. Babu, J.H. Jürgensen, Analys av smartaelnätsteknologier inom kategorin elnätslösningar, Sammordningsrådet för smarta elnät, Stockholm, 2014.

[22] IEEE Standard for Calculating the Current-Temperature of Bare Overhead Conductors, IEEE Power Engineering Society, New York, 2007.

[23] C.J. Wallnerström, J. Setréus, P. Hilber, F. Tong, L. Bertling, in: Model of capacity demand under uncertain weather, 11th International Conference on Probabilistic Methods Applied to Power Systems (PMAPS), Singapore, 2010.

[24] CIROTECH AB, Solar Data of Visby (at Latitude 57.5, Longitude 18.5), Available from: http://www.cirotech.se/62141-visby.pdf. Accessed 20 December 2014.

Further reading

[25] O. o. E. D. & E. Reliability, Recovery Act: Smart Grid Investment Grants, (DOE), U.S. Department of Energy. Available from: http://energy.gov/oe/technology-development/smart-grid/recovery-act-smart-grid-investment-grants. Accessed 8 August 2014.

[26] Special Report by Zpryme's Smart Grid, China: Rise of the Smart Grid, 2011, Available from: https://www.smartgrid.gov/sites/default/files/doc/files/China_Rise_Smart_Grid_201103.pdf.

[27] Swedish Coordination Council, National Action Plan 2015–2030, Available from: http://www.swedishsmartgrid.se/english/the-national-action-plan/. Accessed 8 December 2014.

[28] N. Ekstedt, C.J. Wallnerström, S. Babu, P. Hilber, P. Westerlund, J.H. Jürgensen, T. Lindquist, Reliability Data: A Review of Importance, Use, and Availability, NORDAC, 2014. Stockholm.

[29] Smart Grid Gotland, Available from: http://www.smartgridgotland.com/eng/. Accessed 12 December 2014.

[30] SMHI, Available from: http://www.smhi.se. Accessed 20 February 2014.

[31] C. AB, Soldata för Visby, Available from: http://www.cirotech.se/soldata-visby.htm. Accessed 10 April 2014.

[32] S. C. Council, National Action Plan 2015-2030, Available: http://www.swedishsmartgrid. se/english. Accessed 8 December 2014.

[33] European Commission Joint Research Centre (JRC), Smart Grid Projects Outlook 2014, Available from: http://ses.jrc.ec.europa.eu/sites/ses.jrc.ec.europa.eu/files/u24/2014/report/ld-na-26609-en-n_smart_grid_projects_outlook_2014_-_online.pdf. Accessed 8 August 2014.

[34] M. Kolsjö, J. Sellevy, Öppna data i Sverige, PwC, 2014.

[35] NEPP, Teknik för smarta elnät för själva elnäten - kartläggning och behovsanalys, Samordningsrådet för smarta elnät, Stockholm, 2014.

[36] S. Talpur, C.J. Wallnerström, P. Hilber, C. Flood, Implementation of dynamic line rating in a sub-transmission system for wind power integration, Energy Syst. 6 (2014) 233–249.

[37] C.J. Wallnerström, P. Hilber, P. Söderström, R. Saers, O. Hansson, in: Potential of dynamic rating in Sweden, International Conference on Probabilistic Methods Applied to Power Systems (PMAPS), Durham, 2014.

Part Five

Case Studies

Application of cluster analysis for enhancing power consumption awareness in smart grids

12

Guido Coletta*, Alfredo Vaccaro*, Domenico Villacci*, Ahmed F. Zobaa[†]
*University of Sannio, Benevento, Italy, [†]Brunel University, Uxbridge, United Kingdom

1 Introduction

In recent years, transmission and distribution grids have been subjected to radical changes, which mainly affected their governance and operation criteria. The need for increasing the penetration of renewable power generators, the large-scale deployment of the liberalized electricity markets, and the difficulties in upgrading the grid infrastructure are some of the main challenges that should be faced by system operators. In this complex scenario, the role of grid users is drastically changing because they are evolving from passive customers, which were characterized by inelastic power demand profiles, to the so-called prosumers, which can play an active role by properly dispatching their generation units, modulating their power demand in response to price signals, and offering high-value services to grid operators. The increasing number of these new entities is expected to sensibly improve the efficiency and security of existing electrical grids by supporting their evolution to active, flexible, and self-healing systems composed of distributed and interactive resources. Thanks to these features, the development of new tools aimed at allowing the prosumers to optimally coordinate their consumption/production profiles in function of the grid state and the spot energy price has been recognized as one of the most important foundations of future smart grids.

In this context, recent studies reported in the power system literature have demonstrated that in a smart grid domain, the prosumers, if supported by accurate and reliable information about their actual energy consumption, can achieve electric energy savings up to 5–10%, which mainly derive from the detection of load anomalies and the power demand adaption to the electricity price dynamic. These advanced functions, which are usually referred as demand-side management (DSM) and demand response (DR) techniques, allow smart grid operators to interact with the prosumers in order to modify the load profiles during severe grid contingencies, avoiding the need for activating large-scale load shedding plans while also allowing the prosumers to dynamically change their demand profiles according to optimal energy sourcing policies. Despite these benefits, the adoption of these functions in existing power systems is still at its infancy, and several open problems need to be fixed in order to

Application of Smart Grid Technologies. https://doi.org/10.1016/B978-0-12-803128-5.00012-X

support their large-scale deployment. In particular, DSM and DR techniques frequently operate on data-rich but information-limited domains because the decisions that should be identified derive by the solution of complex optimization problems whose cost and constraints functions depend on a large quantity of uncorrelated and heterogeneous data (i.e., historical load profiles, spot prices, environmental variables, etc.). On the other hand, these decisions should be identified in computation times that should be fast enough for the information to be useful in a short-term operation horizon.

Hence, the development of computing paradigms aimed at converting the power demand data into actionable information, allowing the prosumer to have a full understanding of the available information, represent a timely and relevant issue to address. In the light of this need, a promising research direction is the development of smart techniques for load profile analysis, which allows prosumers to detect irregular demand patterns such as unusual energy usage, decreasing efficiency, erroneous regulations, and load malfunctioning. This strategic information could enhance the energy efficiency, reduce the supplying costs, and increase the situational awareness of prosumers about their actual power demand. These features are extremely useful in complex energy systems, characterized by large and spatially distributed loads, where the sensor data streaming could not allow a full understanding of the actual energy usage [1, 2].

To address this issue, the adoption of principal component analysis (PCA) and data clustering techniques have been recognized as the most promising enabling methodologies for load pattern classification. In particular, in [3–5] the authors proposed a methodology based on cluster analysis for classifying the proper class of homogenous users, designing the most effective tariff paradigm for optimal energy sourcing. In [6] a similar technique based on PCA and hierarchical-based clustering techniques has been proposed to solve the same problem. Also, Ozgonenel et al. [7] compare the performances of Euclidean and Mahalanobis distances in k-mean-based clustering in the task of classifying electricity customers for fault detection, and in Almutairi et al. [8] a repeated k-mean clustering methodology, which improves the ability in consumer classification, is presented.

According to these arguments, in this chapter the employment of self-organizing models based on clustering analysis for power consumption awareness in a smart grid domain is proposed. The main idea is to process the sensor data streaming for classifying the load profiles, correlating them with the endogenous measured variables, and identifying irregularities in energy consumption. The analysis of the statistical correlations between the power demand and the endogenous variables is performed by a fuzzy inference system, which computes the key performance indexes for each load profile class. These indexes are processed by an outlier detection technique, which identifies irregular load patterns originated from devices malfunctions, system faults or incorrect control settings. Experimental results obtained on a real case study are presented and discussed in order to emphasize the benefits deriving from the application of the proposed framework on complex load patterns.

2 Mathematical preliminaries

2.1 Elements of cluster analysis

The self-organizing models for load anomaly detection proposed in this chapter are based on a class of data-clustering techniques that aim to organize data in homogeneous clusters, characterized by high similarity degree. In this context, the main problem to solve is how to classify objects in homogeneous groups starting from multivariate observations, maximizing the similarity of the objects in the same cluster, and maximizing, at the same time, the difference between the different clusters.

2.1.1 Clustering techniques

Among the different approaches that can be adopted for cluster formation, the top-down and the bottom-up paradigms are the most frequently adopted. The first one assumes that the cluster is generated starting from a single group containing all objects, then dividing it in smaller and more homogeneous ones until a termination criteria is satisfied. On the contrary, bottom-up processes associate objects starting from the situation that all of them initially represent distinct clusters.

Another very common classification of data-clustering processes is between:

- Partitioning clustering
- Hierarchical clustering

Partitioning clustering techniques provide a division of data set into nonoverlapping sets. The cluster shapes are the product of predetermined criteria for group development and a cluster is evaluated by comparing each object with respect to a point representative of the cluster, namely the centroid. One of the most diffused partitioning clustering algorithms is represented by k-mean methodology. The mathematical formulation of the partitioning clustering can be expressed through the following definition:

Definition 2.1 (Partitioning clustering). Let $X = \{X_1, X_2, ..., X_N\}$ a set of data. The partitioning clustering provides a subset $C = \{C_1, C_2, ..., C_k\}$ such that:

$$X = \bigcup_{i=1}^{k} C_i$$
$$C_i \cap C_j = \varnothing \text{ for } i \neq j$$

Otherwise, hierarchical clustering is based on the inclusion of smaller clusters in bigger ones, hence allowing the presence of subclusters. The outcome of this algorithm is a hierarchy tree, namely the dendrogram, depicting correlations between different clusters levels. It can be noted that hierarchical clustering can be viewed as a sequence of partitioning clusterings, and that one can obtain the latter by considering one of the levels of the dendrogram. The objective function followed by clustering algorithms' points to highlighting the local structure of the data and maximizing the cluster dissimilarity.

A further categorization concerns the possibility that an object belongs exclusively or not to a specific cluster. In the light of this need, three possibilities can be distinguished, leading to different clustering types:

- Exclusive clustering
- Overlapping (or nonexclusive) clustering
- Fuzzy clustering

It is worth observing that, in the first category, objects belong exclusively to a specific cluster while, in the second one, objects can belong simultaneously to more clusters. In many situations, instead, belonging to a cluster is not true, but there is a probability of belonging to different clusters. In this field of application, fuzzy or probabilistic clustering algorithms are very useful, allowing us to address these problems by considering each object belonging to each cluster with a membership degree between 0 and 1, subject to the constraint that the sum of them for each object must be 1. In a probabilistic sense, that is equal to saying that the sum of probabilities for each object must be 1.

The last classification is between

- Partial clustering
- Complete clustering

A complete clustering assigns all objects to clusters while a partial one does not, allowing the addressing of data issues, that is, outliers, noise or out-of-interest data, among all.

2.1.2   Cluster classification

After giving a classification of the clustering algorithms, let us provide now a classification of cluster types. It is possible to distinguish among five cluster typologies (i.e., well-separated, prototype-based, graph-based, density-based, and conceptual clusters).

Well-separated is characterized by the propriety that the distance between points in different clusters is greater than the distance between any two points in the same cluster.

Prototype-based is characterized by the propriety that the distance between each element in the cluster and the relative prototype is minimal with respect to the protothype of the other clusters. The prototype of the cluster is often a centroid (for continuous attributes data) or a medoid (for categorical attributes data).

Graph-based is useful when data are represented by graphs. In such case a cluster is defined by a connected component, that is, a group of objects interconnected with each other, but that present has no connection with elements outside.

Density-based is defined by a high-density data region surrounded by a low-density data region.

Conceptual clusters is the most general cluster definition and includes all the previous. In particular, a cluster, in such a point of view, is simply defined as a group of objects that shares some proprieties. For example, a prototype-based cluster can be seen as the group of objects that shares the propriety to be the closest to the centroid.

2.1.3 k-Means

The k-mean is an exclusive partitioning clustering technique based on prototype-based clusters and represents one of the simplest and popular approaches to cluster analysis. The k-mean technique allows grouping a data set into k predetermined clusters. Let us show the basic k-mean algorithm: the first step is to fix an initial set of k centroids and to assign each object of the dataset to the closest centroid. Then the centroids are updated by considering the objects of their relative clusters. The procedure repeats until the centroid does not change anymore. So, the algorithm can be summarized as follows:

---

1 Initialization of a number of cluster k and an initial centroid set C_0;
2 while *centroids change over a certain tolerance* **do**
3 | Assign elements to the closest cluster w.r.t. the *i*th-1 centroids;
4 | Compute the *i*th centroid set on the basis of the associated objects;
5 end

The first question could be how to compute the distance between each object and the relative centroid. To answer this question, it is necessary to define appropriate metrics, that is, appropriate distances $d(\mathbf{x}_i, \mathbf{x}_j)$, quantifying the affinity degree of couples of different objects $(\mathbf{x}_i, \mathbf{x}_j)$. Formally, a generic distance or metric is defined as:

Definition 2.2 (Metric). A metric (or distance) is defined as:

$$d(x,y) = \left(|\mathbf{x} - \mathbf{y}|^p\right)^{1/p}$$

abiding by the following proprieties:

1. Nonnegativity
   $$d(\mathbf{x}, \mathbf{y}) \geq 0$$
2. Symmetry
   $$d(\mathbf{x}, \mathbf{y}) = d(\mathbf{x}, \mathbf{y})$$
3. Triangle inequality
   $$d(\mathbf{x}, \mathbf{y}) \leq d(\mathbf{x}, \mathbf{z}) + d(\mathbf{z}, \mathbf{y})$$
4. Identity of indiscernible
   $$d(\mathbf{x},\mathbf{y}) = 0 \Leftrightarrow \mathbf{x} = \mathbf{y};$$

There exist miscellaneous distance definitions, for example:

- Manhattan distance (L_1): $d = |\mathbf{x} - \mathbf{y}|$
- Euclidean distance (L_2): $d = \sqrt{(\mathbf{x} - \mathbf{x})^2}$
- Mahalanobis distance: $d = (\mathbf{x}, \mathbf{y})\mathbf{S}^{-1}(\mathbf{x}, \mathbf{y})$

Formally, once the distance d is defined, the k-means procedure is formalized as:

$$S = \underset{S}{\text{argmin}} \sum_{i=1}^{k} \sum_{\mathbf{x} \in \mathbf{S}_i} \text{dist}(\mathbf{c}_i, \mathbf{x}) \tag{1}$$

where k is the number of clusters, $\mathbf{x}$ is an object, $\mathbf{S}_i$ is the ith cluster and $\mathbf{c}_i$ is the centroid of the ith cluster.

For application domain addressed in this chapter, the most-used distances are the L_1 and, among the others, L_2. It could be demonstrated that for the Manhattan metric, the optimal ith centroid is the median of the object of the cluster. For the particular case of L_2 metric:

$$S = \underset{S}{\text{argmin}} \sum_{i=1}^{k} \sum_{\mathbf{x} \in \mathbf{S}_i} \| \mathbf{x} - \mathbf{c}_i \|_2^2 \tag{2}$$

with

$$\mathbf{c}_i = \frac{1}{\dim(\mathbf{S}_i)} \sum_{\mathbf{x} \in \mathbf{S}_i} \mathbf{x} \tag{3}$$

2.1.4 c-Means

c-Means is a fuzzy-based clustering technique extending the concepts introduced by the k-means algorithm. This approach consists of considering each object belonging to all the clusters with a certain confidence degree.

---

1 Initialization of a number of cluster k and an initial centroid set $\mathbf{C}_0$;
2 while *centroids change over a certain tolerance* **do**
3 | Compute the fuzzy membership matrix, ω, w.r.t. the centroids;
   | Compute the ith centroid set;
4 end

Formally, the c-means procedure is formalized as:

$$S = \underset{S}{\text{argmin}} \sum_{i=1}^{k} \sum_{j=1}^{n} \omega_{i,j}^m \, \text{dist}(\mathbf{c}_i, \mathbf{x}_j) \tag{4}$$

where n is the number of objects in the dataset and $m > 1$ is the fuzzifier coefficient indicating the cluster's fuzziness by determining the belonging degree of each object to each cluster. In the limit of $m = 1$ the weights ω tend to 0 or 1, representing the particular case of the k-means algorithm. Particularizing for the Euclidean metric, as for the previous section, the c-means algorithm can be formalized as:

$$S = \underset{S}{\text{argmin}} \sum_{i=1}^{k} \sum_{j=1}^{n} \omega_{i,j}^m \, \| \mathbf{x}_j - \mathbf{c}_i \|_2^2 \tag{5}$$

with

$$c_i = \frac{\sum\limits_{j=1}^{n} \omega_{i,j}^m \, \mathbf{x}_j}{\sum\limits_{j=1}^{n} \omega_{i,j}^m} \tag{6}$$

and

$$\omega_{i,j} = \frac{1}{\sum\limits_{i=1}^{k} \frac{\| \mathbf{x}_i - \mathbf{c}_j \|_2}{\| \mathbf{x}_i - \mathbf{c}_k \|_2}} \tag{7}$$

3 Detecting load outliers by clustering analysis

The described techniques can play an important role in power consumption awareness by classifying load profiles, identifying the statistical correlations between these and the endogenous variables, and detecting irregular energy usage. In this context, the main idea is to properly classify the hourly demand and the environmental temperature profiles, and identifying their correlation by applying a fuzzy inference system. Fuzzy logic completely revises the classical set theory and the semantic concept of truth, turning upside down the traditional foundation of the classical set theory, because it allows the definition of a belonging degree of an object to a set, that is considering that object belongs to that set with a certain measure. According to this paradigm, an object can belong to more than one set. In the same way, the concept of truth is radically changed; in the classical meaning, for example, if A is true, it cannot be false and vice versa. In fuzzy logic, a statement is associated with a certain degree of truth. These features are particularly useful for the application under study, because they allow the obtaining of a reliable classification of the load and temperature profiles, and an effective assessment of their correlations. A generic fuzzy inference system can be organized into three different conceptual steps: (i) fuzzification, (ii) inference, and (iii) defuzzification. A graphic scheme of the above is provided in Fig. 2, and the overall architecture of the proposed framework is shown in Fig. 1.

Analyzing these figures, it is worth nothing that the first step in applying the proposed framework is to obtain reference patterns for both the electrical and the environmental daily profiles by classifying the sensor data streaming generated by the power meter and the meteorological station. The distances between the measured and the reference profiles is then processed by a fuzzy inference system to discriminate the pattern regularity, namely to detect the so-called load and temperature anomalies. Starting from this estimation, and on the basis of heuristic experiences on thermostatic loads, it could be argued that a load anomaly induced by a temperature anomaly could be considered as a regular load pattern, that is, during summer if the environmental temperature is extremely high, the load is expected to sensibly increase. On the contrary, a load anomaly that does not correspond to a temperature anomaly is a clear indication of an irregular load pattern. These straightforward heuristic rules have been translated in proper fuzzy rules, which represent the knowledge base of the proposed fuzzy inference system.

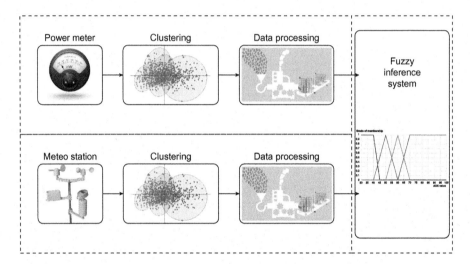

Fig. 1 Proposed algorithm.

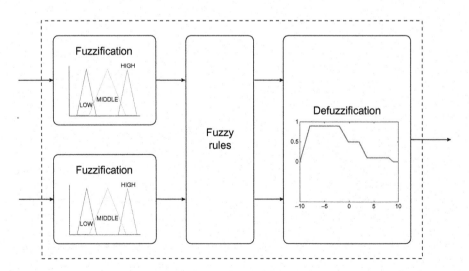

Fig. 2 Fuzzy inference system.

4   Case study

In this section, the proposed framework has been applied in the task of detecting power demand irregularities for a large commercial user located in the south of Italy. The analyzed domain is characterized by complex demand patterns, mainly due to air conditioning, lighting, and medical and technological services, which changes considerably on the basis of the seasons and environmental conditions. The corresponding distribution of the yearly average power demands is reported in Fig. 3.

As shown in Fig. 3, the air conditioning has a bearing of 42% on the total energy consumption. Further, lighting represents another relevant item in the energy computation because many wards need lighting for more than 12–14 h/day. Laundry and kitchen represent another big portion of the energy loads, even though they consist totally of thermal energy requirements.

The electric system under analysis consists of a medium-voltage distribution network connected, by three power lines, to the user substation.

In order to detect the demand irregularities for this complex user, the load profiles and the temperature measurements for 1 year of operation time, reported in Figs. 4 and 5, respectively, have been considered. Analyzing these figures, it is worth noting that the considered yearly profiles have been characterized by complex and heterogeneous patterns, which are characterized by extremely variable electrical and environmental dynamics.

The proposed methodology can be conceptually divided into two main parts: the clustering-based classification of the load and temperature profiles, and the fuzzy-based processing of the computed quantities. The first step consists in classifying the power meter and sensor acquisitions within predetermined clusters, through k-mean- and c-mean-based algorithms. The first are classified into 16 different clusters on the basis of the day (business day/holiday), the loading level (i.e., low and high load), and the season.

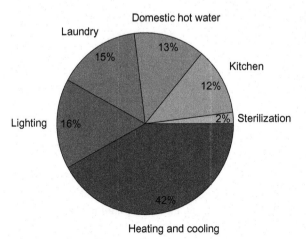

Fig. 3 User energy consumptions.

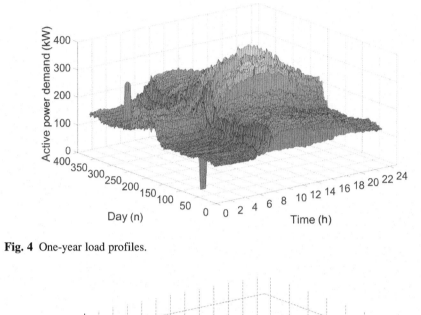

Fig. 4 One-year load profiles.

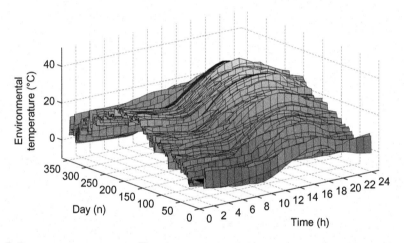

Fig. 5 One-year temperature profiles.

The results of this classification process are shown in Figs. 6–13, which depict the centroids identified by both k-mean and c-mean clustering techniques.

Moreover, the temperature profiles are classified on the basis of the season and the day (hot, middle, and cold), as shown in Figs. 14–21, where, as for the active power load profiles, the cluster centroids for both k-mean and c-mean techniques are represented.

The second conceptual step (i.e., the fuzzy-based outliers detection) computes the distance between each load and temperature profile and the corresponding centroid. When such distance is greater than a threshold value, the corresponding load profile is classified as an anomalous profile. In order to distinguish between endogenous or exogenous effects, the framework correlates demand and temperature profiles: if temperature and demand profiles are both anomalous the anomaly is classified as

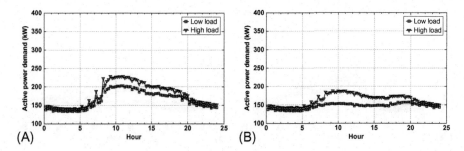

Fig. 6 Wintry load profiles: *k*-mean. (A) Business day; (B) holiday.

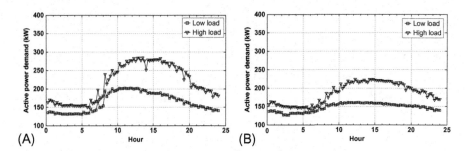

Fig. 7 Spring load profiles: *k*-mean. (A) Business day; (B) holiday.

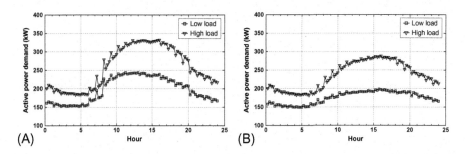

Fig. 8 Summer load profiles: *k*-mean. (A) Business day; (B) holiday.

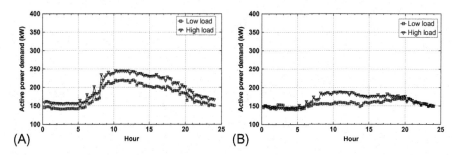

Fig. 9 Fall load profiles: *k*-mean. (A) Business day; (B) holiday.

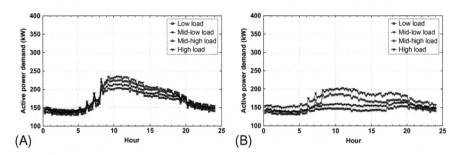

Fig. 10 Wintry load profiles: c-mean. (A) Business day; (B) holiday.

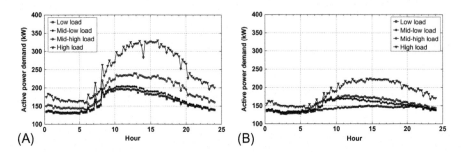

Fig. 11 Spring load profiles: c-mean. (A) Business day; (B) holiday.

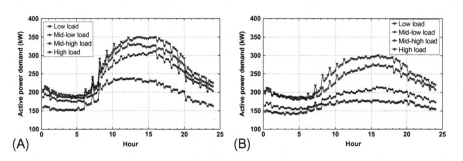

Fig. 12 Summer load profiles: c-mean. (A) Business day; (B) holiday.

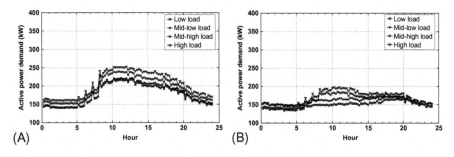

Fig. 13 Fall load profiles: c-mean. (A) Business day; (B) holiday.

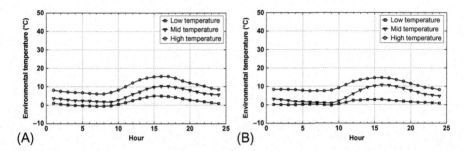

Fig. 14 Wintry temperature profiles: k-mean. (A) Business day; (B) holiday.

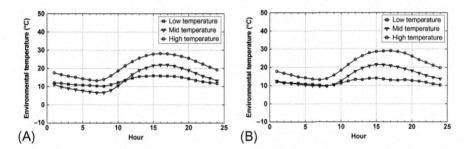

Fig. 15 Spring temperature profiles: k-mean. (A) Business day; (B) holiday.

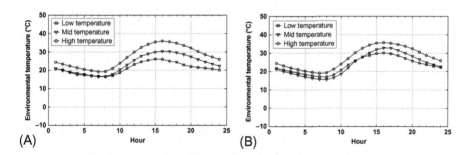

Fig. 16 Summer temperature profiles: k-mean. (A) Business day; (B) holiday.

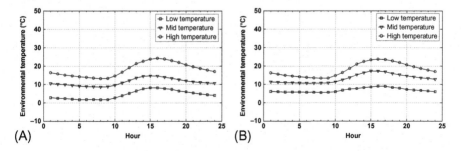

Fig. 17 Fall temperature profiles: k-mean. (A) Business day; (B) holiday.

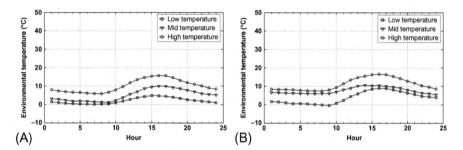

Fig. 18 Wintry temperature profiles: c-mean. (A) Business day; (B) holiday.

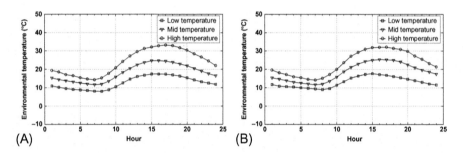

Fig. 19 Spring temperature profiles: c-mean. (A) Business day; (B) holiday.

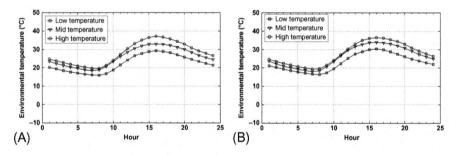

Fig. 20 Summer temperature profiles: c-mean. (A) Business day; (B) holiday.

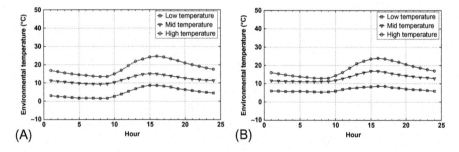

Fig. 21 Fall temperature profiles: c-mean. (A) Business day; (B) holiday.

exogenous while if only the absorption profile is anomalous, it is classified as endogenous, and the considered profile is detected as an outlier.

The results of this process are summarized in Figs. 22–29, where, for each minimum and maximum temperature profile, the temperature and absorption anomalies have been highlighted by denoting them with circles and stars, respectively. Analyzing these results, it is possible to identify the days in which temperature and absorption outliers coexist in the system and, hence, it is possible to identify absorption anomalies that cannot be related to climatic events and that can be reasonably associated with other factors influencing system operation.

Finally, the results of this application can be summarized in tabular format, allowing the performance comparison between k-mean and c-mean data clustering approaches (Tables 1–3).

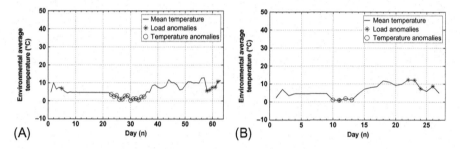

Fig. 22 Wintry anomalies: k-mean. (A) Business day; (B) holiday.

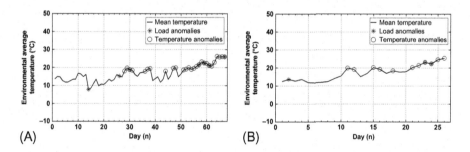

Fig. 23 Spring anomalies: k-mean. (A) Business day; (B) holiday.

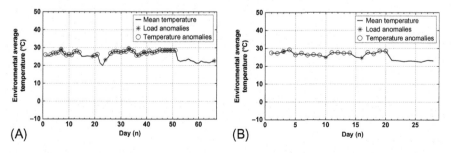

Fig. 24 Summer anomalies: k-mean. (A) Business day; (B) holiday.

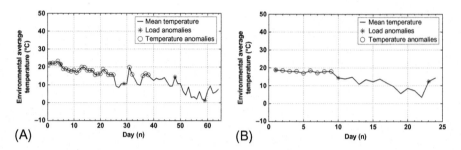

Fig. 25 Fall anomalies: *k*-mean. (A) Business day; (B) holiday.

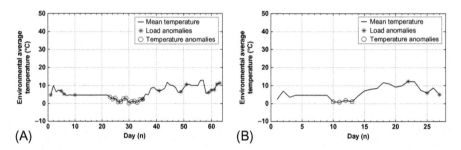

Fig. 26 Wintry anomalies: *c*-mean. (A) Business day; (B) holiday.

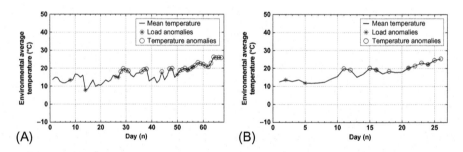

Fig. 27 Spring anomalies: *c*-mean. (A) Business day; (B) holiday.

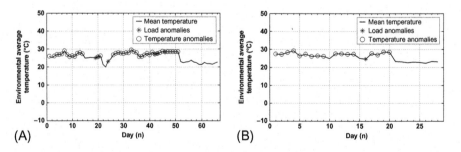

Fig. 28 Summer anomalies: *c*-mean. (A) Business day; (B) holiday.

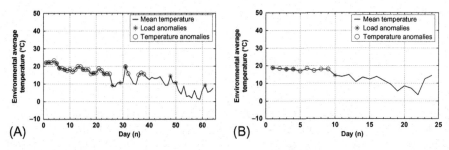

Fig. 29 Fall anomalies: c-mean. (A) Business day; (B) holiday.

Table 1 **Load anomalies**

	k-Mean	c-Means
Winter	6 36 77 78 356 357 359 360 361 362 364	2 6 9 13 48 58 68 72 77 358 359 360 361 362 363 365
Spring	85 100 116 122 159 161 162 163 165 173	85 90 98 100 116 117 134 150 154 156 158 162
Summer	182 184 200 204 207 220 225 228 240 241 242 265	200 201 207 225 233 234 237
Fall	268 272 273 286 292 298 302 303 306 333 348 350	268 269 272 273 275 277 280 281 302 306 310 333 335 352

Table 2 **Temperature anomalies**

	k-Mean	c-Means
Winter	32 33 34 35 36 37 38 39 40 41 42 43 44 45 46 47 48	2 4 5 6 9 10 11 12 13 16 17 18 19 20 23 24 25 26 27 30 31 51 52 55 57 58 59 63 64 67 68 70 71 77 78 356 359 360 361 364
Spring	118 119 120 121 122 123 130 131 132 133 134 141 142 145 146 151 152 153 154 155 156 157 158 159 160 161 162 163 164 165 166 167 168 169 170 171 172 173	96 97 100 102 103 104 107 108 109 110 111 114 115 116 119 120 128 133 134 135 137 138 141 143 149 154 155 161 162 168 169
Summer	174 175 176 178 179 180 181 182 183 184 185 186 187 188 189 190 191 192 193 196 197 201 202 203 209 210 211 212 213 214 215 216 217 218 219 220 221 222 223 226 227 228 229 230 231 232 233 234 235 236 237 238 239 240 241 242 243 244	174 175 176 178 179 180 181 182 183 184 185 186 187 188 189 190 191 192 193 196 197 201 202 203 209 210 211 212 213 214 215 216 217 218 219 220 221 222 223 226 227 228 229 230 231 232 233 234 235 236 237 238 239 240 241 242 243 244
Fall	268 269 270 271 272 273 274 275 276 277 278 279 280 281 282 283 284 285 286 287 288 289 290 291 292 293 294 295 296 297 298 299 300 301 310 311 317 318 319	268 269 270 271 272 273 274 275 276 277 278 279 280 281 282 283 284 285 286 287 288 289 290 291 292 293 294 295 296 297 298 299 300 301 310 311 317 318 319

Table 3 **Endogenous load anomalies**

	k-Means	c-Means
Winter	6 77 78 356 357 359 360 361 362 364	48 72 358 362 363 365
Spring	85 100 116	85 90 98 117 150 156 158
Summer	200 204 207 225 265	200 207 225
Fall	302 303 306 333 348 350	302 306 333 335 352

5 Conclusion

The development of computing paradigms aimed at converting the power demand data into actionable information, allowing the prosumer to have a full understanding of the available information, represents a timely and relevant issue to address. In light of this need, this chapter analyzed the potential role of cluster analysis and fuzzy-based programming for detecting irregular demand patterns, such as unusual energy usage, decreasing efficiency, erroneous regulations, and load malfunctioning. This strategic information could enhance the energy efficiency, reduce the supplying costs, and increase the situational awareness of prosumers about their actual power demand. Experimental results obtained on a real case study have been presented and discussed in order to assess the benefits deriving by the application of the proposed framework in the task of analyzing complex load patterns.

References

[1] Y.-I. Kim, J.-M. Ko, J.-J. Song, H. Choi, Repeated clustering to improve the discrimination of typical daily load profile, J. Electr. Eng. Technol. 7 (3) (2012) 281–287.
[2] S.-L. Yang, C. Shen, et al., A review of electric load classification in smart grid environment, Renew. Sustain. Energy Rev. 24 (2013) 103–110.
[3] G. Chicco, R. Napoli, F. Piglione, Application of clustering algorithms and self organising maps to classify electricity customers, in: 2003 IEEE Bologna Power Tech Conference Proceedings, vol. 1, IEEE, 2003, p. 7.
[4] G. Chicco, R. Napoli, F. Piglione, Comparisons among clustering techniques for electricity customer classification, IEEE Trans. Power Syst. 21 (2) (2006) 933–940.
[5] G. Chicco, R. Napoli, F. Piglione, Load pattern clustering for short-term load forecasting of anomalous days, in: 2001 IEEE Porto Power Tech Proceedings, vol. 2, IEEE, 2001, p. 6.
[6] Y. Li, Y. Cheng, L. Zhang, H. Huang, Application of multivariate statistical analysis to classify electricity customers, in: China International Conference on Electricity Distribution, 2008. CICED 2008, IEEE, 2008, pp. 1–6.
[7] O. Ozgonenel, D.W.P. Thomas, T. Yalcin, I.N. Bertizlioglu, Detection of blackouts by using K-means clustering in a power system, in: 11th International Conference on Developments in Power Systems Protection, 2012. DPSP 2012, IET, 2012, pp. 1–6.
[8] K. Anaparthi, B. Chaudhuri, N.F. Thornhill, B.C. Pal, Coherency identification in power systems through principal component analysis, IEEE Trans. Power Sys. 20 (3) (2005) 1658–1660.

Smart grids and the role of the electric vehicle to support the electricity grid during peak demand

Ahmad Zahedi
James Cook University, Townsville, Australia

1 Introduction

An electricity network is one of the largest, most complicated, and most sophisticated systems in the world. Worldwide, the electricity networks are old, outdated, and not smart enough to meet the 21st century's requirements. Even in developed countries, millions of people are without power for a couple of hours every day. Recently, the number of weather-caused major outages has increased rapidly, mainly because most of the components of an electricity network are outdoor facilities. Even in the countries with a modern power grid, the only way utilities know there is a power outage is when a customer calls and reports. An electricity grid has to be smart enough to be able to handle thousands of generators of different technologies and sizes as well as supplying electricity to millions of customers in a reliable and sustainable way.

A smart phone, for example, means a phone with a computer in it. Similarly, a smart grid means a "fully computerized" electricity grid. One of the key features of the smart grid is automation technology that lets the utility adjust and control millions of devices from a central location.

The concept of a smart grid has been around for many years, but what's important for us is to understand what it really means and how a smart grid can contribute to a reliable, sustainable, and cost-effective power supply. Another key but challenging feature of the smart grid concept is customer involvement through two-way communication between the electricity provider and the customer's electrical appliances.

The future electricity network should also be able to handle a large number of EVs, which will be connected to the grid for charging and discharging. An EV can be integrated into the electricity network easily and conveniently. The EV presents a new demand for electricity during its charging periods and also can play an important role as distributed storage devices that could supply electric power back to the grid.

One of the benefits of an EV is that it can help the electricity network during those periods that the network is facing peak demand. The second benefit is that the EV

Application of Smart Grid Technologies. https://doi.org/10.1016/B978-0-12-803128-5.00013-1

would be able to help shift the grid load from a high-demand time to a low-demand time. The third benefit is that the EV would help smooth variations in power generation caused by variable and intermittent renewable sources such as solar energy and wind power. This benefit is very important because in the near future we will see more renewable and intermittent energy sources connected to the so-called smart grid. The objective of this chapter is to describe the characteristics of a smart grid and its benefit to utilities and customers while also discussing the challenges and evaluating the potential values of the EV's batteries to the electricity grid.

2 Part 1

2.1 What is a smart grid?

There is no uniform definition of a smart grid. According to the European Technology Platform, a smart grid is an electricity network that can intelligently integrate the actions of all users connected to it in order to efficiently deliver sustainable, economic, and secure electricity supplies.

According to the United States Department of Energy, a smart grid is self-healing; enables active participation of consumers; accommodates all generation and storage options; enables introduction of new products, services and markets; optimizes asset utilization; operates efficiently; and provide reliable power quality for the digital economy.

According to the Australian Government (DEWHA), a smart grid combines advanced telecommunications and information technology applications with "smart" appliances to enhance energy efficiency on the electricity power grid, in homes and businesses.

A smart grid includes adding two-way digital communication technology to all devices associated with the grid. A key feature of the smart grid is automation technology that lets the utility adjust and control millions of devices from a central location.

The concept of a smart grid has been around for many years. What's important for us is to understand what a smart grid really means and how it can contribute to national energy policy. The question is, what do we want a smart grid to do cost effectively? One of the key but challenging features of the smart grid concept is customer involvement (customer engagement) in the process of current network structure. The other aspect of a smart grid that is challenging is the use of ICT at all levels.

Fig. 1 shows a conventional (traditional) power system while Fig. 2 shows a future power system.

The traditional power networks can be described as:

- Centralized power generation.
- Controllable generation, uncontrollable loads.
- Generation follows the load.
- Power flows in only one direction.

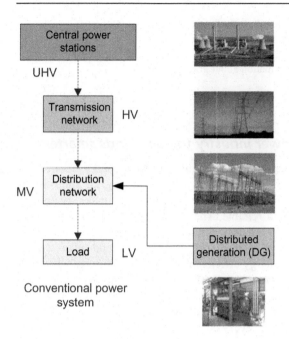

Fig. 1 Traditional power system.

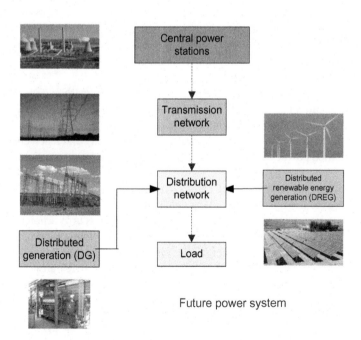

Fig. 2 Future power system.

- Power system operation is mainly based on historical data and experience.
- Overloads in the system are detected by the operators.
- Rerouting of power flow in the case of overload is performed by the operators.
- Utilities do not have sufficient information about grid conditions.
- High power loss.
- Likely events of costly power blackout.

2.2   Innovations in the power industry to make grids smarter

A major transformation in the electric power industry and energy services is underway. This transformation is driven largely by technological innovation, national regulatory rules, the environment, and economics to make electricity cleaner, more affordable, safer, and reliable. The power industry is moving away from high-carbon technologies toward low-carbon technologies in a very short time, adding significant amounts of solar energy, wind power, and other renewable energy as well as increasing energy efficiency. So, in the future we will see a cleaner energy mix while more new and renewable energy technologies will have low carbon emissions. We also will observe that the electricity networks integrate a mix of central power stations and distributed low-voltage generators. By adding more state-of-the-art digital sensors at every corner of the grid, the power industry can better control its interconnected networks. The first digitization move began with smart meters.

Power companies will be able to provide distributed generators, both on the supply side and the demand side. The power industry will also be able to provide a range of customized services to meet a customer's needs and his or her growing expectations (e.g., some commercial customers want 100% renewable energy to meet the company's sustainability goals).

Shifting from high carbon generation technologies to low carbon generation technologies has resulted in lower prices and greater availability. Now, renewable energy is an important part of the world energy mix.

If power companies can deliver electricity more efficiently and teach customers to use it wisely, then power companies can do more to use energy's potential to grow the economy.

With the rollout of the Internet of Things (IoT) at homes and businesses, customers will have more control over all their electrical devices. The IoT will provide the opportunity for customers to control temperature, lighting, home security, choice of energy, etc. The IoT will also provide the opportunity for effective integration of EVs to the grid. Distributed EVs will help stabilize the electric power network.

The evolution of the smart grid means that power industries around the world are in transition to a new environment defined by advanced information and control technologies. This smart grid will provide an opportunity for customer engagement, converting passive customers to active customers while enabling customers to better control their electricity consumption in an efficient and cost effective way. Digital, distributed, modern, and smart power grids will make cities smarter by making the electricity that power companies supply to their customers more secure, reliable, affordable, and clean.

The new generation of power grids will provide an opportunity to customers for the integration of renewable and low-carbon energy sources such as solar and energy storage, hence reducing gas emissions in the electricity generation sector.

Smart grids also deploy new automated technology that improves the resiliency and flexibility of the grid, providing real-time visibility for power system operators in the case of multidirectional power flow. Using intelligent sensors and switches will help systems self-heal by automatically detecting and isolating faults and redirecting the power flow.

2.3 Electricity networks worldwide face a number of challenges

They have aging infrastructure, continuous growth in demand, shifting load patterns, and supply more power using intermittent sources.

Power utilities are facing many challenges, including increasing grid resilience, improving customer engagement, managing the operation of advanced metering infrastructure (AMI), preventing meter malfunctions, optimizing the maintenance of network assets, etc.

Power utilities will be able to solve these challenges by applying big data, cloud computing, and machine learning.

In the near future, we will see that every electrical device, both domestic and commercial, that is connected to the electricity network will have an Internet address. Examples include air conditioning, toasters, the inverter of rooftop solar PV systems, street lighting, EV battery chargers, etc. This will be the so-called IoT, leading an evolution of the grid and all devices connected to the grid to be more digital.

2.4 Some customers demand individualized services

In addition to having reliable, safe, and affordable electricity, some customers demand very high levels of service quality such as stable frequency and voltage, very high reliability, or an especially low carbon energy source. Some customers even want a high degree of engagement with their energy use.

2.5 What factors make an electricity grid smart?

There are a number of factors that make the grid smart:

- Know exactly where a power failure happens and quickly fix it.
- Extend life of aging equipment.
- Detect and minimize outages by sensing potential equipment failures.
- Reduce power loss by using real-time data to match generation and demand.
- Smooth power demand to take advantage of off-peak supply.
- Maintain a sufficient, cost-effective power supply while managing a GHG target.
- Make it easier for consumers to use renewable energy sources.

- Use meters that show consumers their energy use in real time.
- Use variable pricing that allows consumers to choose off-peak energy.
- Help customers to establish a "smart home" that turns appliances on and off.

A smart grid is beneficial for both customers and power utilities.

2.6 Customer side

Because of a lack of visibility, consumers can't be expected to manage what they can't measure or see. The smart grid overcomes this "lack of energy visibility," making real-time usage and pricing data available through:

- In-home displays. This will help consumers to identify energy-intensive appliances that they have and use.
- Online portals.
- Smart phones.

Making customers aware of demand management methods. This will help consumers:

- Reduce energy use during peak times.
- Slow the need for investments in costly generation, transmission. and distribution infrastructure.
- Improve reliability and reduce the overall cost of supply, etc.

2.7 Utility side

The smart grid has benefits for utilities. These include:

- Better managing the grid.
- Offering customers choices.
- Understanding energy usage.
- Reducing cost of electricity.
- Communication with customers.
- Communication with customers' appliances.
- Using more renewable sources of energy.
- Integration of EVs.

One of the important features of a smart grid is so-called self-healing. A self-healing smart grid can provide a number of benefits that lend themselves to a more stable and efficient system. Three of its primary functions include:

- Real-time monitoring and reaction, which allows the system to constantly tune itself to an optimal state.
- Anticipation, which enables the system to automatically look for problems that could trigger larger disturbances.
- Rapid isolation, which allows the system to isolate parts of the network that experience failure from the rest of the system to avoid the spread of disruption, enabling a more rapid restoration.

As a result of these functions, a self-healing smart grid system is able to reduce power outages and minimize their length when they do occur. The smart grid is able to detect abnormal signals, make adaptive reconfigurations, and isolate disturbances, eliminating or minimizing electrical disturbances during storms or other catastrophes.

3 Part 2

3.1 Electric vehicles

An EV is any vehicle that uses electricity for propulsion rather than liquid fuels. Fig. 1 shows the major operational differences between an EV and an ICE vehicle.

The idea of a vehicle propelled by electricity was first conceptualized in the 19th century, and EVs had been manufactured and used in the same century, shortly after the invention of rechargeable lead-acid batteries and electric motors. However, with the development of ICEs, EVs were quickly considered to be an unviable transportation option due to their limited range, power, and durability as well as the large availability of gasoline for ICEs.

Nowadays, with growing concerns for CO_2 emissions and the growing scarcity of fossil fuels, EVs are quickly becoming a viable alternative to the ICE. Because of the environmental benefits associated with EVs, prevalent advancements have been made in the EV industry to influence the uptake of these economically and environmentally beneficial vehicles. These advancements have been seen in battery technologies as well as the integration of energy conservation options, such as regenerative braking.

A major issue that has sparked consumer interest in the ICE over the EV is the difference in range capabilities. Range anxiety is a term commonly associated with the limited range capabilities of EVs compared to ICEs, which has greatly influenced the unpopularity of EVs in the past. To overcome the range issues, advancements in battery technologies, improvements in vehicle efficiency, or the widespread rollout of charging infrastructure are required.

3.2 EV historic timeline

- 1859: Invention of Lead Acid Battery by Gaston Plante.
- 1879: Thomas Edison installs Electric Lights in New York City, popularizing Electricity.
- 1891: First Electric Vehicle is built in the United States.
- 1897: First Commercial US Electric Vehicles. Electrobat Taxis hit the roads in New York.
- 1933–45: German, French, and Dutch automakers sell a small range of EVs, spurred by gas shortages and WWII.
- 1949–51: In Japan, Tama Electric Motorcars sells an EV during a severe gas shortage.
- 1960: Automakers experiment with EVs, though none are widely adopted.
- 1996: General Motors begins leasing the EV1, one of several electric brands rolled out to meet California's Zero Emission rules.
- 2010: Nissan delivers the Leaf, an EV with a 100 mile range, a lithium ion battery, and regenerative braking.
- 2011: The Tesla Roadster electric sports car is offered. It has a range of 245 miles.

Fig. 3 shows the differences between an EV and an ICE.

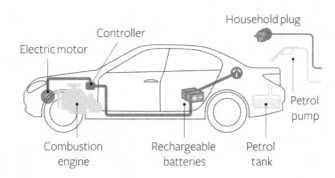

Fig. 3 Differences between an EV and an ICE [1].

3.3 EV types

EVs come in three types:

- Battery electric vehicles (BEV)
- Plug-in hybrid electric vehicles (PHEV)
- Hybrid electric vehicles (HEV)

The BEV operates on electric power only, and for this reason, it does not directly cause any carbon emissions. The BEV's indirect cause of CO_2 emissions is entirely dependent on what means of generation was used to produce the electricity used to charge the vehicle. The PHEV is a hybrid vehicle which is driven by an electric motor but also has a back-up ICE, which is used to recharge the battery while driving. Both the BEV and the PHEV can be plugged into the electricity network to be recharged. The HEV combines an ICE with an electric motor, where the electric motor is used to supplement the combustion engine to provide better fuel economy. The HEV cannot be connected to the grid, where the electric motor is charged from the ICE [1].

3.4 EVs and the smart grid

EVs are becoming an alternative to the conventional ICE vehicle, and will soon become an integrated part of the electricity network. Without applying any control over the charging behavior of EV owners, peak demand could be significantly increased beyond the delivery capacity of electricity networks. Communication between the electricity company and EV owners is made possible with an advanced electricity meter called a smart meter, which can provide the consumer with useful information.

The smart grid is often thought of as a grid that enables closer integration between electricity supply and demand as well as two-way communication between the electricity company and customers via smart meters by using new modern communication technologies. Smart meters can be used to encourage EV owners to charge their EVs outside times of peak demand. Furthermore, EVs can be charged at times when there is excess solar photovoltaic (PV) generated electricity, thus they can aid in the integration of renewable energies. Also, provided that the EV supply equipment is capable of both delivering and receiving energy to and from the EV's battery, consumers can be further encouraged to provide electricity during times of peak demand through vehicle-to-grid (V2G) or vehicle-to-home (V2H) support. An in-vehicle device proposed in this paper is to maximize the consumer's understanding of how they can optimally manage their EV, which would result in a smoother rollout of EVs on a large scale. Providing an in-vehicle display will encourage the EV owners to adopt smart charging strategies by informing them of the benefits that may be gained from the adoption of such strategies. Furthermore, the scope of smart charging can be extended to ensure that users charge when the electricity mix on the network has a large renewable density, meaning that the indirect cause of CO_2 emissions from EV operation is far lower than if the electricity was predominantly generated by, for example, a coal-fired power station.

3.5   EV integration within the smart grid and its impact on electricity networks

The impact that EVs have on electricity networks, given that the appropriate preparation is made, could be positive rather than negative. As stated in Ref. [2], the integration of EVs could present a great opportunity for improving both the efficiency and reliability of networks on a large scale. This is because, if operated properly, EVs can be used to provide support to the network during times of peak demand. This can be achieved by discharging the energy stored in the vehicle's battery to the grid to alleviate much of the stress that the network might be experiencing (a process called vehicle-to-grid (V2G) support).

Furthermore, if the EVs are then charged during times of low demand, such as in the early hours of the morning, much flatter load profiles for the electricity network will be seen, which in turn will dramatically improve the network's efficiency. Because this would result in a decentralized form of generation, improved reliability would also be seen, although this factor would depend on the prevalence of EV deployment. Another benefit associated with EV integration is the improved integration of decentralized renewable energy sources.

3.6   Feature requirements for in-vehicle display unit

The fundamental issues associated with the EV industry include the indirect cause of GHG emissions, integration costs and impacts, potential support to integrating renewable energy sources, and potential support to stabilizing the electricity network. The purpose of the device proposed in this paper is to maximize the consumer's understanding of how they can optimally manage their EV, when driving, charging, or discharging, which will subsequently lead to a smoother rollout of EVs on a large scale. For this reason some of the features that need to be included in the proposed device are discussed here.

- *Battery characteristics*: The key parameters that describe the battery's current condition include the state of charge (SOC), state of health (SOH) (in terms of cycle count), and the calendar life (based on the average number of cycles used per day). Each of these key parameters must be included as features of the display board.
- *Travel information*: An important feature of this device would be a traveling guide. This would include road maps, including charging infrastructure locations. The guide would also require the capability of determining the optimal/shortest route for certain trips, including the shortest additional route required for driving to a charging station part way to the desired destination for that trip. For this reason, the guide would need to predict the energy consumed for the trip, and thus what the SOC would be upon reaching the destination.
- *Electricity network*: Important information describing the electricity network and its current and future status would need to be available to ensure that the consumer would be able to optimize their EV management strategy. This includes information regarding electricity prices, system load (i.e., what percentage of the system's maximum capacity is the current load), and future forecasts for these two parameters (provided either by the distributor or via another means).
- *Charging/discharging schedule*: The optimal scheduling scenario is one where the distributor organizes the schedule for the various consumers. This is because the distributor can receive information on the travel requirements of its respective consumers, and thus organize a schedule that both satisfies these requirements and ensures the system will not be overloaded.

A considerably lengthy low-demand time frame exists from late at night to the early hours of the morning. This time horizon would be the optimal period for EV charging in most cases. Furthermore, in markets that use time of use (TOU) tariffs rather than a real-time pricing scheme, the price of electricity would generally remain unchanged for this entire period.

On the consumer end, information should be provided on the possible charging timeslots for the EV, based on the expected charge duration. The duration itself will be provided to the consumer as well as a comparison of the expected total cost for each possible timeslot. Once a timeslot is selected, this decision would be sent from the display board to the smart meter, where it could then be transmitted to the distributor to organize the command required to trigger charging at the chosen time.

A similar procedure as mentioned above can be applied when choosing the times for providing ancillary support (i.e., V2G) where optional timeslots will be available to the consumer, who can then choose the timeslot they prefer, and information on the expected duration and revenue received will be provided. The charge/discharge management flow chart is shown in Fig. 4. During the charging schedule, it would be important to monitor the energy management (EM) of the power entering the battery to understand the equivalent amount of CO_2 (g/kWh) indirectly emitted from the energy stored in the battery.

- *Economics*: The last and possibly most critical feature of this display board is the information regarding the economics of the EV. The EV economics should describe the overall financial return that consumers can reap from their EV, which effectively classifies the EV and its battery as a means of investment.

The economics should address the following aspects: (i) battery equivalent wear cost; (ii) revenues received and expenses incurred due to charging and discharging; and (iii) other savings or income due to incentives.

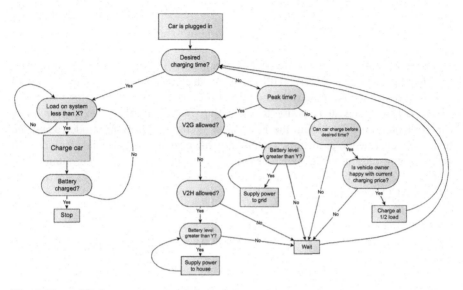

Fig. 4 Charge/discharge management flow chart.

The total profit or loss can be measured over the lifetime of the battery, which will take into account each of the key costs mentioned above. Based on the provision of this information, a financial summary can be compiled at the battery replacement date, stating overall how the consumer profited from the demand side management (DSM) practice that was adopted.

The last feature that can be incorporated is advice or recommendations on how costs can be further minimized and revenues maximized based on the consumer's practice of DSM.

3.7 In-vehicle communication board

The development of an in-vehicle communication board involves establishing the various display features that would provide comprehensive information to the owners of the EV about conditions of the electricity grid, the EV and its battery, pricing options, and charging and discharging schedule options. This is necessary information that EV owners need to know before connecting their EV battery to the electricity grid. Electricity costs at the time of EV-grid connection is an important factor. These costs include the purchase price of electricity for charging, the selling price of electricity for discharging, and the equivalent battery wear costs that are based on the limited energy throughput of the EV's battery.

Having considered these factors mentioned above, the optimal smart charging practices could be obtained by determining those that yielded minimal costs. Smart charging practices typically involve charging when the electricity demand is low and discharging when the demand is high. The adoption of these practices would therefore result in the flattening of network load profiles. The adoption of smart charging practices can yield significant benefits to the operator of the EV, which in turn can benefit the electricity distributor by minimizing the effect that the EV will have on the electricity network.

The disadvantages of EVs have been their high cost, low top speed and short range. Hybrid plug-in electric vehicles (HPEVs) using an electric battery in conjunction with a conventional internal combustion engine have been on the market for more than 25 years. Developed in response to escalating fuel costs, they can be run on a charge-depleting mode (using the battery) or a charge-sustaining mode (using the fuel). Developments in battery technology have enabled auto manufacturers to develop pure plug-in electric vehicles (PEVs), of which an increasing number are on the market. EVs are more energy-efficient than conventional vehicles and dramatically cut CO_2 emissions.

From the grid point of view, the EV is considered as an electrical device representing a new demand for electricity during the periods that EVs need to be charged, but they can play the role of storage device that could supply electric power back to the grid.

Through an effective communication with the grid, the EV battery can be used as a storage device that can make the electricity grid more reliable, especially with large proportion of renewable sources such as grid-connected solar and wind.

3.8 Challenges

- Cost
- Range limitation
- Safety and reliability
- Progress through R&D
- Driving adoption through education

3.9 Opportunities

- Government funding for research will be helpful for technology advancement and cost reduction.
- International R&D cooperation and coordination can help address common areas of need, accelerating technological breakthroughs.
- Consumer education campaigns and clear fuel economy labeling can help enhance public awareness.
- Etc.

3.10 Research questions

- How significant is the impact of charging and discharging of EVs on electricity demand, specifically in regional areas?
- Is the existing distribution network infrastructure capable of handling the increase in demand associated with widespread EV uptake?
- What is the best method of charge management to cope with the increase in electricity demand?

- How will the extended range of new EVs, such as Tesla and the Nissan Leaf, impact customer charging behavior?
- What costs can be expected for consumers charging EVs on existing distribution networks? And as a consequence, how will the consumer's ability to exploit time of use rates affect electricity demand?
- What benefits can "smart" technology provide in reducing the impact of EV charging?
- Understanding vehicle use profiles, EV benefits, and battery life challenges.
- Integrating renewable resources (solar and wind) with vehicle charging.
- Developing and testing grid interoperability standards.
- Exploring grid services technology opportunities.
- Wireless technologies for communications and control of EV systems.
- Monitoring and sense-and-control of charging.
- Software systems for EV energy management.
- Smart charging infrastructure.
- EV fleet management technologies and services.
- V2G and V2H.
- Smart charging infrastructure and scalability.
- Environmental issues and benefits.
- Energy management.
- Role of renewables in EV integration, especially solar and wind.
- Power quality, reliability, and stability effects as a result of EVs.
- Customer adoption, customer behavior and customer response.
- Pricing models for charging stations, roaming across territories.

4 Conclusions

Considering EV batteries as distributed storage in a smart grid environment, there are some technical issues that need to be dealt with:

- Energy management and control strategy in integrating EVs into the grid is the key to using EVs as distributed storage and need to be carefully examined.
- We need to understand the control and management issues when thousands of people plug in their EV to a grid.
- We need to realize that at some point during the day, the local utility might experience a deficit of electricity.
- Before beginning charging, the EV batteries need to communicate with the utility to determine if there is spare capacity in the system to begin charging the batteries, so we need an intelligent inverter.

References

[1] Sparking an Electric Vehicle Debate in Australia, ESAA, 2013. Available from: http://ewp.industry.gov.au/files/Sparking%20an%20Electric%20Vehicle%20Debate%20in%20Australia.pdf. Accessed 19 April 2015.
[2] E. Anderson, Real-time pricing for charging electric vehicles, Electr. J. 27 (9) (2014) 105–111.

Further Reading

[1] AGL, What is a Load Factor?, AGL Energy Sustainability Blog, 2010. Available from: http://aglblog.com.au/2010/05/what-is-a-load-factor/. Accessed 18 September 2015.

[2] Timeline: History of the Electric Car, PBS, 2009. Available from: http://www.pbs.org/now/shows/223/electric-car-timeline.html. Accessed 12 September 2015.

[3] K. Young, C. Wang, L.Y. Wang, K. Strunz, Electric vehicle battery technologies, in: Electric Vehicle Integration, Springer Science + Business Media, New York, 2013, pp. 15–56 (Chapter 2).

[4] S.F. Tie, C.W. Tan, A review of energy sources and energy management system in electric vehicles, Renew. Sust. Energ. Rev. 20 (2013) 82–102.

[5] J. Neubauer, A. Brooker, E. Wood, Sensitivity of battery electric vehicle economics to drive patterns, vehicle range, and charge strategies, J. Power Sources 209 (2012) 269–277.

[6] S. Hickey, Electric Vehicle Fleet Trial Final Report, Ergon Energy, Townsville, 2014.

[7] D.B. Richardson, Electric vehicles and the electric grid: a review of modeling approaches, impacts, and renewable energy integration, Renew. Sust. Energ. Rev. 19 (2013) 247–254.

[8] K. Kraatz, A. Zahedi, in: Energy management and control strategies of electric vehicle integrated into the smart grid, Conference Proceedings, AUPEC 2015.

Measurement-based voltage stability monitoring for load areas

14

Kai Sun, Fengkai Hu
University of Tennessee, Knoxville, TN, United States

1 Introduction

Growth in electrical energy consumption and penetration of intermittent renewable resources would make power transmission systems operate close to their stability limits more often. Among all stability issues, voltage instability due to the inability of the transmission or generation system to deliver the power requested by loads is one of the major concerns in today's power system operations. Voltage stability refers to "the ability of a power system to maintain steady voltages at all buses in the system after being subjected to a disturbance from a given initial operating condition" [1]. In contrast to generator rotors in the angular stability of a power system, the driving forces for voltage instability are mainly loads and the means by which voltage or reactive power are controlled to support loads. Usually, voltage instability initiates from a local bus but may develop to a wide area or even a systemwide instability. In a heavily stressed power system, voltage instability becomes a major concern in system operations and an impacting factor on the power transfer capability of a transmission network. When voltage instability leads to loss of voltage in a significant part of the system, such a process is called voltage collapse, in which many bus voltages in the system decay to an unrecoverable level. As a consequence of voltage collapse, the system may experience a power blackout over a large area. A system restoration procedure would then be undertaken to bring the blackout area back to service.

At present, some electricity utilities use model-based voltage stability assessment (VSA) online software tools to assist operators in foreseeing potential voltage instability and estimating voltage stability margin information. Based on a state estimate on the operating condition, those software tools employ power system models to simulate assumed disturbances such as contingencies and load changes by either steady-state analysis or dynamic analysis. However, such a model-based approach has its limitations: the fidelity of its results highly depends on the accuracy of power system models; and it needs a convergent, accurate state estimate in order to conduct a stability assessment, which may be hard to obtain under stressed system conditions. Real-time, accurate voltage stability margin information is critical for system operators to be aware of any potential or already developing voltage instability. However, it is challenging for model-based VSAs to provide such real-time margin information,

Application of Smart Grid Technologies. https://doi.org/10.1016/B978-0-12-803128-5.00014-3

especially for a large system. Many generators, buses, and transmission lines have to be modeled; also, lots of uncertainties need to be addressed such as load variations and intermittent behaviors of renewable generation.

Many countries are deploying synchronized phasor measurement units (PMUs) on transmission systems to provide wide area measurements for real-time stability monitoring. That leads to more interests in developing measurement-based VSA methods to directly assess voltage stability from measurements on monitored buses [2–23]. A family of measurement-based methods is based on Thevenin's Theorem: local measurements at the monitored load bus or area are used to estimate a Thevenin equivalent (TE) approximating the rest of the system, that is, a voltage source connected through a Thevenin impedance as illustrated by Fig. 1. The power transfer to the monitored load reaches its maximum when that Thevenin impedance $\overline{Z}_{Thevenin}$ has the same magnitude as the load impedance $\overline{Z}_{Load}$ [4,5]. Note: a variable having a bar represents a complex-value variable or a phasor. Based on a TE, voltage stability indices can be obtained [6,7]. Multiple such equivalents may together be applied to a load area [8]. For practical applications, paper [9] demonstrates a TE-based method on a realistic EHV network, and some other works consider load tap changers and overexcitation limiters in their models for better detection of voltage instability [10–12]. Influences from the changes on the external system and measurement errors on TE estimation are concerned in [13]. The equivalent circuit considering HVDC integrated wind energy is studied in [14].

Basically, a TE-based method works satisfactorily on a radially fed load bus or a transmission corridor, and its computational simplicity makes it suitable for real-time application. In recent years, some research efforts have been directed to how to extend the application of the TE to a broader transmission system, for example, the coupled single-port circuit or the Channel Components Transform [15–19]. Some other efforts tested TE-based methods on load areas. References [20,21] apply a TE-based method to a load center area. The method requires synchronized measurements on all boundary buses of the area in order to merge them into one fictitious load bus, for which a TE can be estimated and applied. That method was demonstrated on a real power transmission system in the real-time environment using PMU data [22]. Paper [23] improves the TE estimation for a load area to better tolerate the fluctuations in voltage phase angles and power factors measured at boundary buses. Paper [24] proposes a

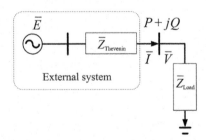

Fig. 1 Thevenin equivalent for a load bus.

measurement- and model-based hybrid approach using the TE to assess voltage stability under $N - 1$ contingencies.

However, as illustrated in [25], a TE-based method may not provide an accurate voltage stability margin for each of the multiple tie lines together feeding a load area if it merges those tie lines, as done in [20,21]. Even when estimating the total transfer limit of multiple tie lines, the TE is accurate only if the boundary buses through which those tie lines feed the load area are strongly connected and the external system is coherent so as to be regarded as a single voltage source. Those are two necessary conditions for merging tie lines and boundary buses and developing a meaningful TE. If connections between boundary buses are weak, when the load of the area gradually increases, tie lines may reach their power transfer limits at different time instants. In other words, voltage instability may initiate near one of the boundary buses and then propagate to the rest of the area. Thus, only monitoring the total power transfer limit of all tie lines using a single TE may delay the detection of voltage instability.

This article introduces an $N+1$ buses equivalent system proposed in [26] for measurement-based voltage stability monitoring that extends and generalizes the TE. This new equivalent has N buses interconnected to represent a load area with N boundary buses and one voltage source representing the external system. In fact, the TE is its special case with $N=1$. By modeling boundary buses separately for a load area, a new method based on that equivalent is developed and demonstrated in this article to calculate power transfer limits at individual boundary buses.

In the rest of this article, Section 2 introduces the new method in detail, including the $N+1$ buses equivalent system, its parameter identification, and analytical solutions of power transfer limits of its tie lines. In Section 3, an online scheme to implement the new method is presented. Section 4 first uses a four-bus power system to illustrate the advantages of the new method over a traditional TE-based method, and then validates the new method on the Northeast Power Coordinating Council (NPCC) 48-machine, 140-bus power system. Finally, conclusions are drawn in Section 5.

2   *N*+1 buses equivalent system

For a load area fed by N tie lines, voltage instability may be a concern with its boundary buses when power flows of those tie lines approach their transfer limits. As shown in Fig. 2, an $N+1$ buses equivalent system is used to model those boundary buses and tie lines while reducing the network details both inside and outside the load area. Assume the external system to be strongly coherent without any angular stability concern. Thus, it is represented by a single voltage source with phasor $\overline{E}$ connected by N branches with impedances $\overline{z}_{E1} \dots \overline{z}_{EN}$ (representing N tie lines) to N boundary buses, respectively. Each boundary bus is monitored and connects an equivalent load with impedance $\overline{z}_{ii}$, modeling the portion of the load seen from that bus. Connection between any two boundary buses i and j is modeled by impedance $\overline{z}_{ij}$. The power transfer limit of each tie line is a function of $N(N-1)/2+2N+1$ complex parameters of that equivalent system, including voltage phasor $\overline{E}$, N tie line impedances $\overline{z}_{Ei}$'s, $N(N-1)/2$ transfer impedances $\overline{z}_{ij}$'s, and N load impedances $\overline{z}_{ii}$'s.

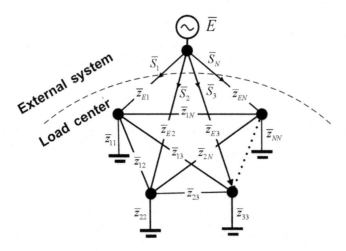

Fig. 2 $N+1$ buses equivalent.

Let $\overline{S}_i = P_i + jQ_i$ denote the complex power fed to boundary bus i and let $\overline{V}_i$ denote the bus voltage phasor. Using synchronized measurements on $\overline{S}_i$ and $\overline{V}_i$, all parameters of the equivalent can be identified online (e.g., every 0.1–1 s) using the latest measurements of a sliding time window. The rest of the section presents the algorithms for estimating the parameters of the external system (i.e., $\overline{E}$ and $\overline{z}_{Ei}$) and the load area (i.e., $\overline{z}_{ii}$ and $\overline{z}_{ij}$) and then derives the analytical solution of each tie line power transfer limit.

2.1 Identification of external system parameters

Assume that the sliding time window contains K measurement points. The external system parameters are assumed to be constant during the time window and hence are estimated by solving the following optimization problem. Nodal power injection Eq. (1) holds at each measurement point k of the time window.

$$\overline{S}_i(k) = \left(\frac{\overline{E} - \overline{V}_i(k)}{\overline{z}_{Ei}}\right)^* \times \overline{V}_i(k), \quad k = 1 \sim K \tag{1}$$

where $\overline{S}_i(k) = P_i(k) + jQ_i(k)$ and $\overline{V}_i(k) = V_i(k)\angle\theta_i(k)$ are, respectively, the received complex power and voltage phasor at boundary bus i at time point k. The magnitude of $\overline{E}$, denoted by E, can be estimated from measurements at each boundary bus i, whose estimation error is

$$e_i^{ex}(k) = E - \left|(P_i(k) - jQ_i(k))(r_{Ei} + jx_{Ei}) + (V_i(k))^2\right|/V_i(k) \tag{2}$$

The optimization problem in Eq. (3) computes the optimal estimates of E, r_{Ei}, and x_{Ei}.

$$\min J^{ex} = \sum_{k=1}^{K} \sum_{i=1}^{N} \frac{\omega_e}{N} \left[e_i^{ex}(k) \right]^2 + \sum_{i=1}^{N} \omega_z \left(\frac{r_{Ei}}{r_{Ep}} - 1 \right)^2 + \sum_{i=1}^{N} \omega_z \left(\frac{x_{Ei}}{x_{Ep}} - 1 \right)^2$$
$$s.t. \quad E > 0, \ r_{Ei} \geq 0 \tag{3}$$

The first term summates the estimation errors of E for all buses and all time points. The second and third terms respectively summate normalized differences in r_{Ei} and x_{Ei} between their estimates for the current and previous time window ("p" indicates the estimate from the previous time window). ω_e and ω_z are weighting factors, respectively, for variances of E and $\overline{z}_{Ei}$ over the time window. For instance, if the network topology of the external system does not change, $\overline{z}_{Ei}$ will be constant. Thus, there should be $\omega_z > \omega_e$ to allow more changes in E. The sequential quadratic programming (SQP) method [27] is used to solve the optimization problem in Eq. (3).

The above optimization problem for the external system is actually a nonconvex problem, so it needs to select good initial values for the external system parameters. As long as the initial values are in a neighborhood of the optimum that is considered as the true solution, the SQP method can make sure to converge to that solution. However, the problem does have multiple local optima. In practice, the initial values of external system parameters for optimization are determined as follows: at the beginning when the new measurement-based method is performed or whenever a major disturbance, for example, a line outage and a generator outage, is detected on the external system, the least square method is applied to the K data points of the current time window to estimate the parameters as new initial values; otherwise, the new method selects initial values from the optimization results of the previous time window. Because of the nonconvex nature of this optimization problem, if the initial values selected at the beginning are far from the true solution, the optimization may converge to a different solution with errors in parameter estimation and, consequently, cause inaccurate transfer limits at the end. Those errors may last until the next time the least square method is performed to recreate new initial values. Later in a case study on the NPCC system, the results from different initial values with errors intentionally added are compared. In practice, the least square method may be performed at a certain frequency, for example, every a few minutes, even if a major disturbance is not detected.

2.2   Identification of load area parameters

Load area parameters include load impedance $\overline{z}_{ii}$ and transfer impedance $\overline{z}_{ij}$ (i and $j = 1, \ldots, N$), whose admittances are $\overline{y}_{ii}$ and $\overline{y}_{ij}$. In each time window, assume constant $\overline{y}_{ij}$ if there is no topology change in the load area, and allow $\overline{y}_{ii}$ to change. From power flow equations,

$$\overline{y}_{ii}(k) = \frac{\overline{S}_i^*(k) - \displaystyle\sum_{j=1,\ldots,N, j \neq i} \left[V_i^2(k) - \overline{V}_i^*(k)\overline{V}_j(k) \right] \overline{y}_{ij}}{V_i^2(k)} \tag{4}$$

A window of K data points has NK values of $\bar{y}_{ii}$ and $N(N-1)/2$ values of $\bar{y}_{ij}$ to be estimated. Thus, there are totally $N^2 - N + 2NK$ real parameters to estimate. K data points can provide $2NK$ nodal power injection equations in the real realm. Because $N^2 - N + 2NK > 2NK$, there are insufficient equations to solve all parameters. In each window, if we may assume a constant power factor for each $\bar{y}_{ii}$, its conductance g_{ii}, and susceptance b_{ii} of $\bar{y}_{ii}$ at two adjacent time points will satisfy

$$\frac{g_{ii}(k-1)}{b_{ii}(k-1)} = \frac{g_{ii}(k)}{b_{ii}(k)}, \quad (k = 2 \sim K) \tag{5}$$

Thus, $K-1$ more equations are added to each bus and the entire load area needs to solve $3NK - N$ equations. From Eq. (6), there is $K = N$.

$$N^2 - N + 2NK = 3NK - N \tag{6}$$

It means that each time window needs to have at least N data points to be able to solve all parameters of the load area. For instance, for load areas with $N = 2, 3$, and 4 boundary buses, we need at least the same numbers of data points to solve 10, 24, and 44 unknowns, respectively.

From Eq. (4), g_{ii} and b_{ii} are both functions of $\bar{y}_{ij}$'s $(i \neq j)$. An error index on Eq. (5) is defined as

$$e_i^{in}(k) = g_{ii}(k-1)b_{ii}(k) - g_{ii}(k)b_{ii}(k-1) \tag{7}$$

The second optimization problem is formulated as Eq. (8) for estimating load area parameters by minimizing three weighted summation terms. g_{ij} and b_{ij} are the estimated conductance and susceptance of $\bar{y}_{ij}$; the second and third terms, respectively, summate their normalized differences from the previous time window ("p" indicates the estimate from the previous time window); ω_{pf} and ω_y are the weighting factors, respectively, for variances of the power factor and $\bar{y}_{ij}$ over the time window.

$$\min J^{in} = \sum_{k=2}^{K} \sum_{i=1}^{N} \frac{\omega_{pf}}{N} \left[e_i^{in}(k) \right]^2 + \sum_{\substack{i,j=1 \\ i \neq j}}^{N} \omega_y \left(\frac{g_{ij}}{g_{ijp}} - 1 \right)^2 + \sum_{\substack{i,j=1 \\ i \neq j}}^{N} \omega_y \left(\frac{b_{ij}}{b_{ijp}} - 1 \right)^2$$

$$s.t. \ g_{ij} > 0 \tag{8}$$

The SQP method is also used to solve all $\bar{y}_{ij}$'s for each time window. Then, calculate load admittance $\bar{y}_{ii}$ directly by Eq. (4).

For this second optimization on the estimation of load area parameters, the initial values of g_{ij} and b_{ij} are also required, which can be obtained directly from the reduced bus admittance matrix about the load area by eliminating all buses except boundary buses.

2.3 Solving the power transfer limit of each tie line

From real-time estimates of equivalent parameters, the active power transfer limit of each tie line can be solved analytically as a function of equivalent parameters.

The admittance matrix of the load area of the equivalent system is given by

$$
\mathbf{Y} = \begin{bmatrix} \overline{Y}_{11} & \cdots & \overline{Y}_{1i} & \cdots & \overline{Y}_{1N} \\ \vdots & \ddots & & & \vdots \\ \overline{Y}_{i1} & & \overline{Y}_{ii} & & \overline{Y}_{iN} \\ \vdots & & & \ddots & \vdots \\ \overline{Y}_{N1} & \cdots & \overline{Y}_{Ni} & \cdots & \overline{Y}_{NN} \end{bmatrix}
\tag{9}
$$

where $\overline{Y}_{ij} = \overline{Y}_{ji} = -\overline{y}_{ij}$ and $\overline{Y}_{ii} = \sum_{j=1}^{N} \overline{y}_{ij}$.

Let $\mathbf{Y_E}$ be a column vector about all admittances $\overline{y}_{Ei} = 1/\overline{z}_{Ei}$ $(i = 1, \ldots, N)$.

$$
\mathbf{Y_E} = \begin{bmatrix} \overline{y}_{E1} & \overline{y}_{E2} & \cdots & \overline{y}_{EN} \end{bmatrix}^T
\tag{10}
$$

Use a superscript "$\mathbf{D}$" to indicate a diagonal matrix whose elements are from a vector, that is,

$$
\mathbf{Y_E^D} = \begin{bmatrix} \overline{y}_{E1} & & 0 \\ & \ddots & \\ 0 & & \overline{y}_{EN} \end{bmatrix}
\tag{11}
$$

A vector about the injected currents satisfies Eq. (12), where $\mathbf{V} = \begin{bmatrix} \overline{V}_1 & \overline{V}_2 & \cdots & \overline{V}_N \end{bmatrix}^T$.

$$
\begin{cases} \mathbf{I} = \mathbf{VY} \\ \mathbf{I} = \overline{E}\mathbf{Y_E} - \mathbf{Y_E^D}\mathbf{V} \end{cases}
\tag{12}
$$

Then, there is

$$
\mathbf{V} = \overline{E}\left(\mathbf{Y} + \mathbf{Y_E^D}\right)^{-1}\mathbf{Y_E} = \overline{E}\frac{\mathrm{adj}\left(\mathbf{Y} + \mathbf{Y_E^D}\right)}{\det\left(\mathbf{Y} + \mathbf{Y_E^D}\right)}\mathbf{Y_E}
\tag{13}
$$

For simplicity, let α_i denote the ith element of $\mathrm{adj}(\mathbf{Y} + \mathbf{Y_E^D})\mathbf{Y_E}$ and let $\gamma = \det(\mathbf{Y} + \mathbf{Y_E^D})$. There is

$$
\overline{V}_i = \frac{\overline{E}\alpha_i}{\gamma}
\tag{14}
$$

The complex power transferred on each tie line is a function of elements of $\mathbf{Y}$, $\overline{E}$ and $\mathbf{Y_E}$.

$$
\mathbf{S} = \begin{bmatrix} \overline{S}_1 & \overline{S}_2 & \cdots & \overline{S}_N \end{bmatrix}^T = (\mathbf{I}^*)^{\mathbf{D}} \times \mathbf{V} = \left(\overline{E}\mathbf{Y_E^D} - \mathbf{V^D Y_E^D}\right)^* \mathbf{V}
\tag{15}
$$

If changes on the external system are ignored, each complex power is a function of load admittances, that is,

$$\overline{S}_i(\overline{y}_{11}, \ldots, \overline{y}_{NN}) = (\overline{E}^* \overline{y}_{Ei}^* - \overline{y}_{Ei}^* \overline{V}_i^*) \overline{V}_i = |\overline{E}|^2 \left(\overline{y}_{Ei}^* - \overline{y}_{Ei}^* \frac{\alpha_i^*}{\gamma^*} \right) \frac{\alpha_i}{\gamma} \tag{16}$$

$$P_i(\overline{y}_{11}, \ldots, \overline{y}_{NN}) = |\overline{E}|^2 \operatorname{Re} \left[\left(\overline{y}_{Ei}^* - \overline{y}_{Ei}^* \frac{\alpha_i^*}{\gamma^*} \right) \frac{\alpha_i}{\gamma} \right] \tag{17}$$

Based on the aforementioned constant power factor assumption over a time window for each load, P_i is a function of all load admittance magnitudes, that is, y_{jj} $(j = 1, \ldots, N)$. Its maximum $P_{i,j}^{\text{Max}}$ with respect to the change of y_{jj} at bus j is reached when

$$\frac{\partial P_i(y_{11}, \ldots, y_{NN})}{\partial y_{jj}} = 0 \quad i, j = 1 \ldots N \tag{18}$$

If an analytical solution of y_{jj} is obtained from Eq. (18) as a function of equivalent parameters, an analytical expression of $P_{i,j}^{\text{Max}}$ can be derived by plugging the solution of y_{jj} into Eq. (17). $P_{i,j}^{\text{Max}}$ with $i = j$ represents the maximum active power transfer to bus i with only the local load at bus i varying; $P_{i,j}^{\text{Max}}$ with $j \neq i$ represents the maximum active power transfer to bus i with only the load at another bus j varying.

For a general $N + 1$ buses equivalent, the analytical solution of $P_{i,j}^{\text{Max}}$ can be obtained by solving a quadratic equation. The proof based on Eq. (18) is given in the Appendix of [26].

This paper assumes that each y_{jj} may change individually, so there are N such transfer limits $P_{i,1}^{\text{Max}} \cdots P_{i,N}^{\text{Max}}$ for each tie line. By estimating the real-time pattern of load changes, the limit that best matches that pattern will be more accurate and selected. The voltage stability margin on a tie line is defined as the difference between the limit and the real-time power transfer.

3 Online scheme for implementation

Compared with the traditional TE-based approach, this new measurement-based method uses a more complex $N + 1$ buses equivalent to model more details about the boundary of a load area. Synchronized measurements on all boundary buses are needed from either state estimation results if only steady-state voltage instability is concerned or PMUs for real-time detection of voltage instability. The parameters of the equivalent are identified using real-time measurements over a recent time window for several seconds. In a later case study on the NPCC system, results from 5 to 10 s time windows will be compared. If the scheme uses PMUs in order to speed up online parameter identification and filter out noise or dynamics irrelevant to voltage stability, the original high-resolution measurements (e.g., at 30 Hz) may be downgraded to a low-sampling rate f_s (e.g., 2 Hz) by averaging the raw data over a time internal of $1/f_s$ (i.e., 0.5 s for 2 Hz).

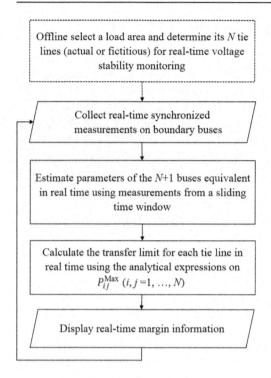

Offline select a load area and determine its N tie lines (actual or fictitious) for real-time voltage stability monitoring

Collect real-time synchronized measurements on boundary buses

Estimate parameters of the $N+1$ buses equivalent in real time using measurements from a sliding time window

Calculate the transfer limit for each tie line in real time using the analytical expressions on P_{ij}^{Max} ($i, j = 1, ..., N$)

Display real-time margin information

Fig. 3 Flowchart of the $N+1$ buses equivalent-based method.

As shown by the flowchart in Fig. 3, the new method first identifies all branches connecting the load area with the rest of the system, which comprise a cut set partitioning the load area. Those branches are assumed to be coming from the same voltage source $\overline{E}$. Then, any branches coming to the same boundary buses are merged. Assume that M branches are yielded. The proposed method is able to calculate the transfer limit for each branch using an $M+1$ buses equivalent. However, if M is large, it will result in huge computational burdens in estimating $M(M-1)/2 + 2M + 1$ parameters and consequently calculating M limits. Different from the TE that merges all M branches to one fictitious tie line, this new approach may group some branches across the boundary and only merge each group to one fictitious tie line. The criteria of grouping are: the boundary buses in one group are tightly interconnected; the branches in one group reach limits almost at the same time; and it is not required to monitor the branches within a group individually. For the fictitious tie line representing a group of branches, only its total limit is calculated. Thus, after merging some groups of branches to fictitious tie lines, the final number of tie lines becomes $N < M$. A simpler $N+1$ buses equivalent is used, which still keeps characteristics of the load area regarding voltage stability. The TE-based approach is a special case of this new method with $N = 1$.

The next two sections will test the new method on a four-bus power system and the NPCC 140-bus system using data generated from simulation. All computations involved in the algorithms of the new method are performed in MATLAB on an Intel Core i7 CPU desktop computer.

4   Demonstration on a four-bus power system

This section demonstrates the new method on a four-bus power system with one constant voltage source supporting three interconnected load buses representing a load area. The system is simulated in MATLAB with gradual load increases at three load buses. The simulation results on three load buses are treated as PMU data and fed to the new method. The system represents a special case of the system in Fig. 1 with $N = 3$. Let $\overline{E} = 1.0\angle 5°$ pu and three tie lines have the same impedance $\overline{z}_{E1} = \overline{z}_{E2} = \overline{z}_{E3} = 0.01 + j0.1$ pu. At the beginning, three load impedances $\overline{z}_{11} = \overline{z}_{22} = \overline{z}_{33} = 1 + j1$ pu. Consider two sets of transfer impedances in Table 1, respectively, for weak and tight connections between three boundary buses.

Keep the impedance angle of $\overline{z}_{33}$ unchanged but gradually reduce its modulus by 1% every 2 s until active powers on all tie lines meet limits. As shown in Fig. 4, three PV (power-voltage) curves (P_i vs V_i) are drawn for two groups of transfer impedances. For Set A, the transfer impedances between boundary buses are not ignorable compared with the tie line and load impedances, so three PV curves are distinct. However, when those impedances decrease to the values in Set B, three PV curves basically coincide, which is the case where a TE can be applied.

Simulation results are recorded at a 1 Hz sampling rate. The new method is performed every 1 s using the data of the latest 10 s time window. All computations of the new method on each time window are finished within 0.05 s in MATLAB. The new method gives each tie line three limits for three extreme load increase assumptions. For two sets of impedances, Fig. 5 shows P_1 to P_3, the total tie line flow $P_\Sigma = P_1 + P_2 + P_3$, and their limits calculated by the new method. The limits from solutions of $\partial P_1/\partial y_{33}$ to $\partial P_3/\partial y_{33}$ match the actual load increase, so their sum is defined as the total tie line flow limit $P_\Sigma^{\text{Max(New)}}$. For comparison, the TE-based method in [20,21] is also performed to give the total tie line flow limit as $P_\Sigma^{\text{Max(TE)}}$ in Fig. 5.

Tests on those two groups of data show that when a TE-based method is applied to a load area, voltage instability is detected only when the total tie line flow meets its limit. However, due to the uneven increase of load, one tie line may be stressed more to reach its limit earlier than the others, which can successfully be detected by the new method. Another observation on Fig. 5 is that the curve of $P_\Sigma^{\text{Max(New)}}$ is flatter than that of $P_\Sigma^{\text{Max(TE)}}$, and $P_\Sigma^{\text{Max(TE)}}$ is more optimistic and less accurate than $P_\Sigma^{\text{Max(New)}}$ when the system has a distance to voltage instability.

For Set A with a weak interconnection between boundary buses, Fig. 5A shows that three individual tie line flows P_1, P_2, and P_3 meet their limits at $t = 680$ s, 676 s, and

Table 1   Values of transfer impedances

Set	$\overline{z}_{12}$ (pu)	$\overline{z}_{13}$ (pu)	$\overline{z}_{23}$ (pu)
A	$0.01 + j0.1$	$0.015 + j0.15$	$0.005 + j0.05$
B	$0.0005 + j0.005$	$0.0008 + j0.0075$	$0.0003 + j0.0025$

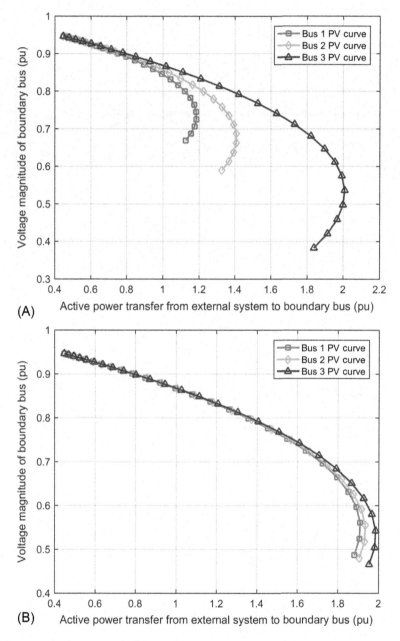

Fig. 4 PV curves for weak and tight connections between boundary buses. (A) Set A (weak); (B) Set B (tight).

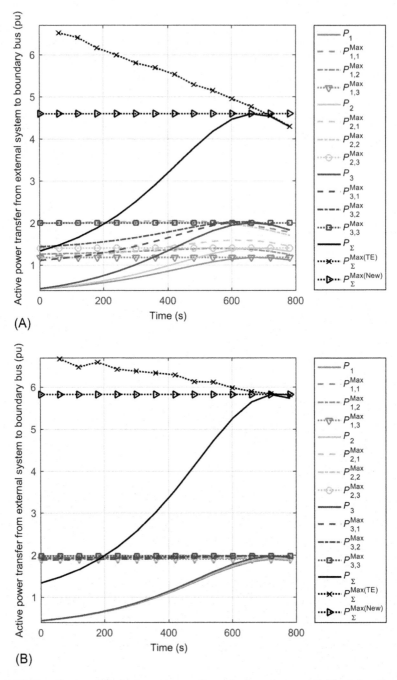

Fig. 5 Tie line flows and limits for two sets of transfer impedances. (A) Set A (weak connection); (B) Set B (tight connection).

666 s, respectively. The total tie line flow P_Σ meets $P_\Sigma^{\text{Max(New)}}$ at $t = 680$ s. $P_\Sigma^{\text{Max(TE)}}$ from the TE-based method is not as flat as $P_\Sigma^{\text{Max(New)}}$, and it is met by P_Σ after $t = 700$ s. Fig. 6 gives details of Fig. 5A on each tie line flow and its three limits, calculated according to Eq. (18) for three load increase assumptions. The limit from the solution of $\partial P_i / \partial y_{33}$ matches the actual load increase, and its curve is flat and met by P_i earlier than the other two. By using the new method, zero margin is first detected at $t = 666$ s on the third tie line and then on the other two tie lines as well as the total tie line flow. However, the TE-based method detects zero margin much later because, first, it only monitors the total tie line flow limit and second, the limit curve is not as flat as that given by the new method.

For Set B with a tight connection between boundary buses, Fig. 5B shows that all tie line flows reach their limits at $t = 732$ s. Also, P_Σ meets both $P_\Sigma^{\text{Max(New)}}$ and $P_\Sigma^{\text{Max(TE)}}$ at that same time.

5 Case studies on the NPCC test system

The proposed new method is tested on a NPCC 48-machine, 140-bus system model [28]. As highlighted in Fig. 7, the system has a Connecticut Load Center (CLC) area supported by power from three tie lines, that is, 73-35, 30-31, and 6-5. Line 73-35 is from the NYISO region and the other two are from north of the ISO-NE region. Powertech's TSAT is used to simulate four voltage instability scenarios about the CLC area:

- Scenario A: Generator trip followed by load increase leading to voltage instability.
- Scenario B: Generator trip followed by a tie line tip causing voltage instability.
- Scenario C: Two successive tie line trips causing voltage instability
- Scenario D: Shunt switching to postpone voltage instability.

The load model at each load bus adopts the default load model setting in TSAT, that is, 100% constant current for real power load and 100% constant impedance for reactive power load. Simulation results on the voltages at boundary buses 35, 31, and 5 and the complex powers of the three tie lines are recorded at 30 Hz, that is, the typical PMU sampling rate. The raw data are preprocessed by an averaging filter over 15 samples to be downgraded to 2 Hz. The processed data are then fed to the new method for estimating the external and load area parameters and calculating transfer limits. That data preprocessing improves the efficiency of two optimizations for parameter estimation while keeping the necessary dynamics on voltage stability in the data. The new method is performed every 0.5 s on data of the latest 5 s time window.

5.1 Scenario A: Generator trip followed by load increase leading to voltage instability

To create a voltage collapse in the CLC area, all its loads are uniformly increased by a total of 0.42 MW per second from its original load of 1906.5 MW with constant load power factors. At $t = 360$ s, the generator on bus 21 is tripped, which pushes the system

Fig. 6 Line flows and limits for Set A. (A) P_1 and limits; (B) P_2 and limits; (C) P_3 and limits.

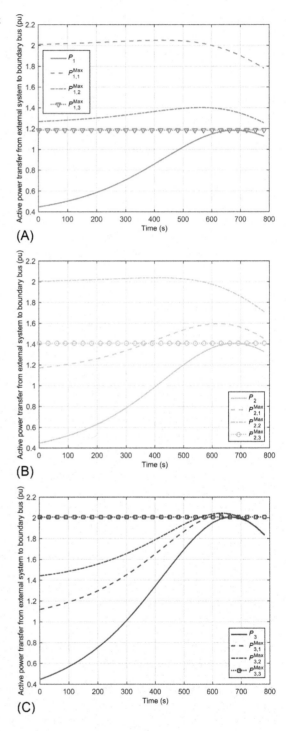

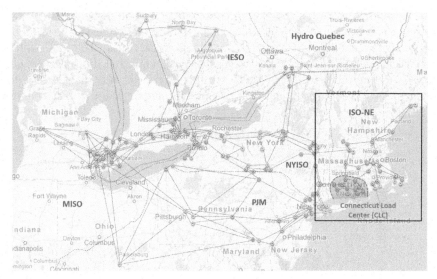

(A)

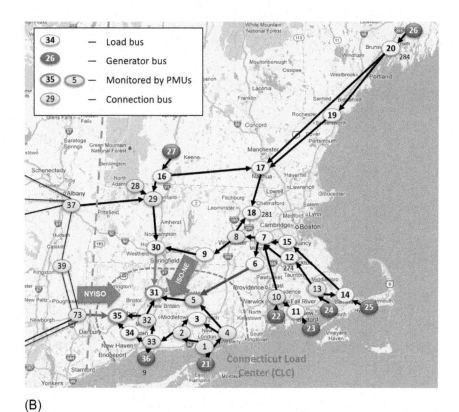

(B)

Fig. 7 Map of NPCC system and CLC area. (A) System topology; (B) CLC area.

to be close to the voltage stability limit. Shortly after a slight load increase, voltage collapse happens around $t = 530$ s, as shown in Fig. 8 on three boundary bus voltages. Fig. 9 indicates the PV curves monitored at three boundary buses. To better illustrate the PV curves, the figure is drawn using the data sampled at 25 s intervals until $t = 500$ s to filter out transient dynamics on the curves right after the generator trip and the voltage collapse at the end. Note that the generator trip causes a transition from

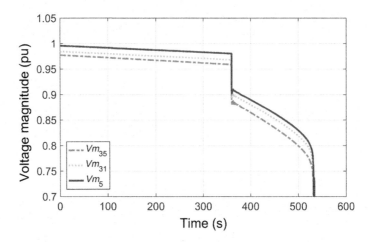

Fig. 8 Voltage magnitudes at CLC boundary buses (Scenario A).

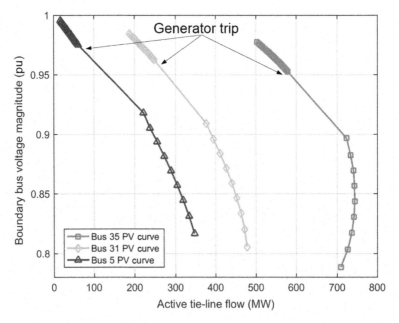

Fig. 9 PV curves monitored at the CLC boundary buses.

the precontingency PV curves to the postcontingency PV curves with a more critical condition of voltage stability. Bus 35 is the most critical bus because the "nose point" of its postcontingency PV curve is passed.

When the external system has strong coherency, the proposed new method can be applied to reduce it to one voltage phasor $\overline{E}$ connected with boundary buses of the load area by N branches, respectively. Fig. 10 shows the voltage magnitudes and angles of all buses in Fig. 7B that are outside the CLC load center, indicating a strong coherency of the external system. Therefore, the proposed method is valid and may adopt a 3 + 1 buses equivalent to calculate the power transfer limits separately for three tie lines.

Fig. 11 gives estimates of three load impedance magnitudes seen at the boundary buses. The figure shows that the load seen from bus 5 changes more significantly than the others. Hence, $P_{35,5}^{Max}$, $P_{31,5}^{Max}$ and $P_{5,5}^{Max}$ calculated from $\partial P_i/\partial y_5 = 0$ are more accurate and used to calculate the stability margin.

To compare with the new method, Fig. 12 gives the total transfer limit estimated using a TE-based method. The margin stays positive until the final collapse of the entire system.

Fig. 13 gives the results from the new method and each tie line has three transfer limits. Before the generator trip, all lines have sufficient margins to their limits. After the trip, more power is needed from the external system, so the active powers of the three tie lines all increase significantly to approach their limits. In Fig. 13A, P_{35} of 73-35 reaches the limit $P_{35,5}^{Max}$ at $t = 473.5$ s. From Fig. 13B and C, the other two lines keep positive margins until the final voltage collapse. It confirms the observation from Fig. 9 that voltage collapse will initiate from bus 35. If the limit and margin information on an individual tie line is displayed for operators in real time, an early remedial action may be taken before voltage collapse. However, such information is not available from a TE-based method.

Once parameters of the $N+1$ buses equivalent are estimated, transfer limits are directly calculated by their analytical expressions. The major time cost of the new method is with online parameter estimation. Fig. 14 gives the probability density about the times spent respectively on estimating the external system and the load area over one time window. According to the figure, parameter estimation over one time window can be accomplished within 0.02–0.1 s. The average total time cost for each cycle of the new method's online procedure (i.e., steps 2–5 in Fig. 3) is 0.0614 s, which includes 0.0221 s for external parameter estimation, 0.0271 s for load area parameter estimation, and 0.0122 s for transfer limits calculation. The times on measurements input and margin display are ignorable. The test results indicate that the new method can be applied in an online environment.

Fig. 15 compares the estimated $P_{35,5}^{Max}$ using a 5 s sliding time window with that using a 10 s sliding window. The two results match very well, which indicates that the new method is not very sensitive to the length of the sliding time window.

To test the results of the new method using inaccurate initial values in the external system parameter estimation, Fig. 16 compares the tie line power limits for 73-35 with three different sets of initial values: 110%, 100%, and 90% of the estimates from a least square method on the first time window at the beginning. From the figure, a

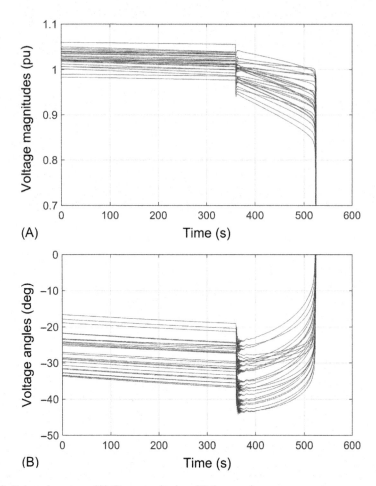

Fig. 10 External system. (A) Bus magnitudes; (B) bus angles.

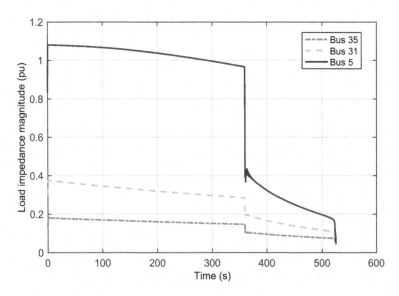

Fig. 11 Real-time estimation of load impedance magnitudes.

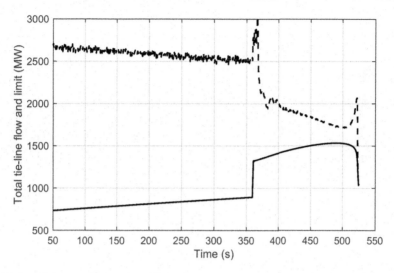

Fig. 12 Result from a TE-based method.

10% error will cause a less than 5% error in the transfer limit estimation before the generator trip and about a 2% error in the limit after the generator trip. If the least square method is used to reestimate the initial values when the generator trip is detected, that 2% error can be eliminated and these three limits will merge to one limit associated with accurate initial values. Considering that in the real world, small errors in the initial values cannot be avoided completely, a small positive threshold rather than zero may be defined for the transfer margin as an alarm of voltage instability.

5.2 Scenario B: Generator trip followed by a tie line tip causing voltage instability

Before $t = 400$ s, this scenario is the same as Scenario A. At $t = 400$ s, tie line 73-35 is tripped to cause immediate voltage collapse. The voltage magnitudes of three boundary buses are shown in Fig. 17.

Fig. 18 shows the tie line power flows and their limits for this scenario. After the generator trip, all tie line flows become closer to their limits, and tie line 73-35 carries 795 MW, which is higher than the total margin of 667 MW on tie lines 31-30 and 6-5. When tie line 73-35 is tripped at $t = 400$ s, its flow is transferred to the other two tie lines to cause them to meet limits. That explains why voltage collapse happens following that tie line trip. If the above tie line margin information is presented to the system operator before $t = 400$ s, the operator will be aware that the system following the generator trip cannot endure such a single tie line trip contingency and may take a control action.

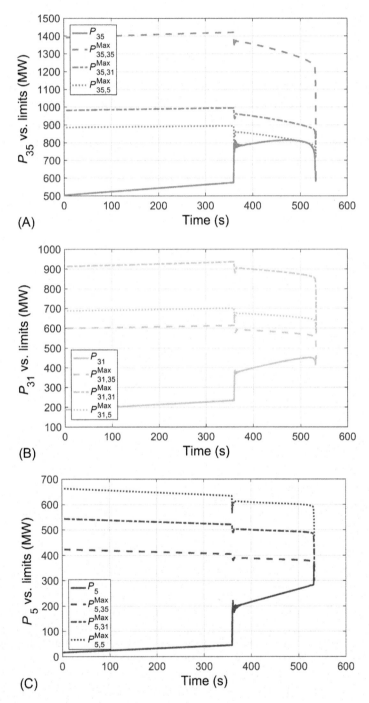

Fig. 13 Transfer limits of each tie line calculated by the new approach (Scenario A).
(A) P_{35} versus its limits; (B) P_{31} versus its limits; (C) P_5 versus its limits.

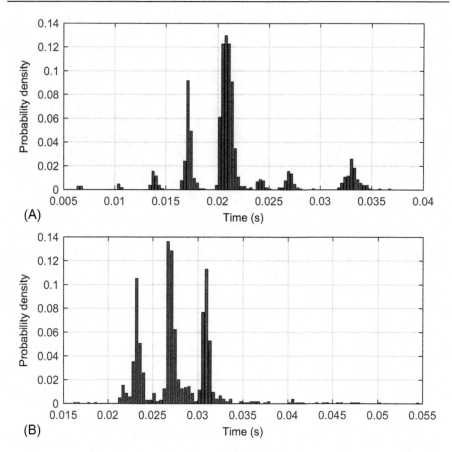

Fig. 14 Time performances on online parameter estimation. (A) Time for estimating external system parameters; (B) time for estimating load area parameters.

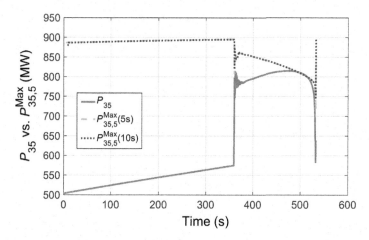

Fig. 15 Comparison of different optimization time windows.

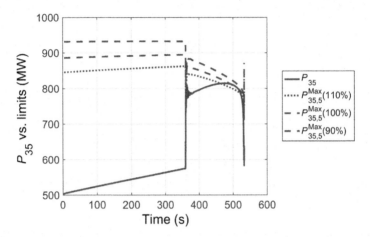

Fig. 16 Comparison of initial values with errors.

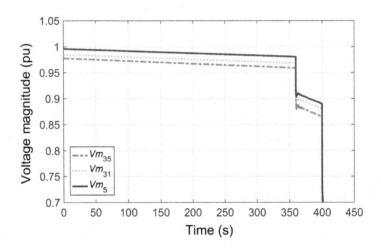

Fig. 17 Voltage magnitudes at CLC boundary buses (Scenario B).

5.3 Scenario C: Two successive tie line trips causing voltage instability

In this scenario, two successive tie line trips on 31-30 and 6-5 are simulated to test the adaptability of the new method to "$N-1$" and "$N-2$" conditions. During $t=0$–100 s, all loads in the CLC area are uniformly increased by 0.43 MW/s from its original load of 1906.5 MW with constant power factors unchanged. At $t=100$ s, the tie line 31-30 is tripped, and thus the voltages of three boundary buses drop significantly. During $t=100$–200 s, loads keep increasing at a lower speed equal to 0.37 MW/s. At $t=200$ s, the tie line 6-5 is tripped, causing voltage collapse as shown in Fig. 19.

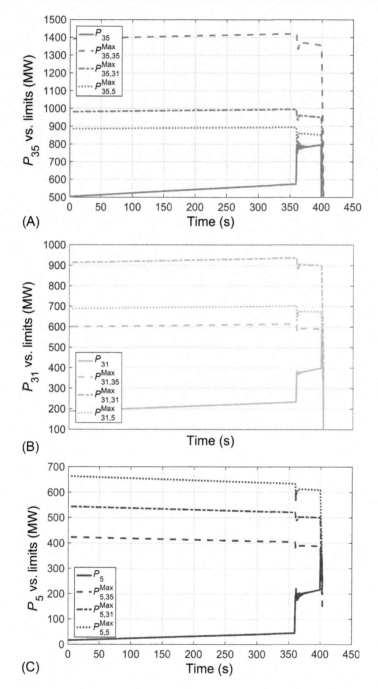

Fig. 18 Transfer limits of each tie line calculated by the new approach (Scenario B). (A) P_{35} versus its limits; (B) P_{31} versus its limits; (C) P_5 versus its limits.

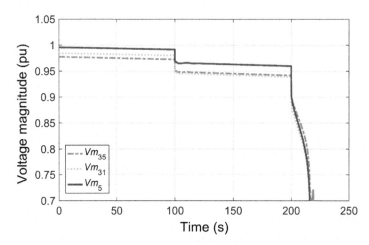

Fig. 19 Voltage magnitudes at CLC boundary buses (Scenario C).

Fig. 20 gives the results on each tie line and the transfer limits. Before tie line 31-30 is tripped at $t = 100$ s, all tie lines have sufficient transfer margin. After that trip, $P_{31} = 0$ is captured from measurements and the new method sets the corresponding $y_{E31} = 0$ to adapt to the new "$N - 1$" condition. Then, the transfer limit on tie line 73-35 drops to 677 MW. Before the next tie line 6-5 is tripped at $t = 200$ s, tie lines 73-35 and 6-5 totally transfer 733 MW to the CLC area, which is higher than the limit 677 MW of tie line 73-35. Therefore, the second tie line trip causes zero margin on tie line 73-35, followed by a voltage collapse.

5.4 Scenario D: Shunt switching to postpone voltage instability

This scenario considers switching in a capacitor bank located in the CLC area to postpone voltage collapse. Everything of this scenario during $t = 0$–440 s is the same as Scenario A. At $t = 440$ s when the transfer margin on the most critical tie line 73-35 drops to 3%, a 50 MVAR capacitor bank at bus 33 is switched in. Due to its additional VAR support, the voltage of the CLC area increases, as shown in Fig. 21, about voltage magnitudes of three boundary buses. Fig. 22 shows the transfer limits on tie line 73-35 for this scenario. Both the tie line flow and limits increase after that switch. The slight tie line flow increase is caused by voltage-sensitive loads in the load area, which is smaller than the increase of the limit. Therefore, zero margin happens at $t = 496.5$ s, that is, 23 s later than the 473.5 s of Scenario A. This scenario demonstrates that adding VAR support in the load area will increase the voltage stability margin, which is correctly captured by the new method.

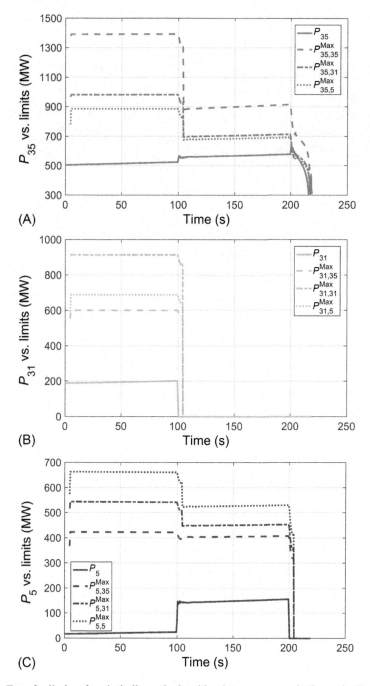

Fig. 20 Transfer limits of each tie line calculated by the new approach (Scenario C).
(A) P_{35} versus its limits; (B) P_{31} versus its limits; (C) P_5 versus its limits.

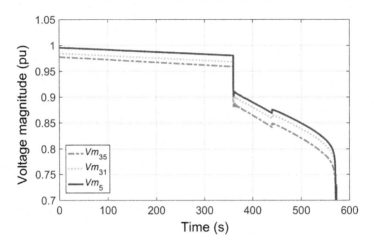

Fig. 21 Voltage magnitudes at CLC boundary buses (Scenario D).

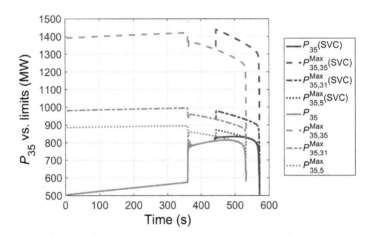

Fig. 22 P_{35} versus its limits.

6 Discussion and conclusions

This article has introduced a new measurement-based method proposed in [26] for real-time voltage stability monitoring of a load area fed by multiple tie lines. The new method is based on an $N+1$ buses equivalent system whose parameters are estimated directly from synchronized measurements obtained at the boundary buses of the load area. For each tie line, the method calculates the transfer limit and margin against voltage instability analytically from that estimated equivalent system. The new method has been demonstrated in detail on a four-bus system and then tested by case studies on a 140-bus NPCC system model.

Compared to a traditional TE-based method for measurement-based voltage stability monitoring, the new method has two apparent advantages. First, the new method offers detailed limit and margin information on individual tie lines so as to identify the tie line and boundary bus with the smallest margin as the location where voltage instability more likely initiates. Second, as demonstrated on the four-bus system, before the voltage collapse point, the total tie line flow limit from the new method is more accurate than the limit from the TE-based method. The latter fluctuates more and is not as flat as the former because the TE-based method does not model the weak or strong connection between boundary buses. The above second advantage makes the new method more suitable for online monitoring and early warning of voltage instability, and the first advantage can help the system operator to identify the location where voltage instability more likely initiates and accordingly choose more effective control resources, for example, those having shorter electrical distances to the tie line with the smallest margin. Recent studies in [29] demonstrate that voltage stability margin information for an anticipated "$n-1$" condition can also be provided by sensitivity analysis studies on the $N+1$ buses equivalent.

References

[1] P. Kundur, J. Paserba, V. Ajjarapu, G. Andersson, A. Bose, C. Canizares, N. Hatziargyriou, D. Hill, A. Stankovic, C. Taylor, T. Van Cutsem, V. Vittal, Definition and classification of power system stability (IEEE/CIGRE joint task force on stability terms and definitions), IEEE Trans. Power Syst. 19 (3) (2004) 1387–1401.

[2] T. Van Cutsem, L. Wehenkel, et al., Decision tree approaches to voltage security assessment, IEE Proc. C Gener. Transm. Distrib. 140 (3) (1993) 189–198.

[3] R. Diao, K. Sun, V. Vittal, et al., Decision tree-based online voltage security assessment using PMU measurements, IEEE Trans. Power Syst. 24 (2) (2009) 832–839.

[4] K. Vu, M.M. Begovic, D. Novosel, et al., Use of local measurements to estimate voltage-stability margin, IEEE Trans. Power Syst. 14 (3) (1999) 1029–1035.

[5] I. Smon, G. Verbic, F. Gubina, Local voltage-stability index using tellegen's theorem, IEEE Trans. Power Syst. 21 (3) (2006) 1267–1275.

[6] B. Milosevic, M. Begovic, Voltage-stability protection and control using a wide-area network of phasor measurements, IEEE Trans. Power Syst. 18 (1) (2003) 121–127.

[7] B. Leonardi, V. Ajjarapu, Development of multilinear regression models for online voltage stability margin estimation, IEEE Trans. Power Syst. 26 (1) (2011) 374–383.

[8] M. Parniani, et al., "Voltage stability analysis of a multiple-infeed load center using phasor measurement data." IEEE PES Power Systems Conference and Exposition, November 2006.

[9] S. Corsi, G.N. Taranto, A real-time voltage instability identification algorithm based on local phasor measurements, IEEE Trans. Power Syst. 23 (3) (2008) 1271–1279.

[10] C. D. Vournas, and N. G. Sakellaridis, "Tracking maximum loadability conditions in power systems." in Proc. 2007 Bulk Power System Dynamics and Control-VII, Charleston, SC, August 2007.

[11] M. Glavic, T. Van Cutsem, Wide-area detection of voltage instability from synchronized phasor measurements. Part I: principle, IEEE Trans. Power Syst. 24 (3) (2009) 1408–1416.

[12] C. Vournas, C. Lambrou, M. Glavic et al., "An integrated autonomous protection system against voltage instability based on load tap changers." 2010 iREP Symposium, August 2010.

[13] S.M. Abdelkader, D.J. Morrow, Online Thévenin equivalent determination considering system side changes and measurement errors, IEEE Trans. Power Syst. 30 (99) (2014) 1–10.

[14] L. He, C.-C. Liu, Parameter identification with PMUs for instability detection in power systems with HVDC integrated offshore wind energy, IEEE Trans. Power Syst. 29 (2) (2014) 775–784.

[15] W. Li, et al., Investigation on the Thevenin equivalent parameters for online estimation of maximum power transfer limits, IET Gener. Transm. Distrib. 4 (10) (2010) 1180–1187.

[16] Y. Wang, I.R. Pordanjani, et al., Voltage stability monitoring based on the concept of coupled single-port circuit, IEEE Trans. Power Syst. 26 (4) (2011) 2154–2163.

[17] J.-H. Liu, C.-C. Chu, Wide-area measurement-based voltage stability indicators by modified coupled single-port models, IEEE Trans. Power Syst. 29 (2) (2014) 756–764.

[18] W. Xu, et al., A network decoupling transform for phasor data based voltage stability analysis and monitoring, IEEE Trans. Smart Grid 3 (1) (2012) 261–270.

[19] I.R. Pordanjani, et al., Identification of critical components for voltage stability assessment using channel components transform, IEEE Trans. Smart Grid 4 (2) (2013) 1122–1132.

[20] P. Zhang, L. Min, J. Chen, Measurement-based voltage stability monitoring and control, US Patent 8,126,667, 2012.

[21] K. Sun, P. Zhang, L. Min, Measurement-based voltage stability monitoring and control for load centers, EPRI Report No. 1017798, 2009.

[22] F. Galvan, A. Abur, K. Sun, et al., in: Implementation of synchrophasor monitoring at Entergy: tools, training and tribulations, IEEE PES General Meeting, San Diego, CA, 23–26 July, 2012.

[23] K. Sun, F. Hu, N. Bhatt, "A new approach for real-time voltage stability monitoring using PMUs." 2014 IEEE ISGT-Asia, Kuala Lumpur, 20–23 May 2014.

[24] H. Yuan, F. Li, Hybrid voltage stability assessment (VSA) for N-1 contingency, Electr. Power Syst. Res. 122 (2015) 65–75.

[25] F. Hu, K. Sun, et al., in: An adaptive three-bus power system equivalent for estimating voltage stability margin from synchronized phasor measurements, IEEE PES General Meeting, National Harbor, MD, 27–31 July, 2014.

[26] F. Hu, K. Sun, A. Del Rosso, E. Farantatos, N. Bhatt, Measurement-based real-time voltage stability monitoring for load areas, IEEE Trans. Power Syst. 31 (4) (2016) 2787–2798.

[27] J. Nocedal, S.J. Wright, Numerical Optimization (Springer Series in Operations Research and Financial Engineering), Springer, 2006.

[28] J.H. Chow, R. Galarza, P. Accari, et al., Inertial and slow coherency aggregation algorithms for power system dynamic model reduction, IEEE Trans. Power Syst. 10 (2) (1995) 680–685.

[29] F. Hu, K. Sun, D. Shi, Z. Wang, "Measurement-based voltage stability assessment for load areas addressing n-1 contingencies," IET Gener. Transm. Distrib. 11 (15) (2017) 3731–3738.

Index

Note: Page numbers followed by *f* indicate figures, and *t* indicate tables.

Printed in the United States
By Bookmasters